PETROLEUM REFINING

1

CRUDE OIL

PETROLEUM PRODUCTS

PROCESS FLOWSHEETS

- Catalytic Cracking of Heavy Petroleum Fractions.
  D. DECROOCQ

- Applied Heterogeneous Catalysis. Design. Manufacture.
  Use of Solid Catalysts.
  J.-F. LE PAGE

- Methanol and Carbonylation.
  J. GAUTHIER-LAFAYE and R. PERRON

- Chemical Reactors. Design. Engineering. Operation.
  P. TRAMBOUZE, H. VAN LANDEGHEM and J.-P. WAUQUIER

- Petrochemical Processes. Technical and Economic Characteristics.
  A. CHAUVEL and G. LEFEBVRE

  Volume 1. Synthesis-Gas Derivatives and Major Hydrocarbons.
  Volume 2. Major Oxygenated, Chlorinated and Nitrated Derivatives.

- Resid and Heavy Oil Processing.
  J.-F. LE PAGE, S. G. CHATILA and M. DAVIDSON

- Scale-Up Methodology for Chemical Processes.
  J.-P. EUZEN, P. TRAMBOUZE and J.-P. WAUQUIER

- Industrial Energy Management.
  V. KAISER

- Computational Fluid Dynamics Applied to Process Engineering.

- Industrial Water Treatment.
  F. BERNÉ and J. CORDONNIER

Proceedings of Symposiums

- Characterization of Heavy Crude Oils and Petroleum Residues.
  Caractérisation des huiles lourdes et des résidus pétroliers.
  Symposium international, Lyon, 1984.

- International Symposium on Alcohol Fuels.
  Symposium international sur les carburants alcoolisés.
  VIIe Symposium international, Paris, 1986.

Institut Français du Pétrole Publications

PETROLEUM REFINING

CRUDE OIL

PETROLEUM PRODUCTS

PROCESS FLOWSHEETS

Edited by

Jean-Pierre Wauquier

Institut Français du Pétrole

Translated from the French
by David H. Smith

1995

ÉDITIONS TECHNIP   27 RUE GINOUX 75737 PARIS CEDEX 15

Translation of
"Le raffinage du Pétrole.
Tome 1. Pétrole brut. Produits pétroliers.
Schémas de fabrication", J.-P. Wauquier
© 1994, Éditions Technip, Paris.

ISBN 2-7108-0685-1

Series ISBN 2-7108-0686-X

Printed in France
by Imprimerie Chirat, 42540 Saint-Just-la-Pendue

The authors contributing to this volume, which was edited and coordinated by J.-P. Wauquier, *Institut Français du Pétrole,* are the following persons:

Raymond BOULET   Chapters 1, 2, and 3
Institut Français du Pétrole (IFP)

Sami G. CHATILA   Chapter 8
Institut Français du Pétrole (IFP)

Edouard FREUND   Preface
Institut Français du Pétrole (IFP)

Jean-Claude GUIBET   Chapter 5
Institut Français du Pétrole (IFP)

Gérard HEINRICH   Chapter 10
Institut Français du Pétrole (IFP)

Henri PARADOWSKI   Chapter 4 and Appendix 1
Technip

Jean-Claude ROUSSEL   Chapters 1, 2 and 3
Institut Français du Pétrole (IFP)

Bernard SILLION   Chapter 9
Institut Français du Pétrole (IFP)

Bernard THIAULT   Chapters 6, 7 and Appendix 2
Bureau de Normalisation du Pétrole (BNPét)

David H. SMITH   Translation from the original
Institut Français du Pétrole (IFP)   French version

Presentation of the Series

"PETROLEUM REFINING"

The series "**Petroleum Refining**" will comprise five volumes covering the following aspects of the petroleum refining industry:

- Crude oil. Petroleum products. Process flowsheets.
- Separation processes.
- Conversion processes.
- Materials and equipment.
- Refinery operations and management.

The series is designed for the engineers and technicians who will be operating the refineries at the beginning of the next century. By that time, solutions will necessarily have been found for a number of problems: increasingly severe product specifications and, more especially, environmental protection. The series will provide those in the refining industry with the essentials of petroleum refining as well as information on the special technologies they will be using.

A group of eminent specialists was placed in charge of writing the series, with those involved being listed at the beginning of each relevant volume. We would like to thank all of them for their dynamic and even enthusiastic approach to this project.

The series is the long-awaited revision of the two-volume "*Le pétrole. Raffinage et Génie Chimique*" that was first published in 1965 under the direction of Pierre Wuithier. Revision of the original work, which was a highly successful publication, had been under consideration for several years. Jacqueline Funck, then Director of Information and Documentation at the *Institut Français du Pétrole*, was instrumental in this endeavor.

In 1990, at the request of the Director of the Center for Refining, Petrochemistry and Engineering at the *École Nationale Supérieure du Pétrole et des Moteurs*, the late Jean Durandet initiated the conceptual foundations for the new series that he was unable to complete. We would like to pay a tribute to him here.

Jean-Pierre Wauquier has accepted to take over for us as editor of the first two volumes. Pierre Leprince, followed by other well known figures will then carry on and complete this comprehensive series.

Michel VERWAERDE
Director of Publications
Institut Français du Pétrole

Foreword

Edouard Freund

Refining has the function of transforming crude oil from various origins into a set of refined products in accordance with precise specifications and in quantities corresponding as closely as possible to the market requirements.

Crude oils present a wide variety of physical and chemical properties. Among the more important characteristics are the following:

- distillation curve, which leads to a first classification of light crude oils having high distillate yields as opposed to heavy or extra heavy crudes

- sulfur content (crudes having low or high contents)

- chemical composition, this is used only to characterize particular crudes (paraffinic or naphthenic).

As a whole, a given crude is generally used to make products most of which have positive added values. This is particularly the case for motor fuels and specialty products. However, some of the products could have negative added values, as in the case of unavoidable products like heavy fuels and certain petroleum cokes.

The products could be classified as a function of various criteria: physical properties (in particular, volatility), the way they are created (primary distillation or conversion). Nevertheless, the classification most relevant to this discussion is linked to the end product use: LPG, premium gasoline, kerosene and diesel oil, medium and heavy fuels, specialty products like solvents, lubricants, and asphalts. Indeed, the product specifications are generally related to the end use. Traditionally, they have to do with specific properties: octane number for premium gasoline, cetane number for diesel oil as well as overall physical properties such as density, distillation curves and viscosity.

Chemical composition does not generally come into play, except for the case where it is necessary to establish maximum specifications for undesirable compounds such as sulfur, nitrogen, and metals, or even more unusually, certain compounds or families of compounds such as benzene in premium gasolines. By tradition, the refiner supposedly possesses numerous degrees of freedom to generate products for which the properties but not the composition are specified.

Nevertheless, we are witnessing most recently an important development in petroleum product specifications regarding two main factors:

- product quality improvements, e.g., lubricant bases having a very high viscosity index
- environmental limits on composition established to reduce emissions overall or to modify them to reduce their impact. This constraint concerns principally the motor and heating fuels.

From complex cuts characterized in an overall manner, there is a transition towards mixtures containing only a limited number of hydrocarbon families or even compounds. This development has only just begun. It affects for the moment only certain products and certain geographical zones. It is leading gradually to a different view of both refining and the characterization of petroleum products.

Simple conventional refining is based essentially on atmospheric distillation. The residue from the distillation constitutes heavy fuel, the quantity and qualities of which are mainly determined by the crude feedstock available without many ways to improve it. Manufacture of products like asphalt and lubricant bases requires supplementary operations, in particular separation operations and is possible only with a relatively narrow selection of crudes (crudes for lube oils, crudes for asphalts). The distillates are not normally directly usable; processing must be done to improve them, either mild treatment such as hydrodesulfurization of middle distillates at low pressure, or deep treatment usually with partial conversion such as catalytic reforming. The conventional refinery thereby has rather limited flexibility and makes products the quality of which is closely linked to the nature of the crude oil used.

The following are found in the flowsheet of a refining complex:

- conversion processes capable of transforming a part of the atmospheric residue, such as catalytic cracking of vacuum distillate or (partially) atmospheric residue, hydrocracking of vacuum distillate and thermal processes such as visbreaking
- processes for treating distillates from conversion units which are more severe than those utilized for distillates from primary distillation, as well as conversion feedstock pretreatment processes
- processes of a new kind, that modify or even synthesize components or narrow cuts derived from conversion process effluents.

We cite isomerization of C_5–C_6 paraffinic cuts, aliphatic alkylation making isoparaffinic gasoline from C_3–C_5 olefins and isobutane, and etherification of C_4–C_5 olefins with the C_1–C_2 alcohols. This type of refinery can need more hydrogen than is available from naphtha reforming. Flexibility is greatly improved over the simple conventional refinery. Nonetheless some products are not eliminated, for example, the heavy fuel of marginal quality, and the conversion product qualities may not be adequate, even after severe treatment, to meet certain specifications such as the gasoline octane number, diesel cetane number, and allowable levels of certain components.

These problems are solved in the refinery of the future, the refinery beyond 2000, with the arrival of deep conversion processing such as residue hydrocracking, carbon rejection and gasification processes which can lead to the elimination of heavy fuel production if needs be, supplementary processes for deep treatment of distillates coming from conversion or deep conversion, and synthesis of compounds from light ends of the same conversion processes which led to the advanced flow schemes themselves. This type of refinery approaches that of a petrochemical complex capable of supplying the traditional refining products, but meeting much more severe specifications, and petrochemical intermediates such as olefins, aromatics, hydrogen and methanol.

This brief description of past and present refining developments leads to a certain number of important remarks. First of all, we are observing a gradual, continuous evolution. It could hardly be otherwise, considering the large time factors —it takes several years to build a refinery— the capital investment, and the tightness of the product specifications. Moreover, refining evolves around successive modifications to a basic flow scheme containing a limited number of processes. These processes have been greatly improved over the past twenty years from the technological point of view and, for catalytic processes, the level of performance of the catalysts in service. On the other hand, very few new processes have appeared: as early as 1970 one could almost have built the refinery of the year 2000 but with much lower performance with regard to energy, economics, and product quality. Among the truly new processes, one can name selective oligomerization, light olefin etherification and very low pressure reforming with continuous catalyst regeneration.

The preceding discussion on the role of refining and the development of flow schemes shows clearly the importance attributed to the characterization of crude oils and petroleum products.

First of all, one should note that refining a low cost raw material into low or medium added value products requires extremely delicate optimization. It is out of the question to give them much more than the specifications require; thus highlighting the importance of being able to predict the various product yields and qualities that a given crude oil can supply. A profound understanding of crude oils appears therefore indispensable. That is the role of crude oil analysis, an operation traced in part to refining, with the development of distillation and other separation processes and which remains largely empirical. Progress in this field will remain necessarily slow, considering the highly complex crude oil chemistry which is practically impossible to break down without preliminary separations. That is not the case for characterizing cuts obtained from distillation or even residues following operations such as deasphalting or dewaxing, even though the heavy cuts remain still very complex and impossible to analyze in detail. The problem of characterizing these cuts is similar in approach to that for the products.

Product characterization aims at defining their end-use properties by means of conventional standard measurements related as well as possible — and in any case, being the object of a large consensus — to end-use properties. We cite for example that octane numbers are supposed to represent the resistance of gasoline to knocking in ignition engines.

Whatever the development of knowledge in the fields of chemical analysis and structure-property relationships, the characterization by determination of conventional properties of usage and other values related empirically to properties of usage will remain mandatory and unavoidable, as a minimum because it is required with regard to specifications.

However, this conventional method presents a certain number of limitations. In the first place, the traditional end-use property itself can be difficult to determine. Consider the cetane number for example: is it a good characterization of diesel fuel with respect to its behavior in commercial diesel engines? In the second place, concern for protecting the environment imposes new specifications which are often specifications linked to the composition of products; very low content of certain contaminants, reduced levels of certain families of compounds, or even a specific compound as already discussed.

These new specifications impose, at least partially, a new approach that aims to establish composition-property relationships, indispensable for the choice and operation of processes controlling product composition. Of course, this type of chemical approach presents its own interest which is independent of new limits introduced as needed by new specifications. One can thus foresee that this new approach to characterizing products will gradually become reality, all the while in keeping with the idea that it can probably never completely replace the conventional methods, notably the in the case of complex cuts that are mainly heavy fractions. The development of analytical and characterization methods will have in turn a great impact on refining by providing a more rational approach to planning refineries and by playing an important role in optimizing the operation of the various units.

We should therefore conclude that refining will witness a very important evolution, without revolution, but which will affect both the processes and procedures utilized, the objective being to produce "clean" products in a "clean", energy-efficient manner.

Contents

**Chapter 3   Characterization of Crude Oils
and Petroleum Fractions**

**Chapter 4   Methods for the Calculation
of Hydrocarbon Physical Properties**

Chapter 5   **Characteristics of Petroleum Products for Energy Use (Motor Fuels – Heating Fuels)**

Nomenclature

The nomenclature used in Volume 1 is based on the recommendations of the IUPAC (International Union of Pure and Applied Chemistry) for the system of units utilized as well as for their symbols. The reference is entitled,

"Quantities, Units and Symbols in Physical Chemistry"

prepared by I. Mills, T. Cvitas et al. edited by Blackwell Scientific Publications, Oxford, UK, 1993.

Any deviations result from a deliberate choice, either to conform to current usage in the profession, or to avoid ambiguity in the interpretation of symbols.

In addition to fundamental units from the SI system, i.e., m, kg, s, mol, K, A, and cd, multiples and sub-multiples of these units as well as derived or combined units are also used and indicated in parentheses.

Symbols

a	coefficient of RKS equation of state (energy parameter)	$m^6 \cdot bar/mol^2$
a	absorptivity coefficient in the infrared	$(l/(g \cdot cm))$
A	degrees API	$-$
A	absorbance	$-$
$\mathring{A}$	Angström	$\left(10^{-10}\,m\right)$
B_o	magnetic field intensity	$T = kg/\left(A \cdot s^2\right)$
b	coefficient of RKS equation of state (covolume)	m^3/mol
c	speed of light	$\left(3 \cdot 10^8\,m/s\right)$
c, C	concentration	$mol/m^3, kg/m^3, (kg/l), (g/l)$

C	accompanied by a number: hydrocarbon whose carbon number is equal to the number	—
C_p	isobaric molar or mass specific heat	$J/(mol \cdot K), J/(kg \cdot K)$
d	specific gravity	—
d	diameter	m
D	diffusivity, diffusion coefficient	m^2/s
E	energy	J
e	electron charge	$1.602 \ 10^{-19} \ A \cdot s$
eV	electron volt	$1.602 \ 10^{-19} \ J$
F	Helmholtz molar free energy	J/mol
f	fugacity	Pa, bar
G	Gibbs free energy or Gibbs molar free energy	J, J/mol
G	molar flow of gas phase	mol/s
g	acceleration of gravity	$9.81 \ m/s^2$
H	enthalpy, molar enthalpy, weight enthalpy	J, J/mol, J/kg
H	Henry's constant	bar
h	Planck's constant	$6.626 \ 10^{-34} \ J \cdot s$
h	height	m
Hp	horsepower	(746 W)
I	radiation intensity	cd
J	molar flux	$mol/(s \cdot m^2)$
K_W	Watson characterization factor	—
k, K	constant	variable
L	molar liquid flow rate	mol/s
l	length	m
m	prefix for milli (one-thousandth)	(10^{-3})
m	constant for the Soave equation	—
M, m	mass	kg
m/e	mass/charge	$kg/(A \cdot s)$
M	molecular weight	kg/mol, kg/kmol
M	blending index	—
M_o	magnetization	$T = kg/(A \cdot s^2)$
N	normal (0°C, atmospheric pressure)	—
N_C	number of carbon atoms	—
n	prefix for nano	(10^{-9})
n	quantity of matter	mol, kmol
n	refractive index	—
P	pressure	Pa, bar, (mm Hg, torr)

p	partial pressure	Pa, bar
Pa	parachor	$10^3 \, (\text{mN/m})^{1/4} \, (\text{m}^3/\text{kmol})$
ppb	parts per billion (billion = thousand million)	−
ppm	parts per million	−
p	prefix for pico	(10^{-12})
r	stoichiometric ratio (for combustion)	−
R	number of rings (in a chemical formula)	−
R	ideal gas constant	$0.083 \cdot \text{m}^3 \cdot \text{bar}/(\text{K} \cdot \text{kmol})$
		$8.31 \, \text{J}/(\text{mol} \cdot \text{K})$
r, R	radius	m
S	entropy, molar entropy	J/K, J/(mol·K)
S	weight % sulfur	−
S	standard specific gravity, $d \, ^{60°F}_{60°F}$	−
T	temperature	K
T	transmittance	−
t	time	s
U	internal energy, molar internal energy	J, J/mol
V	potential	$V = \text{m}^2 \cdot \text{kg}/(\text{A} \cdot \text{s}^3)$
V	volume	m^3
V	molar volume	m^3/mol
W	weight content	−
x	mole fraction in liquid phase	−
x, y, z	Cartesian coordinates	m
y	mole fraction in vapor phase	−
Z	atomic number	−
z	compressibility factor	−
z	mole fraction in feed	−
z	Rackett's parameter (see Ra index)	−
γ	activity coefficient	−
δ	solubility parameter	$(\text{J}/\text{m}^3)^{1/2}$
λ	electrical conductivity	(S/m)
λ	thermal conductivity	$\text{W}/(\text{m} \cdot \text{K})$
μ	reduced mass $= (m_1 \cdot m_2)/(m_1 + m_2)$	kg
μ	dynamic or absolute viscosity	Pa·s
μ	prefix for micro	(10^{-6})
ν	kinematic viscosity	m^2/s
ν	frequency	s^{-1}, Hz
$\bar{\nu}$	wave number	(cm^{-1})

π	vapor pressure	bar, Pa
ρ	density	kg/m^3
σ	interfacial tension	N/m
φ	equivalence ratio (fuel mixture)	—
ϕ	fugacity coefficient	—
ω	acentric factor	—

Indices

A	relative to component A
A	aromatic
b	bubble, at the boiling point
B	relative to component B
c	critical, pseudocritical
e	estimated
e	relative to the flash point
e	effective
f	relative to a petroleum fraction
f	formation
f	fusion
g	gas
gp	ideal gas
h	hydrates
H	hydrogen
HC	hydrocarbon
i	initial
i	relating to a component i
j	relating to a component $j \neq i$
l	liquid
m	mixture
m	molar
m	mass
N	naphthenic
n	degree of polymerization
p	relative to pressure (partial or total), at constant pressure
P	paraffinic

r	reaction
R	in rings (carbon)
Ra	Rackett
ref	reference
r	reduced
s	at saturation pressure
s	solid
T	at temperature T
T	at the triple point
v	vapor, vaporization
v	volume
w	weight, mass
4	water at 4°C
15	at 15°C
298	at 298 K
100	at 100°F
210	at 210°F
$\bar{\nu}$	for the wave number defined (see symbol $\bar{\nu}$)
$+$	higher than, in the case of a hydrocarbon chain (e.g., C_{5+} signifying all hydrocarbons having 5 or more carbon atoms)

Superscripts

$+$	carbonium ion (carrier of a positive charge)
$+$	greater than the indicated temperature (e.g., $375°C^+$ or 375^+)
15	at 15°C
$°$	total (fugacity)
	in the pure state (solubility in water)
	at infinite dilution (diffusivity)
	of reference (pressure)
	at standard state (pressure = 1 bar)
	in the ideal gas state
	as saturated liquid
1	at temperature T_1
2	at temperature T_2

Other Symbols

π	3.1416
Π	product operator
Σ	summation operator
Δ	difference operator
d	differential operator
∂	partial differential operator
$\int$	integral operator
exp	exponential
/	division sign
$\cdot$	multiplication sign
log	logarithm base 10, common logarithm
ln	Naperian logarithm, natural logarithm
$\in$	belongs to
$\simeq$	about equal to
$\sim$	proportional to
$\leq$	equal to or less than
$\geq$	equal to or greater than
$\neq$	different from or unequal to
$=$	equal
$+$	plus
$-$	minus
atm	atmosphere (1 atm = 101,325 Pa)
abs	absolute (pressure)
rel	relative (pressure)

Temperature Scale Relations

$$T\ (^\circ F) = 1{,}8\ T\ (^\circ C) + 32$$
$$T\ (K) = T\ (^\circ C) + 273.15$$
$$T\ (^\circ R) = 1{,}8\ T\ (K)$$

Abbreviations and Acronyms

AFNOR	Association Française de Normalisation
AFTP	Association Française des Techniciens du Pétrole
AIChE	American Institute of Chemical Engineers
ANSI	American National Standards Institute
AP	aniline point
API	American Petroleum Institute
AR	atmospheric residue
ARDS	atmospheric residue hydroconversion
ASTM	American Society for Testing and Materials
ASVAHL	Association pour la Valorisation des Huiles Lourdes
AVL	Anstalt für Verbrennungskraftmaschinen List
BMCI	Bureau of Mines Correlation Index
BrN	bromine number
BNPét	Bureau de Normalisation du Pétrole
BP	British Pharmacopœia
BP	boiling point
BP	British Petroleum Corporation
BS	bright stock (heavy lubricating stock)
BSI	British Standards Institution
BTS	low sulfur content (*basse teneur en soufre*)
BTH	benzothiophenes
BTX	benzene – toluene – xylenes
BWR	Benedict-Webb-Rubin (equation of state)
CCAI	calculated carbon aromaticity index

CCE/CEC	Commission des Communautés Européennes/ Commission of the European Communities
CCI	calculated cetane index
CCS	cold cranking simulator
CEE/EEC	Communauté Économique Européenne/ European Economic Community
CEC	Coordinating European Council for the Development of Performance Tests for Lubricating Oils and Motor Fuels
CEI/IEC	Commission Électrotechnique Internationale/ International Electrotechnical Commission
CEN	Comité Européen de Normalisation/ European Committee for Standardization
CENELEC	Comité Européen de Normalisation des Industries Electriques
CEP	car efficiency parameter
CFPP	cold filter plugging point
CFR (motor)	Cooperative Fuel Research
CI	cetane index
CN	cetane number
CII	calculated ignition index
CLO	clarified oil (catalytic cracking)
CNOMO	Comité de Normalisation des Moyens de Production
CONCAWE	Conservation of Clean Air and Water in Europe
COSTALD	Corresponding States Liquid Density
CPDP	Centre Professionnel du Pétrole
CRC	Coordinating Research Council
CSR	Chambre Syndicale du Raffinage
Cx	aerodynamic coefficient (automobile)
DAB	dialkylbenzenes
DAO	deasphalted oil
DBTH	dibenzothiophenes
DEA	diethanolamine
ΔR100	RON–MON of the fraction distilled at 100°C
DHYCA	Direction des Hydrocarbures et des Carburants (French Ministry of Industry)
DI	diesel index
DII	diesel ignition improvers
DIN	Deutsches Institüt für Normung

DIPPR	Design Institute for Physical Property Data
DMF	dimethylformamide
DMSO	dimethylsulfoxide
DON	distribution octane number
ECE (cycle)	Economic Commission for Europe
EEC (CEE)	European Economic Community (Communauté Économique Européenne) (has become "European Union" at the end of 1993)
EFTA (AELE)	European Free Trade Association (Association Européenne de Libre Échange)
EN	European Norm
ENSPM-FI	École Nationale Supérieure du Pétrole et des Moteurs Formation Industrie
EP	end point (or FBP – final boiling point)
EPS	electrostatic precipitation
ETBE	ethyl tertiary butyl ether
EU	European Union
EUDC	extra-urban driving cycle
E 70-100-180-210	volume fraction distilled at 70-100-180-210°C
FAM	Fachausschuss Mineralöl-und-Brennstoff-Normung
FCC	fluid catalytic cracking
FDA	Food and Drug Administration
FEON	front end octane number
FIA	fluorescent indicator adsorption
FID	flame ionization detector
FOD	*fuel oil domestique* (domestic fuel oil or home-heating oil)
FP	flash point
FVI	fuel volatility index
GFC	Groupement Français de Coordination pour le développement des essais de performance des lubrifiants et des combustibles pour moteurs
GHSV	gas hourly space velocity
GHV	gross heating value (GCV = gross calorific value)
GPC	gas phase chromatography
GPC	gel permeation chromatography
LPG	liquefied petroleum gas
HC	hydrocarbon
HCO	heavy cycle oil (heavy gas oil from catalytic cracking)

HDC	hydrocracking
HDN	hydrodenitrogenation
HDS	hydrodesulfurization
HMN	heptamethylnonane
HPLC	high pressure liquid chromatography
H/C	hydrogen/carbon ratio
IFP	Institut Français du Pétrole
IP	initial point (or IBP – initial boiling point)
IP	Institute of Petroleum (UK)
IR	infrared
ISO	International Organization for Standardization
JFTOT	jet fuel thermal oxidation tester
JP	jet power, jet-propelled, jet propulsion
JPI	Japanese Petroleum Institute
LCO	light cycle oil (gas oil from catalytic cracking)
LHSV	liquid hourly space velocity
LL	low lead (content)
M	blending index for gasoline
MAV	maleic anhydride value
MEA	monoethanolamine
MEK	methylethylketon
MON	motor octane number
MS	mass spectroscopy
MSEP	microseparometer surfactants
MTBE	methyltertiarybutylether
MW	molecular weight
NATO	North Atlantic Treaty Organization
NF	Norme Françaiṡe (AFNOR)
NF	National Formulary
NHV	net heating value (NCV = net calorific value)
NMHC	non-methane hydrocarbons
NMP	N-methylpyrrolidone
NMR	nuclear magnetic resonance
NPD	nitrogen phosphorus thermionic detector
NPRA	National Petroleum Refiners Association
N/S	nitrogen/sulfur ratio

OCVP	ethylene-propylene copolymers
ON	octane number
PAO	poly alpha-olefins
PCB	polychlorinated biphenyls
PIBSA	polyisobutylene succinic anhydride
PONA	paraffins, olefins, naphthenes, aromatics
PNA	polynuclear aromatics
PSA	pressure swing adsorption (gas separation by adsorption)
PVC	polyvinyl chloride
RBT	ring-and-ball temperature
RCC	residue catalytic cracking
RF	radio frequency
RF	residual flow (diesel injectors)
RKS	Redlich-Kwong-Soave equation of state
RON	research octane number
RTFOT	rolling thin film oven test
RUFIT	rational utilization of fuels in private transport
RVP	Reid vapor pressure
S	sensitivity (= RON − MON)
SAE	Society of Automotive Engineers
SARA	saturates-aromatics-resins-asphaltenes
SBPs	special boiling point spirits
S/C	sulfur/carbon ratio
SCD	sulfur chemiluminescence detector
SD	simulated distillation
SE	specific energy
SEC	steric exclusion chromatography
SHED	sealed housing for evaporation determination
SIA	Société des Ingénieurs de l'Automobile
SR	straight run
SRK	Soave-Redlich-Kwong equation of state
SSU	Saybolt Seconds Universal
TAME	tertiary amyl methyl ether
TBA	tertiary butyl alcohol
TBN	total basic number
TBP	true boiling point

TBTS	very low sulfur content (*très basse teneur en soufre*)
TDC	top dead center
TDS	Technical Data Services
TEL	tetraethyl lead
THC	total hydrocarbons (see Table 5.27)
TML	tetramethyl lead
TMS	tetramethyl silane
TR	jet power (JP), turbojet
TVP	true vapor pressure
UV	ultra-violet
UOP	Universal Oil Products
VABP	volume average boiling point
VI	viscosity index
VD	vacuum distillate
VLS	very low sulfur
V/L	volatility criterion, vapor-liquid (ratio)
VLI	volatility index
VR	vacuum residue
VRDS	vacuum residue hydroconversion
WS	white spirit
WSIM	water separation index modified
XR	X-ray
XRD	X-ray diffraction
XRF	X-ray fluorescence
100N	100 neutral (lube bases)
200N	200 neutral (lube bases)
350N	350 neutral (lube bases)
600N	600 neutral (lube bases)

Composition of Crude Oils and Petroleum Products

Jean-Claude Roussel

Raymond Boulet

Crude oils have physical and chemical characteristics that vary widely from one production field to another and even within the same field.

The roughest form of characterization, but nevertheless one that has great economic consequence, is the classification of "light" and "heavy" crude. Because crude oil is composed essentially of hydrocarbon molecules, its specific gravity varies inversely with its H/C atomic ratio. Specific gravities for various crude oils will range from 0.7 to 1.0; they are often expressed in degrees API[1] (American Petroleum Institute) which will vary between 70 and 5. As we will see, it is clear that this variable gravity reflects compositions of chemical families that are very different from each other.

Figure 1.1 illustrates the diversity of products derived from petroleum classified according to their distillation ranges and number of carbon atoms. From one crude to another, the proportions of the recovered fractions vary widely. A good illustration is the gasoline fraction (one of the most economically attractive); a crude from Qatar gives about 37 per cent by volume whereas a Boscan crude oil only yields 4.5%.

All these differences influence the conditions of production, transport, storage and refining adapted to the crude and its derived products: hence the necessity for knowing the composition as precisely as possible.

1.1 Pure Components

Crude oils are mixtures of pure components, but these are extremely numerous and the difficulty to describe the different fractions increases with the number of carbon atoms.

---

[1] $°API = \dfrac{141.5}{\text{Standard specific gravity, S}} - 131.5$ (see 4.1.2.2).

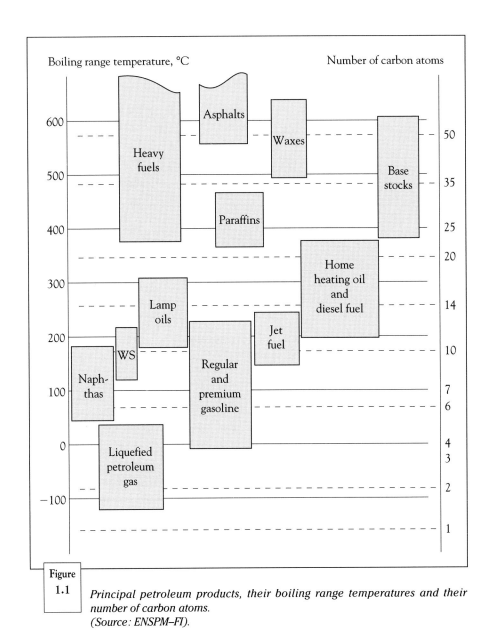

Figure 1.1   *Principal petroleum products, their boiling range temperatures and their number of carbon atoms.*
(Source: ENSPM–FI).

We will recall briefly the different families of hydrocarbons without attempting to give a summary of a course in organic chemistry. The reader is particularly encouraged to refer to general reference works for information concerning the nomenclature and properties of these compounds (Lefebvre, 1978).

1.1.1 **Hydrocarbons**

Hydrocarbons constitute the essential components of petroleum; their molecules contain only carbon and hydrogen; they are grouped into many chemical families according to their structure. All structures are based on the quadrivalency of carbon.

The carbon–carbon molecule chains can be as follows:

- either linked by a single bond (given the suffix "–ane") $-\overset{|}{\underset{|}{C}}-\overset{|}{\underset{|}{C}}-$

- or linked by multiple bonds,

$$\text{double,} \quad \overset{\diagdown}{\diagup}C=C\overset{\diagup}{\diagdown} \quad \text{(suffix "–ene")}$$

$$\text{or triple,} \quad -C\equiv C- \quad \text{(suffix "–yne")}$$

that are said to be unsaturated. The same molecule can contain several multiple bonds; for two sets of double bonds, the suffix is "–diene"; the multiple bonds are said to be conjugated when two of them are separated by a single bond.

1.1.1.1 **Saturated Aliphatic Hydrocarbons or Alkanes or Paraffins**

These consist of a chain of carbon atoms each carrying 0 to 3 hydrogen atoms except for the simplest molecule, methane: CH_4. Each carbon atom is linked to four other atoms which can be either carbon or hydrogen. Their general formula is:

$$C_nH_{2n+2}$$

They can be structured as straight chains as are the normal paraffins or *n*-alkanes and they are described by the following formula:

$$CH_3-\left(CH_2\right)_n-CH_3$$

Their boiling points increase with the number of carbon atoms. For molecules of low carbon numbers, the addition of a carbon increases the boiling point about 25°C. Further additions result in a smaller increase. The density increases with the molecular weight: 0.626 kg/l for pentane which has 5 atoms of carbon, 0.791 kg/l for pentacosane which has 25 carbon atoms, but the density is always much lower than 1.

One or more hydrogen atoms can be substituted by a carbon atom or chain of hydrocarbons, in which case they are called isoparaffins or isoalkanes.

Example: isopentane

$$CH_3-CH-CH_2-CH_3$$
$$\underset{CH_3}{\overset{|}{}}$$

Branching can take place at different locations in the chain, giving the possibility of, for equal numbers of carbon atoms, different molecules called isomers.

Examples:

2 methyl-hexane $CH_3-CH-CH_2-CH_2-CH_2-CH_3$
$\qquad\qquad\qquad\quad |$
$\qquad\qquad\qquad CH_3$

3 methyl-hexane $CH_3-CH_2-CH-CH_2-CH_2-CH_3$
$\qquad\qquad\qquad\qquad\quad |$
$\qquad\qquad\qquad\qquad CH_3$

Isoparaffins have boiling points lower than normal paraffins with the same number of carbon atoms. Table 1.1 presents some physical properties of selected paraffins

	Overall formula	Structural* formula	Molecular weight	Boiling. point, °C (1 atm)	Specific gravity d_4^{15} (liquid)
Methane	CH_4	C	16.0	−161.5	0.260
Ethane	C_2H_6	C−C	30.1	−88.6	0.377
Propane	C_3H_8	C−C−C	44.1	−42.1	0.508
n-Butane	C_4H_{10}	C−C−C−C	58.1	−0.5	0.585
Isobutane	C_4H_{10}	$\begin{array}{c} \text{C} \\ | \\ \text{C−C−C} \end{array}$	58.1	−11.7	0.563
n-Pentane	C_5H_{12}	C−C−C−C−C	72.1	+36.1	0.631
n-Heptane	C_7H_{16}	C−C−C−C−C−C−C	100.2	98.4	0.688

Table 1.1	*Physical constants of selected alkanes.* * Hydrogen atoms are omitted to simplify illustration.

1.1.1.2 Saturated Cyclic Hydrocarbons or Cycloparaffins or Naphthenes

These hydrocarbons contain cyclic (or ring) structures in all or part of the skeleton. The number of carbon atoms in the ring thus formed can vary. Refer to Table 1.2.

Their boiling points and densities are higher than alkanes having the same number of carbon atoms.

The rings most frequently encountered in crude oils are those having five or six carbon atoms. In these rings, each hydrogen atom can be substituted by a paraffinic "alkyl" chain that is either a straight chain or branched.

	Overall formula	Structural* formula	Molecular weight	Boiling. point, °C (1 atm)	Specific gravity d_4^{15} (liquid)
Cyclopentane	C_5H_{10}		70.1	49.3	0.750
Methylcyclopentane	C_6H_{12}		84.2	71.8	0.753
Cyclohexane	C_6H_{12}		84.2	80.7	0.783
Methylcyclohexane	C_7H_{14}		98.2	100.9	0.774

Table 1.2   *Physical constants of selected cycloparaffins.*
* Hydrogen atoms are omitted to simplify illustration.

The general formula for cycloparaffins having a single ring is C_nH_{2n}. There also exist cycloparaffins with two, three, or four, etc. rings attached. Hence decalin has two rings attached to each other and the general formula is C_nH_{2n-2}.

In cycloparaffins having four and five rings, one finds hydrocarbons having retained in part the structure of living matter at the petroleum's origin (steranes and hopanes). They serve as biochemical markers.

1.1.1.3 **Aromatic Hydrocarbons**

Cyclic and polyunsaturated, aromatics are present in high concentrations in crude oils. The presence in their structure of at least one ring containing three conjugated double bonds accounts for some remarkable properties. In fact, the first three, benzene, toluene, and the xylenes, are basic raw materials for the petrochemical industry. They also are large contributors to the octane number of a gasoline. On the other hand, the higher homologs are generally considered a nuisance because they cause environmental and public health problems and they impair catalyst activity by coke deposition.

The basic pattern common to all aromatics is the benzene ring as illustrated in Kekule's formula:

The structural formula is often represented by:

The hydrogen atom attached to each carbon atom in the hexagon has been omitted by convention.

The general formula is the following:

$$C_nH_{2n-6}$$

Hydrogen atoms can be substituted in the following ways:

- Either by alkyl groups which are designated by the letter R, which is equivalent to $C_{n'}H_{2n'+1}$, to give alkyl-aromatics. The prefixes *ortho, meta* and *para* are used to show positions of substitutes on the ring. Therefore, C_8H_{10} or dimethyl benzenes (xylenes), are represented in the following manner:

| orthoxylene | metaxylene | paraxylene |

- Or by other aromatics.

  In this instance a second aromatic ring can be substituted for two adjacent hydrogen atoms giving condensed polynuclear aromatics, for example:

| naphthalene | anthracene |

phenanthrene

- Or by a naphthenic ring which can also be substituted for two adjacent hydrogen atoms forming a naphthene aromatic such as tetralin or tetrahydronaphthalene.

Table 1.3 summarizes data for these aromatic hydrocarbons.

	Overall formula	Structural* formula	Molecular weight	Boiling. point, °C (1 atm)	Specific gravity d_4^{15} (liquid)
Benzene	C_6H_6		78.1	80.1	0.884
Toluene	C_7H_8		92.1	110.6	0.871
Ethylbenzene	C_8H_{10}		106.2	136.2	0.871
o-xylene	C_8H_{10}		106.2	144.4	0.884
m-xylene	C_8H_{10}		106.2	139.1	0.868
p-xylene	C_8H_{10}		106.2	138.4	0.865

Table 1.3	*Physical constants of selected aromatics*.* * The hydrogen atoms have been omitted for the sake of simplification.

1.1.1.4 **Unsaturated Aliphatic Hydrocarbons or Olefins or Alkenes**

In this group certain carbon atoms are linked only to three atoms, which implies the existence of one or more double bonds between carbon atoms.

Taking into account the double bond, an olefin situation is encountered that is much more complex than that of the preceding families. For example, the C_4H_8 butene isomers have many arrangements:

- 1-butene $\qquad\qquad CH_2=CH-CH_2-CH_3$

- cis 2-butene
$$CH_3-\underset{\underset{\textstyle H}{|}}{C}=\underset{\underset{\textstyle H}{|}}{C}-CH_3$$

- trans 2-butene
$$CH_3-\underset{\underset{\textstyle H}{|}}{C}=\overset{\overset{\textstyle H}{|}}{C}-CH_3$$

- isobutene
$$\begin{array}{c} CH_3 \diagdown \\ \diagup \\ CH_3 \end{array} C=CH_2$$

The terms *cis* and *trans* signify, respectively, that the two hydrogen atoms are on the same side or on the opposite side with respect to the perpendicular plane of the double bond.

There are little or no olefins in crude oil or "straight run" (direct from crude distillation) products but they are found in refining products, particularly in the fractions coming from conversion of heavy fractions whether or not these processes are thermal or catalytic. The first few compounds of this family are very important raw materials for the petrochemical industry: e.g., ethylene, propylene, and butenes.

Table 1.4 gives the physical properties of selected olefins.

1.1.1.5 **Other Hydrocarbons**

Normally absent or in trace amounts in crude oil, products of conversion processes such as diolefins, acetylenes, etc., are encountered. Table 1.4 gives the physical properties of some of them. Noteworthy is 1-3 butadiene:

$$CH_2=CH-CH=CH_2 \quad \text{as well as isoprene:} \quad CH_2=C(CH_3)-CH=CH_2$$

the basic monomers for a number of polymers.

1.1.2 **Non-Hydrocarbon Compounds**

Molecules in this category contain atoms other that carbon and hydrogen. The distinction is made between organic and organometallic compounds.

	Overall formula	Structural* formula	Molecular weight	Boiling. point, °C (1 atm)	Specific gravity d_4^{15} (liquid)
Ethylene	C_2H_4	C=C	28.0	−103.7	
Propylene	C_3H_6	C−C=C	42.1	−47.7	0.523
1-Butene	C_4H_8	C=C−C−C	56.1	−6.3	0.601

Cis 2-butene	C_4H_8	C C
		  \ /
		   C=C
			56.1	+3.7	0.627

Trans 2-butene	C_4H_8	C
		  \
		   C=C
		        \
		         C
			56.1	0.8	0.610

Isobutene	C_4H_8	C−C=C
		  |
		  C
			56.1	−6.9	0.601

1-Pentene	C_5H_{10}	C=C−C−C−C	70.1	+30.0	0.646
1,3-Butadiene	C_4H_6	C=C−C=C	54.1	−4.4	0.627

Isoprene	C_5H_8	C=C−C=C
		  |
		  C
			68.1	34.1	0.686

Cyclopentadiene	C_5H_6	C−C // \\\\ C C \ / C	66.0	40.0	

Table 1.4	*Physical constants of selected unsaturated hydrocarbons.* * The hydrogen atoms have been omitted to simplify illustration.

1.1.2.1 Heteroatomic Organic Compounds

Sulfur Compounds

Sulfur is the heteroatom most frequently found in crude oils (see Table 1.5). Sulfur concentrations can range from 0.1 to more than 8 weight percent; moreover, this content is correlated with the gravity of the crude oil and, therefore, its quality (light or heavy).

Sulfur might be present in inorganic forms: elemental S, hydrogen sulfide H_2S, carbonyl sulfide COS, or positioned within organic molecules as in the following:

- sulfides, where sulfur is positioned as part of a saturated chain:

$$CH_3-CH_2-CH_2-S-(CH_2)_4-CH_3 \text{ (propyl pentyl sulfide)}$$

  or as cyclic sulfides having 4 or 5 carbon atoms in a ring

Crude oil	Origin	Visc. mm^2/s	Asph. wt. %	O wt. %	N wt. %	S wt. %	Ni ppm	V ppm
Batiraman	Turkey	1180	22.1	0.53	0.49	7.04	99	153
Boscan	Venezuela	595	14.1	0.79	0.74	5.46	125	1220
Lacq. sup.	France	81.7	13.2	0.57	0.42	4.94	19	29
Chauvin Source	Canada	28	6.0	0.48	0.66	2.80	35	67
Bellshill Lake	Canada	7.9	2.2	0.34	< 0.3	1.97	11	18
Emeraude	Congo	113	1.7	1.10	0.65	0.57	64	9
Anguille	Gabon	14.1	1.2	0.92	0.26	0.82	115	14
Duri	Sumatra	51	0.7	0.65	0.47	< 0.10	39	1.5
Pematang	Sumatra	10.2	0.1	0.51	0.26	< 0.10	15	0.6
Edjeleh	Algeria	5.3	0.1	0.73	0.34	< 0.10	1.5	2.3
Hassi Messaoud	Algeria	2.32	0.1	1.93	0.38	< 0.10	< 0.2	< 0.2

| Table 1.5 | *Some characteristics of 250°C$^+$ fractions.* |

- disulfides having the general formula $R-S-S-R'$, often present in light fractions
- thiols or mercaptans, $C_nH_{2n+1}SH$ found in low boiling fractions wherein the hydrogen bonded to the sulfur has acid characteristics
- thiophenes and their derivatives, often present in fractions boiling over 250°C, constitute an important group of sulfur compounds. The sulfur atom is positioned in the aromatic rings.

Examples:

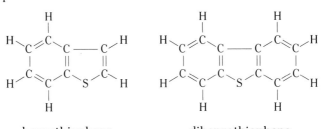

benzothiophene   dibenzothiophene

A knowledge of these compounds is important because they often have undesirable attributes, e.g., unpleasant odor, the SO_2 formed by combustion, catalyst poisoning. There are a number of refining processes to eliminate sulfur compounds.

Oxygen Compounds

Crude oils generally contain less oxygen than sulfur (Table 1.5). Even though it is not abundant, oxygen can play a consequential role; in particular, it is responsible for petroleum acidity. Oxygen is found in the following compounds:

- Phenols formed by the substitution of a hydrogen atom by a hydroxyl group, OH, in an aromatic ring:

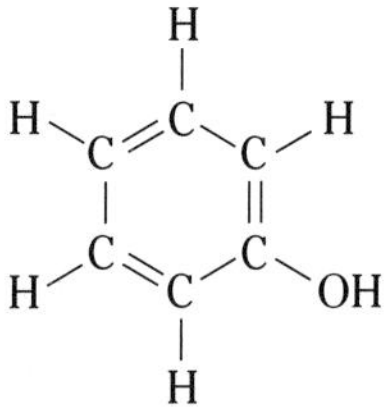

- Furanes and benzofuranes in which an oxygenated ring is condensed into one or more aromatic rings.

  Example: dibenzofurane

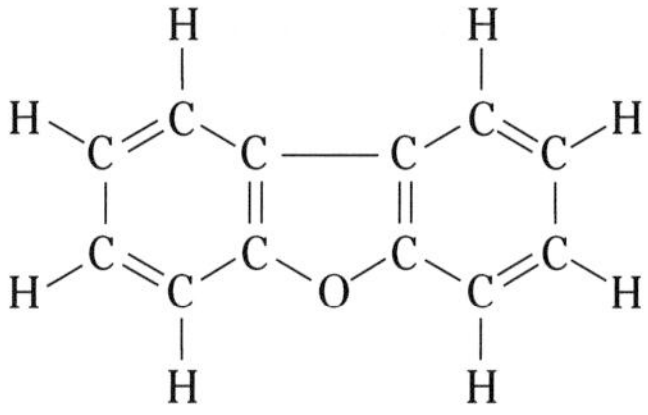

- Carboxylic acids: $R-COOH$

  where R is an alkyl radical, an aromatic ring, or a saturated ring. In this case, they are naphthenic acids for which the carboxylic group is bonded to either a cyclopentane or cyclohexane molecule. Abundant in certain crude oils, naphthenic acids cause corrosion problems.

- Esters: $R-COO-R'$

  where R and R' are alkyl radicals or aromatics.

Nitrogen Compounds

In crude oil, nitrogen is found mostly in fractions boiling over 250°C and is particularly concentrated in resins and asphaltenes. Nitrogen takes the following forms:

- In saturated or aromatic amides:

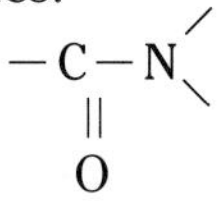

- As amines:

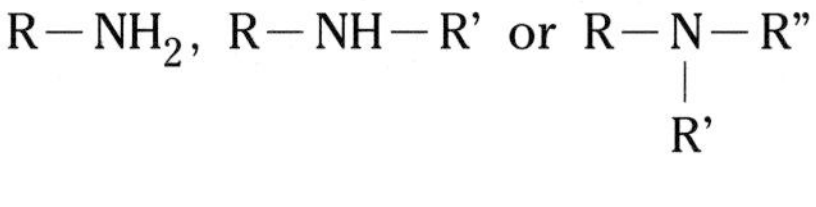

$$R-NH_2, \quad R-NH-R' \quad or \quad R-N-R''$$
$$\qquad\qquad\qquad\qquad\qquad\qquad |$$
$$\qquad\qquad\qquad\qquad\qquad\qquad R'$$

- As carbazoles, where a ring containing nitrogen is condensed with one or more aromatic rings, forming neutral compounds.

  For example, the compound dibenzopyrrole:

- As pyridines, with basic nitrogen:

  Nitrogen is incorporated in a hexagonal ring having three double bonds. The compounds in this family are those which can give a basic character to petroleum products and are thus a poison to acid catalysts.

  Examples:

2-methyl pyridine   acridine

Following certain refining processes like catalytic cracking, sizeable amounts of nitrogen can appear in light cuts and cause quality problems such as instability in storage, brown color, and gums.

1.1.2.2 Organometallic Compounds

In the heaviest fractions such as resins and asphaltenes (see article 1.2), metal atoms such as nickel and vanadium are found. They belong in part to molecules in the porphyrine family where the basic pattern is represented by four pyrrolic rings, the metal being at the center of this complex in the form Ni^{++} or VO^{+}:

1.2 **Compounds whose Chemistry is Incompletely Defined**

When it comes to the heaviest of petroleum fractions, modern analytical methods are not able to isolate and characterize the molecules completely. In the absence of something better, the analyst separates the heavy fractions into different categories, which leads merely to definitions that are workable but are no longer in terms of exact structure.

1.2.1 **Asphaltenes**

Asphaltenes are obtained in the laboratory by precipitation in normal heptane. Refer to the separation flow diagram in Figure 1.2. They comprise an accumulation of condensed polynuclear aromatic layers linked by saturated chains. A folding of the construction shows the aromatic layers to be in piles, whose cohesion is attributed to π electrons from double bonds of the benzene ring. These are shiny black solids whose molecular weight can vary from 1000 to 100,000.

The portion that is soluble in normal heptane is given the term maltenes.

According to the nature of the solvent employed, the yields and constitutions of the asphaltenes are different. In the United States, asphaltenes are obtained by precipitation from normal pentane.

In industry, the elimination of asphaltenes from oil involves using propane or butane. The utilization of a lighter paraffin results in the heavier paraffins precipitating along with the asphaltenes thereby diminishing their aromatic character. The oil removed from its asphaltene fraction is known as deasphalted oil or DAO. The precipitated portion is called asphalt.

Asphaltenes have high concentrations of heteroelements: sulfur, nitrogen, nickel and vanadium. Their content varies widely in petroleum oils (Table 1.5). They cause a number of problems throughout the petroleum industry.

In oil bearing formations, the presence of polar chemical functions of asphaltenes probably makes the rock wettable to hydrocarbons and limits their production. It also happens that during production, asphaltenes precipitate, blocking the tubing. The asphaltenes are partly responsible for the high viscosity and specific gravity of heavy crudes, leading to transport problems.

The refining industry generally seeks either to eliminate asphaltenes or to convert them to lighter materials because the presence of heteroelements cause pollution problems, e.g., sulfur and nitrogen, catalyst poisoning, and corrosion (formation of metal vanadates during combustion).

All the problems briefly described above justify the large effort to characterize asphaltenes by techniques seldom found elsewhere in the petroleum industry. One of these is to analyze asphaltenes by steric exclusion chromatography, a widespread technique for studying polymers. In this method, the components are eluted by a solvent travelling through a column having microporous packing. According to whether the particles to be separated have larger or smaller sizes than those of the micropores, they can or can not penetrate into the micropores, leading to a process of separation by size (in reality, by hydrodynamic volume), the largest particles are the first to appear. This chromatographic method is then utilized to characterize asphaltenes by determining their mass distribution, the system generally being calibrated using polymer standards. One observes that during the visbreaking process, the size and the polydispersity of asphaltenes is greatly reduced.

Because they contain many islets of condensed aromatics, the carbon-rich asphaltenes can begin to acquire the spatial organization of graphite layers.

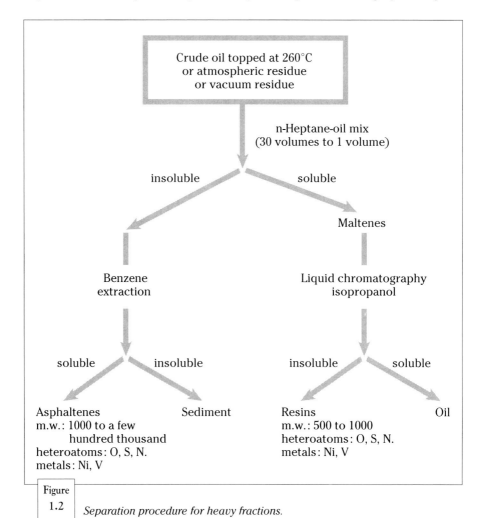

Figure 1.2   *Separation procedure for heavy fractions.*

This is the reason for numerous studies that have been conducted with X-ray diffraction, a method not described here. Suffice it to say that X-ray diffraction is useful in providing values of aromaticity, the distances between layers and between aliphatic chains, the thickness of particles, etc.

Other techniques such as X-ray diffusion or small angle neutron diffusion are also used in attempts to describe the size and form of asphaltenes in crude oil. It is generally believed that asphaltenes have the approximate form of very flat ellipsoids whose thicknesses are on the order of one nanometer and diameters of several dozen nanometers.

1.2.2 **Resins**

If maltenes are subjected to liquid chromatography (see 2.1.2.4) the components eluted by the more polar solvents are called resins. Their composition, once again, depends on the procedure used.

Resins are generally molecules having aromatic characteristics and contain heteroatoms (N, O, S, occasionally Ni and V); their molecular weight ranges from 500 to 1000.

In English language publications, resins are frequently called "polar compounds" or "N, S, O compounds".

Fractionation and Elemental Analysis of Crude Oils and Petroleum Cuts

Jean-Claude Roussel

Raymond Boulet

Crude oils form a continuum of chemical species from gas to the heaviest components made up of asphaltenes; it is evidently out of the question, given the complexity of the mixtures, to analyze them completely. In this chapter we will introduce the techniques of fractionation used in the characterization of petroleum as well as the techniques of elemental analysis applied to the fractions obtained.

Because of their diversity and complexity as well as the gradual internationalization of the different standards, it has proven necessary to standardize the methods of sample preservation, handling, fractionation, and analysis throughout the chain of separation and treatment. All these stages are the object of precise protocols established by official national and international organizations. They describe in as minute detail as possible the procedures employed not only for each analysis but very often giving different procedures for the same analysis in different matrices. These are the standards or standardized methods discussed in Chapter 7.

2.1 Preparatory and Analytical Fractionations

2.1.1 Distillation

One distinguishes preparatory distillations that are designed to separate the fractions for subsequent analysis from non-preparatory analytical distillations that are performed to characterize the feed itself. For example, the distillation curve that gives the recovered volume or weight as a function of the distillation temperature characterizes the volatility of the sample.

Along the same lines, a distillation can be simulated by gas phase chromatography. As in a refinery, distillation in the laboratory is very often the first step to be carried out, because it gives the yields in different cuts: gasoline, kerosene, etc., and makes further characterization of the cuts possible.

2.1.1.1 **Preparatory Laboratory Distillation**

This is the ASTM D 2892 test method and corresponds to a laboratory technique defined for a distillation column having 15 to 18 theoretical plates and operating with a 5:1 reflux ratio. The test is commonly known as the TBP for *True Boiling Point*.

The procedure applies to stabilized, i.e., debutanized, crudes, but can be applied to any petroleum mixture with the exception of liquefied petroleum gas, very light naphtha, and those fractions having boiling points over 400°C.

The sample can vary from 0.5 to 30 liters enabling the following:

• recovery of liquefied gas, distillation cuts and residue
• determination of both weight and volume yields for the cuts
• generation of a distillation curve giving temperature as a function of weight or volume recovered.

The apparatus can be used for either atmospheric distillations or low pressure distillations as low as 2 mm of mercury (0.266 kPa).

Beyond 340°C in the reboiler, the residue begins to crack thermally. If the distillation is stopped at this point, the residue is called the atmospheric residue. In order to continue, the distillation is conducted under a low pressure, "vacuum", so as to reduce the temperature in the reboiler.

The distillation continues to a boiling point corresponding an atmospheric boiling point of about 535°C. The remaining material is called *vacuum residue*.

Commercial equipment is available which automatically switches from atmospheric distillation to vacuum distillation and calculates the distillation curve as temperatures under atmospheric pressure conditions as a function of weight or volume per cent recovery.

2.1.1.2 **Non-Preparatory Distillation**

Material having a boiling range from 0 to 400°C

Corresponding to ASTM D 86 (NF M 07-002), this method applies to gasolines, kerosenes, heating oils, and similar petroleum products.

Sample size is 100 ml and distillation conditions are specified according to the type of sample. Temperature and volume of condensate are taken simultaneously and the test results are calculated and reported as boiling temperature as a function of the volume recovered as shown in Table 2.1.

Distilled volume, %	Temperature, °C
IP	33.0
5	42.5
10	45.5
20	50.0
30	55.0
40	62.5
50	75.0
60	99.0
70	131.0
80	151.5
90	167.5
95	177.5
EP	183.5
% distillate	97.8
% residue	0.8
% losses	1.4

Table 2.1

Typical results for an ASTM D 86 distillation of a gasoline.

Heavy Fractions

As stated above for the TBP distillation, petroleum cannot be heated above 340°C without its molecules starting to crack. Because of this, analytical distillation of heavy fractions is done according to the ASTM D 1160 method for petroleum materials that can be partially or completely vaporized at a maximum temperature of 400°C at pressures from 50 to 1 mm of mercury (6.55 to 0.133 kPa).

The sample is distilled at predetermined and precisely controlled temperatures under conditions that give a fractionation equivalent to about one theoretical plate.

The results are presented as a distillation curve showing the boiling temperature (corrected to atmospheric pressure) as a function of the distilled volume.

2.1.1.3 **Distillation Simulated by Gas Chromatography**

Gas phase chromatography is a separation method in which the molecules are split between a stationary phase, a heavy solvent, and a mobile gas phase called the carrier gas. The separation takes place in a column containing the heavy solvent which can have the following forms:

- either impregnated on an inert support having a specific surface of 1 to 10 m² per gram, in the form of a powder whose particle size ranges from 100 to 200 μm. These are called packed columns; their length is on the order of a few meters and their diameter is from 2.5 to 4 mm

- or a tube wall coating covering the inside surface of a tube of a few dozen meters in length and having a diameter of from 0.1 to 0.5 mm. These are called capillary columns. Another similar technique is to use a quartz tube for which the stationary phase, often a polysiloxane, is cross-linked onto the inside surface.

Used in virtually all organic chemistry analytical laboratories, gas chromatography has a powerful separation capacity. Using distillation as an analogy, the number of theoretical plates would vary from 100 for packed columns to 10^5 for 100-meter capillary columns as shown in Figure 2.1.

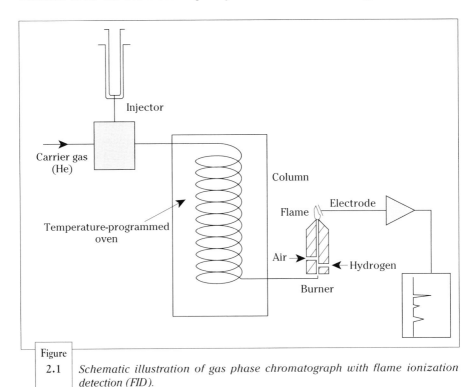

| Figure 2.1 | *Schematic illustration of gas phase chromatograph with flame ionization detection (FID).* |

The column is swept continuously by a carrier gas such as helium, hydrogen, nitrogen or argon. The sample is injected into the head of the column where it is vaporized and picked up by the carrier gas. In packed columns, the injected volume is on the order of a microliter, whereas in a capillary column a flow divider (split) is installed at the head of the column and only a tiny fraction of the volume injected, about one per cent, is carried into the column. The different components migrate through the length of the column by a continuous succession of equilibria between the stationary and mobile phases. The components are held up by their attraction for the stationary phase and their vaporization temperatures.

The large number of stationary phases can be classified in two groups:

- Apolar stationary phases having no dipolar moments, that is their center of gravities of their positive and negative electric charges coincide. With this type of compound, the components elute as a function of their increasing boiling points. The time difference between the moment of injection and the moment the component leaves the column is called the retention time.

- Polar stationary phases which have a polar moment. These phases interact with the dipolar moments of polar components themselves and those components capable of induced polarization such as aromatics.

There is a detector at the column outlet. Of the many varieties, the two most widely used are described below:

- Thermal conductivity or katharometric detectors have two platinum wires through which a constant electric current passes. One of these wires is swept by carrier gas only and the other is swept by the column effluent carrier gas plus the eluted components. The two wires are integrated into two of the branches in a Wheatstone bridge circuit at steady state. As a new component appears, the thermal conductivity of the environment surrounding the wire changes, as does its temperature and, as a result, its resistance. It is the imbalance thus created in the Wheatstone bridge circuit that is measured as the chromatographic signal. The katharometer is not very sensitive and is seldom used in capillary columns where the mass of the eluted components is very small. Furthermore, owing to its principle of operation, its response varies with the thermal conductivity of the solute: it is not, then, a universal detector and it needs to be calibrated.

- A Flame Ionization Detector or FID operates on the following principle. At the column outlet, the carrier gas and the eluted components pass through a burner fed by hydrogen and air. When a component arrives, its combustion produces ions that are collected by an electrode positioned near the flame. The resulting direct current is amplified and constitutes the chromatographic signal. The detector is very sensitive and is well-adapted for use with capillary columns; its response is approximately proportional to the weight of carbon present in the solute, a quality that greatly simplifies quantitative analysis. Refer to Figure 2.1.

The simulated distillation method uses gas phase chromatography in conjunction with an apolar column, that is, a column where the elution of components is a function of their boiling points. The column temperature is increased at reproducible rate (programed temperature) and the area of the chromatogram is recorded as a function of elution time.

A correlation between retention times and boiling points is established by calibration with a known mixture of hydrocarbons, usually normal paraffins, whose boiling points are known (see Figure 2.2). From this information, the distribution of boiling points of the sample mixture is obtained.

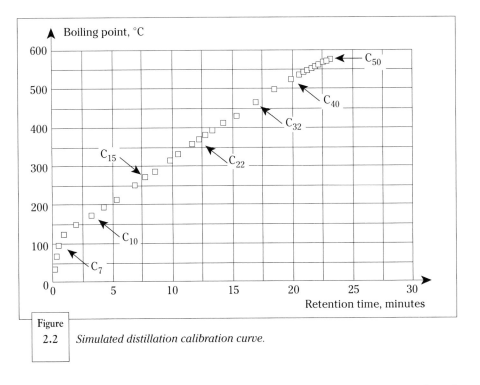

Figure 2.2 *Simulated distillation calibration curve.*

Figure 2.3 shows a chromatogram for a vacuum gas oil simulated distillation curve. The ordinate axis, in this case the retention time, is related to a temperature scale using a calibration curve. The chromatogram is divided into horizontal bands corresponding the desired separations. The detector used here is a flame ionization detector so that area percentages correspond directly to weight percentages.

The advantages of this method are its rapidity (maximum duration, 70 min), the small quantity of sample required (1 µl), and its ease of automation.

Two ASTM methods described below are based on this technique:

- D 2887, applies to products and petroleum fractions whose final boiling points are equal to or below 538°C (1000°F), and have boiling points above 38°C (100°F). The results obtained are equivalent to those obtained from the TBP distillation, ASTM D 2892.

- D 3710, applies to products and petroleum fractions whose final boiling points are equal to or less than 260°C (500°F).

In practice, simulated distillation by gas phase chromatography is used for the following objectives:

- either as a guide to control a TBP distillation by evaluating the recovered volumes of each fraction

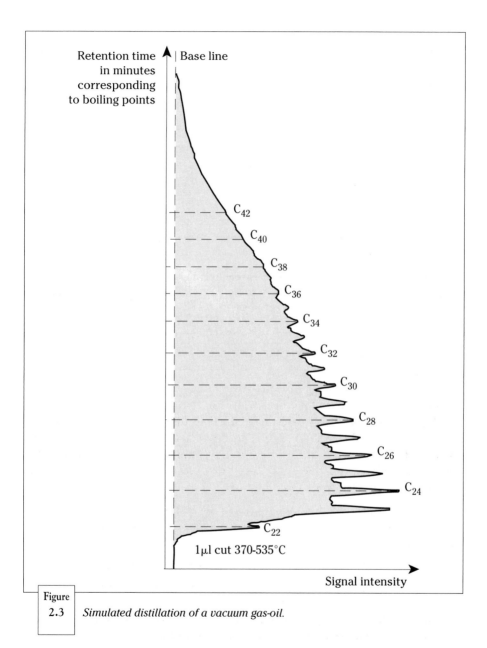

Retention time in minutes corresponding to boiling points

| Base line

C_{42}
C_{40}
C_{38}
C_{36}
C_{34}
C_{32}
C_{30}
C_{28}
C_{26}
C_{24}
C_{22}

1 μl cut 370-535°C

Signal intensity

Figure
2.3   *Simulated distillation of a vacuum gas-oil.*

• or for a very quick estimation of the yields of the light fractions from conversion processes.

Finally, other methods are used to obtain simulated distillation by gas phase chromatography for atmospheric or vacuum residues. For these cases, some of the sample components can not elute and an internal standard is added to the sample in order to obtain this quantity with precision.

An example of the good correlation between TBP and simulated distillation is given in Figure 2.4, where it is shown that 71% of a Kuwait crude distils below 535°C.

Recently, chromatographs and their associated columns have been able to elute components with boiling points up to 700°C under atmospheric pressure.

Backflushing techniques, well-known to analysts, are also extremely valuable tools.

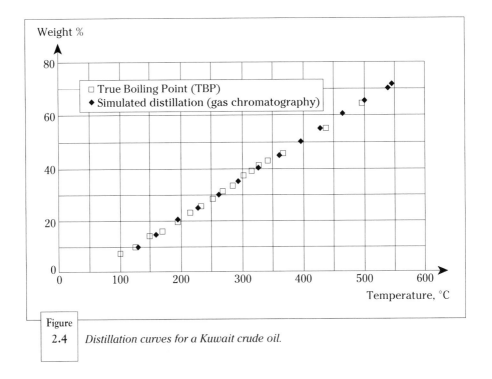

Figure
2.4   Distillation curves for a Kuwait crude oil.

2.1.2 **Other Separations**

In order to simplify the analysis of petroleum and its fractions, other preliminary separation techniques are employed, aiming generally to separate certain classes of components.

2.1.2.1 **Solvent Extraction**

This technique is based on the selectivity of a solvent for different families or individual components in a mixture. Solvent extraction can be either analytical or preparatory in function.

The powerful solvent properties of dimethylsulfoxide (DMSO), for example, are used to dissolve selectively the polynuclear aromatics found in oils and paraffins. The procedure is shown in Figure 2.5.

These methods can be written into standards such as the following:

- analysis of polynuclear aromatics in petroleum oils (IP 346)
- analysis of aromatics in products for human consumption, as used by the Food and Drug Administration, etc.

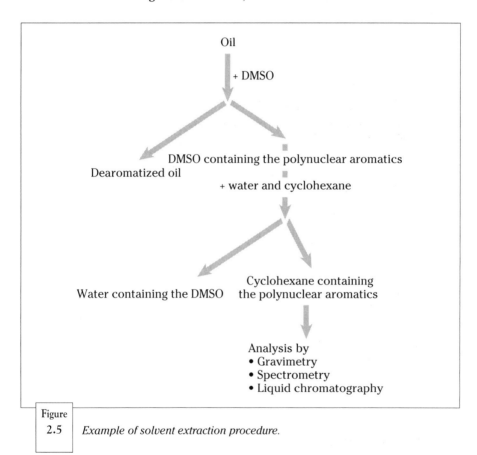

| Figure 2.5 | *Example of solvent extraction procedure.* |

2.1.2.2 **Precipitation**

This technique is used to quantify one or more components in a mixture, i.e., extracting them from mixtures to facilitate their final analysis. An example is that for the asphaltenes, already described in the definition of these components in article 1.2.1.

2.1.2.3 **Subtractive Methods for a Chemical Family**

These methods are grouped as either physical or chemical processes.

For physical processes, two examples are the elimination of normal paraffins from a mixture by their adsorption on 5 Å molecular sieves or by their selective formation of solids with urea (clathrates)

For chemical processes, some examples are the elimination of aromatics by sulfonation, the elimination of olefins by bromine addition on the double bond (bromine number), the elimination of conjugated diolefins as in the case of the maleic anhydride value (MAV), and the extraction of bases or acids by contact with aqueous acidic or basic solutions.

2.1.2.4 **Liquid Chromatography**

Liquid chromatography is a separation technique based on the selective adsorption on a solid, silica or alumina for example, or a mixture of the two, of the different components of a liquid mixture.

Liquid chromatography, having a resolving power generally less than that of gas phase chromatography, is often employed when the latter cannot be used, as in the case of samples containing heat-sensitive or low vapor-pressure compounds.

The field of application for liquid chromatography in the petroleum world is vast: separation of diesel fuel by chemical families, separation of distillation residues (see Tables 3.4 and 3.5), separation of polynuclear aromatics, and separation of certain basic nitrogen derivatives. Some examples are given later in this section.

The adsorbent, the stationary phase, fills a column of a few decimeters in length and 5 to 10 mm in diameter. The column is swept continually by a solvent or mixture of solvents (the liquid phase).

The mixture to be studied is injected by syringe into the head of the column and the molecules comprising the mixture are adsorbed in varying degrees by the stationary phase and desorbed by the liquid phase. At the end of this succession of equilibria, the components of the mixture, more or less separated from each other, leave the column with the solvent.

The higher the eluting force of the solvent, the more easily the components are desorbed. This force is represented by the solvent's polarity which results in mainly three types of interaction between the solvent and the solute:

- dispersive interactions due to London forces which are particularly strong in polarizable components such as aromatics
- dipole-dipole interactions, the stronger the components' dipolar moments, the stronger are these interactions
- hydrogen bond interactions.

Taken together, these solvent-solute interactions make up the solvent polarity, which is represented well by Hildebrand's solubility parameter (1950).

Added to these interactions are the electrostatic forces related to the dielectric constants and which are important when it is necessary to separate ionic components.

It follows that, when it is required to elute all the components with reasonable volumes of solvent, it is usually necessary to change the solvent in order to displace the adsorption equilibrium. For example, one can begin the separation with heptane as the solvent, then, gradually, toluene and finally going to solvents with increasingly higher polarities such as methylene chloride and methanol. It is common practice to change the solvent composition gradually rather than suddenly. This is called the method of solvent gradients.

At the column outlet, either of two detection methods are employed:

- physical measurement such as refractive index where the pure solvent index is compared with that of the solvent containing the solute, using a differential refractometer
- by light absorption or fluorescence at certain characteristic wavelengths, mainly in the ultraviolet spectrum.

Note that in liquid phase chromatography there are no detectors that are both sensitive and universal, that is, which respond linearly to solute concentration regardless of its chemical nature. In fact, the refractometer detects all solutes but it is not very sensitive; its response depends evidently on the difference in refractive indices between solvent and solute whereas absorption and UV fluorescence methods respond only to aromatics, an advantage in numerous applications. Unfortunately, their coefficient of response (in ultraviolet, absorptivity is the term used) is highly variable among individual components.

Some techniques have been described that are based on the concept of flame ionization used in gas chromatography. The results are generally unsatisfactory because it is necessary to evaporate the solvent prior to introducing the mixture into the detector.

Liquid phase chromatography can use a supercritical fluid as an eluent. The solvent evaporates on leaving the column and allows detection by FID. At present, there are few instances in the petroleum industry using the supercritical fluid technique.

2.2 **Elemental Analysis**

The determination of the elemental composition of a petroleum cut is of prime importance because it provides a quick means of finding out the quality of a given cut or determining the efficiency of a refining process. In fact, the quality of a cut generally increases with the H/C ratio and in all cases, with a decrease in hetero-element (nitrogen, sulfur, and metals) content.

2.2.1 **Some Definitions**

In this section, frequent mention will be made of the following terms:

- **sensitivity:** the ratio of the change in measured value corresponding to the concentration of the element to be analyzed
- **detectable limit:** (or threshold of sensitivity) the lowest detectable value of the variable for which the method can confirm that the value is not zero. It is considered as equal to two or three times the standard deviation obtained for a solution in which the concentration is higher than the detectable value but close to the values of concentration of the blank test solution. See the French standard NF X 07001
- **accuracy (or error):** the closeness of agreement between result of a measurement and the true value (by convention) of the magnitude of the measured quantity.

2.2.2 **Sampling**

The complexity of petroleum products raises the question of sample validity; is the sample representative of the total flow? The problem becomes that much more difficult when dealing with samples of heavy materials or samples coming from separations. The diverse chemical families in a petroleum cut can have very different physical characteristics and the homogeneous nature of the cut is often due to the delicate equilibrium between its components. The equilibrium can be upset by extraction or by addition of certain materials as in the case of the precipitation of asphaltenes by light paraffins.

Before withdrawing a sample it is necessary to agitate it, even if it is a gas, and eventually heat the sample being careful to stay below temperatures which could cause evaporation of the lighter components.

If agitation and heating are not practical as in the cases of large volumes, it is better to withdraw samples from various levels in order to get an average sample.

The procedures are described in the standards ASTM D 270, D 4057 and NF M 07-001.

2.2.3 **Analysis of Carbon, Hydrogen and Nitrogen**

The analysis of carbon, hydrogen and nitrogen is most often based on a method of combustion drawn from the Dumas method. About 1 mg of the sample is placed in an oven heated to $1050°C$ and purged by a mixture of helium and oxygen. The combustion products pass over an oxidizing agent such as Cr_2O_3 or Co_3O_4 where they are converted to CO_2, H_2O, and nitrogen oxides, NO_x, and then carried off by an inert gas. The mixture passes across copper heated to $650°C$ which reduces the NO_x to N_2. The gases are separated in a gas chromatograph column equipped with a katharometer.

Other methods for analyzing combustion products can be substituted for chromatography. Gravimetry can be used, for example, after a series of absorption on different beds, as in the case of water absorption in magnesium perchlorate or CO_2 in soda lime; infra-red spectrometry can be used for the detection of CO_2 and water.

The acceptable limits of this kind of analysis are shown in Table 2.2.

	C	**H**	**N**
Range, weight %	$0.2 - 100$	$0.2 - 100$	$0.2 - 100$
Error	± 0.2	± 0.1	± 0.1

Table 2.2   *Elemental analysis of carbon, hydrogen, and nitrogen.*

A few remarks concerning this table are necessary:

- With regard to hydrogen, the accuracy is deemed insufficient for obtaining the hydrogen balance in a refining process. A rather cumbersome answer used at times is to determine the content by "macro" analysis. The hydrogen content in approximately one gram of sample is calculated by weighing the water formed. More recently, a totally different technique has appeared, hydrogen analysis by nuclear magnetic resonance. Refer to article 3.2.2.2. The order of magnitude of the absolute error is 0.05%.

- For nitrogen, the sensitivity threshold of the Dumas method is largely insufficient. The feedstock specifications for process units and the evaluation of hydrodenitrification processes require measurements covering a range of 0.5 to 2000 ppm weight. For this concentration range, the following methods can be applied:

  - Kjeldahl Method

    By means of an inorganic reaction with H_2SO_4 in the presence of a catalyst, the nitrogen components are converted to $(NH_4)_2SO_4$. The ammonia is displaced by a strong base and analyzed. By increasing the sample size, it is possible to analyze for as low as 1 ppm of nitrogen. The method is long and laborious but remains the method of reference in spite of its tedious procedure.

  - Detection by Chemiluminescence

    The sample is burned in oxygen at 1000°C. Nitrogen oxide, NO, is formed and transformed into NO_2 by ozone, the NO_2 thus formed being in an excited state NO_2^*. The return to the normal state of the molecule is accompanied by the emission of photons which are detected by photometry. This type of apparatus is very common today and is capable of reaching detectable limits of about 0.5 ppm.

– Reduction

The sample is reduced in a hydrogen stream at $800°C$ in the presence of a nickel catalyst. The ammonia formed is detected by coulometry and the test sensitivity is on the order of one part per million.

In conclusion, it is important to understand that the various methods of analysis for trace quantities of nitrogen are not entirely satisfactory. The determination of nitrogen content suffers from mediocre accuracy and inadequate agreement between the different methods. The reason commonly given for these equivocal results is the resistance to combustion — or reduction — of certain nitrogen bonds

$$= CH - N \overset{/}{\underset{\backslash}{}} \text{ and heterocyclics } -N = N -$$

2.2.4 **Oxygen Analysis**

Oxygen is present only in small quantities in petroleum as illustrated in Table 1.5, and its concentration is usually determined by subtracting the combined carbon, hydrogen, and nitrogen total from 100.

Incidentally, numerous petroleum products, particularly those coming from conversion processes, are unstable with respect to oxidation and oxygen analysis is meaningful only if great precautions are taken during sample withdrawal and storage.

The analytical method employed was described by Unterzaucher in 1940.

The material to be analyzed is pyrolyzed in an inert gas at $1100°C$ in the presence of carbon; the carbon monoxide formed, if any, is either analyzed directly by chromatography or analyzed as carbon dioxide after oxidation by CuO. The CO_2 is detected by infra-red spectrometry or by gas phase chromatography.

The carbon is often replaced by a catalyst such as platinized or nickelized carbon or a mixture of carbon, platinum and nickel.

The presence of sulfur, nitrogen, halogens, etc. can interfere with the test. After pyrolysis, it is necessary to eliminate the following components:

• acid gases, by absorption in soda lime
• sulfur compounds, either by cold separation with liquid nitrogen, or by fixation on reduced copper at $900°C$.

This method can measure concentrations on the order of 5000 ppm.

For trace quantities of less than 100 ppm, the most successful method — and the most costly — is neutron activation. The sample is subjected to neutron bombardment in an accelerator where oxygen 16 is converted to unstable nitrogen 16 having a half-life of seven seconds. This is accompanied by emission of β and γ rays which are detected and measured. Oxygen concentrations as low as 10 ppm can be detected. At such levels, the problem is to find an acceptable blank sample.

2.2.5 **Sulfur Analysis**

Knowledge of sulfur content in petroleum products is imperative; the analytical methods are numerous and depend on both the concentration being measured and the material being analyzed.

The methods most generally used are combustion and X-ray fluorescence.

2.2.5.1 **Sulfur Analysis by Combustion**

All these methods begin with combustion of the sample resulting in the sulfur being oxidized to SO_2 and SO_3. Table 2.3 summarizes the different analytical methods with references to the corresponding standards.

	Method	ASTM	NF	Range	Material
Combustion	Coulometry		M 07-052	100 ppm − 100%	All
	Microcoulom.	D 3120		3 −100 ppm	26° < BP < 274°C
	Wickbold	D 2774	T 60-142	1 ppm − 5%	All
	Induction furnace	D 1552	M 07-025	0.05 − 100%	BP > 175°C
	Lamp	D 1266	M 07-031	0.01 − 0.4%	
	Quartz tube		T 60-108	0.05 − 100%	BP > 175°C
	Bomb	D 129	T 60-109	0.1 − 100%	Non-volatile
	UV fluorescence		M 07-059	0.5 ppm − a few %	Liquid
	Hydrogenolysis	D 4045		0.02 − 10 ppm	30° < BP < 371°C
	Non-disp. XF	D 4294	M 07-053	0.01− 5%	

Table 2.3	*Analytical methods for sulfur content.*

A distinction must be made between the different methods with respect to the temperature attained during combustion because the reaction:

$$SO_2 + 1/2\ O_2 \longrightarrow SO_3 \qquad \Delta H^0_r = -95,5 \text{ kJ/mol.}$$

does not favor conversion to SO_3 at high temperatures. Above 1200°C, only SO_2 is stable. The table below gives equilibrium constants from the equation:

$$K_p = \frac{P_{SO_3}}{P_{SO_2^*}\sqrt{P_{O_2}}}$$

in the temperature zone most frequently encountered in commercial equipment.

$T\ °C$	900	950	1000	1100
K_p	1.88	1.20	0.60	−0.5

Methods Giving only SO$_2$

These are called high temperature induction furnace methods which differ only as to the kind of furnace used and employ the same ASTM procedure. The sample is heated to over 1300°C in an oxygen stream and transformed to SO$_2$ which is analyzed with an infra-red detector.

Methods Giving SO$_2$ and SO$_3$ but that Measure Total Sulfur

In these methods, the sulfur oxides produced during combustion are, before detection, either converted into sulfuric acid by bubbling in a hydrogen peroxide-water solution or converted into sulfates.

- Wickbold Method: the sample passes into a hydrogen-oxygen flame having a large excess of oxygen. The resulting sulfur oxides are converted to sulfuric acid by contact with hydrogen peroxide solution.

- Bomb Method: the sample is burned in a bomb under oxygen pressures of 30 bar. The sulfur contained in the wash water is analyzed via gravimetry as barium sulfate.

- Lamp Method: the sample is burned in a closed system in an atmosphere of 70% CO$_2$ and 30% oxygen in order to avoid formation of nitrogen oxides. This method was to have been abandoned as it takes three hours to carry out, but remains officially required for jet fuel sulfur analysis.

- Quartz Tube Method: the sample is burned in a quartz tube and a stream of purified air carries the combustion gases into a hydrogen peroxide solution.

For these different methods, the detection is either by volume measurement, gravimetry, conductimetry or coulometry.

Methods Measuring SO$_2$ Only

The sample is pyrolyzed in an 80/20 mixture of oxygen and nitrogen at from 1050 to 1100°C; the combustion gases are analyzed by iodine titration or by UV fluorescence. Up to 20% of the sulfur can escape analysis, however.

2.2.5.2 **Sulfur Analysis by Hydrogenolysis**

Another method which should be cited apart from the others is to pyrolyze the sample in a hydrogen atmosphere. The sulfur is converted to H$_2$S which darkens lead-acetate-impregnated paper. The speed of darkening, measured by an optical device, provides the concentration measurement. This method attains sensitivity thresholds of 0.02 ppm.

In all these methods, the accuracy depends on the sulfur concentration; in relative value, the error is estimated to be about 0.1%.

2.2.5.3 **Sulfur Analysis by X-Ray Fluorescence**

We will begin by a brief review of the concept of the X-ray fluorescence analytical method widely used in the petroleum industry for studying the whole range of products and for analyzing catalysts as well.

The sample is irradiated by primary X-rays as illustrated in Figure 2.6. Under the effect of the radiation, some electrons orbiting in internal shells are torn away. The "holes" created by the departing electrons are filled by electrons coming from higher levels. This electronic rearrangement is accompanied by the emission of what are called secondary X-rays whose wavelengths are characteristic of the element being bombarded. An element can give emissions of several different wavelengths depending on the electron levels of departure and arrival. For example, a series of K_i rays indicates that the electron transitions end up at the K shell, the index i (α, β, γ) referring respectively to the origins of the transition at 1, 2, or 3 levels above; therefore the emission K_α results from the movement of the electron arriving at the K shell and coming from the L shell.

The spectrum of the secondary emission, that is, the intensity of X-ray radiation as a function of wavelength is established using a crystal analyzer based on Bragg's law.

Sulfur is analyzed on the K_α emission at 5.573 Angstroms.

This method can attain, depending on the sample, concentrations on the order of 10 ppm wt. with an error on the order of 20%.

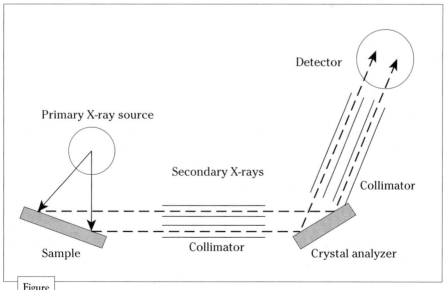

<table>
<tr><td>Figure
2.6</td><td>*Concept of analysis by X-ray fluorescence.*</td></tr>
</table>

The analysis can be performed in either of the following ways:

- by means of a dispersion apparatus of the kind shown in Figure 2.6 for very general use which allows a study of the wavelength and analysis of all elements having an atomic number higher than 11 (sodium)
- or by using less sophisticated equipment, whose X-ray source and monochromator are replaced respectively by a radioactive isotope and the crystal analyser by a filter, isolating the fluorescent wavelength to be analyzed. This type of equipment is generally dedicated for analysis of one or a few elements only, but on the other hand it is simple to use. ASTM D 4294 is the standard employed.

2.2.6 **Analysis of Metals in Petroleum Cuts**

The petroleum industry faces the need to analyze numerous elements which are either naturally present in crude oil as is particularly the case for nickel and vanadium or those elements that are added to petroleum products during refining.

In this section we will discuss only the analytical techniques that are in very general usage without presenting the older chemical methods.

Essentially three methods are used:

- X-ray fluorescence
- atomic absorption
- argon plasma emission.

2.2.6.1 **Using X-Ray Fluorescence for Analysis of Metals**

The method has been described previously. The metals most frequently analyzed are the following:

- nickel Ni, vanadium V, iron Fe, lead Pb as well as metalloids, phosphorus, sulfur, and chlorine in petroleum products
- copper Cu, magnesium Mg, and zinc Zn, in lubricating oil additives.

The detectable limits for a dispersion apparatus are a few $\mu g/g$, and vary according to the environment around from a few $\mu g/g$ for heavy elements in light matrices to a few mg/g for light elements.

2.2.6.2 **Using Atomic Absorption for Analysis of Metals**

We will briefly review the concept of the method whose apparatus is illustrated schematically in Figure 2.7.

The sample to be analyzed can be dissolved in an organic solvent, xylene or methylisobutyl ketone. Generally, for reasons of reproducibility and because of matrix effects (the surroundings affect the droplet size and therefore the effectiveness of the nebulization process), it is preferable to mineralize the sample in H_2SO_4, evaporate it and conduct the test in an aqueous environment.

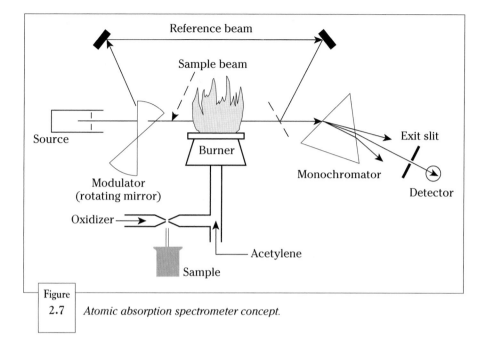

Figure
2.7 | *Atomic absorption spectrometer concept.*

The solution is nebulized, then atomized:

- either in an air/acetylene flame (most often) or in a nitrogen protoxide-acetylene flame (for the most refractory materials)
- or in a graphite furnace for analysis of trace quantities.

In both cases the aerosol is subjected to temperatures exceeding 2300°C. At these temperatures, the molecules or salts are dissociated into their elemental atomic components. The atoms have the capacity to absorb energy carried by the photons provided they have a well-defined frequency that corresponds to the energy needed to cause a peripheral electron to travel from its base energy level to its excited state energy level. This energy level, E_i, is supplied by photons whose frequency, ν_i, is such that $E_i = h\nu_i$. As each atom has the ability to absorb or emit well-defined frequencies, the right kind of energy will be derived from a source made up of the element to be analyzed; hence the terms, "nickel" or "sodium" lamps.

A photon emitted by a source comprising the element to be analyzed will be absorbed by the same atoms if they are present in their free atomic state in the flame or in the furnace.

A comparison between the beam intensity before and after the flame provides a measurement of the quantity of photons absorbed and therefore the concentration of the atom being analyzed. The comparison can be made directly by a double beam analyzer. See Figure 2.7 in which the beam is divided into 2 branches one of which traverses the flame, the other serving as a reference. These beams are then recombined by a semi-transparent mirror. A detector, synchronized to the rotating mirror, responds to the energy difference between the two beams.

Absorption is related to concentration by the Beer-Lambert relation which we will examine further in article 3.2.2.1.

This is a monoelemental method and it requires a tight calibration per element and a different source for each one. There are however, di- or tri-atomic cathodes available.

As shown in Table 2.4, atomic absorption is extremely sensitive. It is particularly suited to the analyses of arsenic and lead in gasolines, for sodium in fuel oils (where it is the only reliable method) and for mercury in gas condensates.

	X-ray fluorescence		Atomic absorption*		Plasma emission*
	Detectable limit, $\mu g/g$	Error, $\mu g/g$	**Flame** Detectable limit, $\mu g/l$	**Furnace** Detectable limit, $\mu g/l$	Detectable limit, $\mu g/l$
Ni	2	1	200	2**	50
V	2	1		2**	20
Fe	2	1	200	1	50
Pb			250	10	200
Cu	3	1	200		20
Mg			50		10
Zn	2	1	200	0.5	20
As	4	1		5	500
Na			50		
S					200
Hg			0.5**		5000

* The detectable limits are given for samples such as they are introduced into the apparatus; they should be previously diluted in order to be nebulized. It thereby is useful to apply a dilution coefficient, usually at least 10. The dilution depends on the sample viscosity.

** After mineralization.

Table 2.4   *Metal analysis in petroleum cuts.*

Taking into account the range of wavelength and the intensity of emission beams, certain elements cannot be determined by atomic absorption, such as C, H, O, N, S, and the halogens.

2.2.6.3 **Using Argon Plasma Atomic Emission for Analysis of Metals**

The sample should be liquid or in solution. It is pumped and nebulized in an argon atmosphere, then sent through a plasma torch: that is, in an environment where the material is strongly ionized resulting from the electromagnetic radiation produced by an induction coil. Refer to the schematic diagram in Figure 2.8.

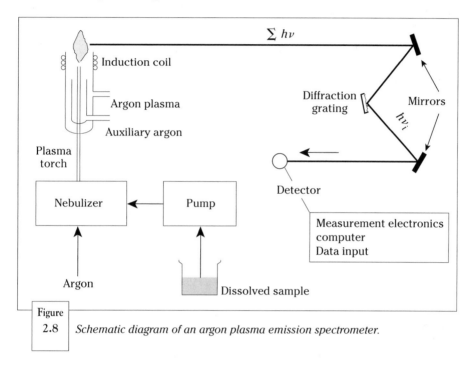

Figure 2.8 *Schematic diagram of an argon plasma emission spectrometer.*

The plasma comprises free positive and negative ions generally in equilibrium at high temperature. It has a high electrical and thermal conductivity and emits photons whose frequency is characteristic of the atoms present. The polychromatic rays are analyzed by a monochromator that gives the plasma's spectrum. The elements are identified by their wavelengths and the signal intensity is proportional to the quantity of ions present in the arc. The response is linear for 5 or 6 orders of magnitude, something which gives a large advantage to the plasma technique over atomic absorption. Furthermore, the method is multi-elemental; it permits the analysis of about 70 trace elements without having to change the device.

With the exception of alkalis, the sensitivity is generally higher than that of atomic absorption (at least flame atomic absorption). Refer to Table 2.4.

This method has a very general application range: analysis for metals in crude oils, in their various distillation cuts, and in their residues as well as for metals contained in spent lubricating oils, water, lubricants, etc.

The choice between X-ray fluorescence and the two other methods will be guided by the concentration levels and by the duration of the analytical procedure; X-ray fluorescence is usually less sensitive than atomic absorption, but, at least for petroleum products, it requires less preparation after obtaining the calibration curve. Table 2.4 shows the detectable limits and accuracies of the three methods given above for the most commonly analyzed metals in petroleum products. For atomic absorption and plasma, the figures are given for analysis in an organic medium without mineralization.

Characterization of Crude Oils and Petroleum Fractions

Jean-Claude Roussel

Raymond Boulet

Although distillation and elemental analysis of the fractions provide a good evaluation of the qualities of a crude oil, they are nevertheless insufficient. Indeed, the numerous uses of petroleum demand a detailed molecular analysis. This is true for all distillation fractions, certain crude oils being valued essentially for their light fractions used in motor fuels, others because they make quality lubricating oils and still others because they make excellent base stocks for paving asphalt.

Furthermore, molecular analysis is absolutely necessary for the petroleum industry in order to interpret the chemical processes being used and to evaluate the efficiency of treatments whether they be thermal or catalytic. This chapter will therefore present physical analytical methods used in the molecular characterization of petroleum.

3.1 Characterization of Crude Oils According to Dominant Characteristics Based on Overall Physical Properties

Because of the differences existing between the quality of different distillation cuts and those resulting from their downstream processing, it is useful to group them according to a major characteristic. That is, they are grouped into the three principal chemical families which constitute them: paraffins, naphthenes and aromatics. From a molecular point of view, their chemical reactivities follow this order:

paraffins < naphthenes < aromatics.

It is important to define the above in both molecular and atomic terms.

Consider the molecule below. From an atomic point of view, an atom common to two structures, aromatic and naphthenic, or aromatic and paraffinic or still further naphthenic and paraffinic will be considered first of all aromatic, then naphthenic, then paraffinic.

Carbon atoms 1, 2, 3, 4, 5 and 6 are aromatic, atoms 7, 8 and 9 are naphthenic, and atoms 10, 11, 12 and 13 are paraffinic.

It is important to keep these differences in mind because, according to the characterization methods employed, one speaks in terms of either the percentage of the kind of molecule or the percentage of the kind of atom. A molecule is said to be aromatic if it has at least one benzene ring as in the case of the molecule shown above; if not it is considered naphthenic if it has at least one naphthenic ring. Finally, with neither an aromatic ring nor a naphthenic ring, we will have a paraffinic molecule. Thus the above molecule is considered to be 100% aromatic even though the atomic carbon fractions are 6/13 for the aromatic carbons, 3/13 for the naphthenic carbons and 4/13 for the paraffinic carbons.

Experience has shown that certain carefully selected physical properties could be correlated with the dominant composition of a petroleum cut or crude oil.

3.1.1 Characterization of Crude Oils Using Specific Gravities of a Light Fraction and a Heavy Fraction

Eleven different groups of crude oils have been defined according to the densities of their heavy gasoline cuts (100–200°C) and their residues with boiling points above 350°C as shown in Table 3.1.

The specific gravity of a pure hydrocarbon is linked to its H/C ratio, the specific gravity decreasing as the H/C ratio increases. Table 3.2 illustrates this variation for hydrocarbons having 14 carbon atoms.

3.1.2 Characterization Factor K_{UOP}, or Watson Factor K_W

The characterization factor K_{UOP} was introduced by research personnel from the Universal Oil Products Company.

It is based on the observations that the specific gravities of hydrocarbons are related to their H/C ratios (and thus to their chemical character) and that their boiling points are linked to the number of carbon atoms in their molecules.

Crude oil base	Specific gravity of heavy gasoline cut, d_4^{15}	Specific gravity of residue (BP > 350°C) d_4^{15}
Paraffinic	Below 0.760	Less than 0.930
Intermediary Paraffinic	Less than 0.760	Between 0.930 and 0.975
Asphaltic Paraffinic	Less than 0.760	Greater than 0.975
Paraffinic Intermediary	Between 0.760 and 0.780	Less than 0.930
Intermediary	Between 0.760 and 0.780	Between 0.930 and 0.975
Asphaltic Intermediary	Between 0.760 and 0.780	Greater than 0.975
Paraffinic Naphthenic	Between 0.780 and 0.800	Less than 0.930
Intermediary Naphthenic	Between 0.780 and 0.800	Between 0.930 and 0.975
Paraffinic Aromatic	Greater than 0.800	Less than 0.930
Aromatic	Greater than 0.800	Between 0.930 and 0.975
Asphaltic	Greater than 0.780	Greater than 0.975

Table 3.1 *Estimating the general nature of crude oils by measurement of two specific gravities.*

	H/C atomic ratio	Specific gravity d_4^{15}
Tetradecane $C_{14}H_{30}$	2.10	0.763
Octylcyclohexane $C_{14}H_{28}$	2.00	0.817
Octylbenzene $C_{14}H_{22}$	1.57	0.858
Butylnaphthalene $C_{14}H_{16}$	1.04	0.966

Table 3.2 *Specific gravity compared with H/C ratio for pure hydrocarbons.*

From these observations, the characterization factor K_{UOP} (or K_W) was defined for pure components using only their boiling points and their densities:

$$K_{UOP} = \frac{(1.8\,T)^{1/3}}{S}$$

T being the boiling temperature (Kelvin) and S being the standard specific gravity (15.6°C/15.6°C). Refer to Chapter 4.

The K_{UOP} values for the pure hydrocarbons investigated are as follows:

- 13 for paraffins
- 12 for hydrocarbons whose chain and ring weights are equivalent
- 11 for pure naphthenes
- 10 for pure aromatics.

To extend the applicability of the characterization factor to the complex mixtures of hydrocarbons found in petroleum fractions, it was necessary to introduce the concept of a mean average boiling point temperature to a petroleum cut. This is calculated from the distillation curves, either ASTM or TBP. The volume average boiling point (VABP) is derived from the cut point temperatures for 10, 20, 50, 80 or 90% for the sample in question. In the above formula, VABP replaces the boiling point for the pure component.

The following temperatures have been defined:

- for a crude oil using its TBP distillation (given as volume),

volume average boiling point: $\quad T = \dfrac{T_{20} + T_{50} + T_{80}}{3}$

- for a petroleum cut using its ASTM distillation curve,

volume average boiling point: $\quad T = \dfrac{T_{10} + 2T_{50} + T_{90}}{4}$

where T_i is the temperature at which $i\%$ of the sample has been distilled.

In this manner, the K_{UOP} of a petroleum cut can be calculated quickly from readily available data, i. e., the specific gravity and the distillation curve. The K_{UOP} value is between 10 and 13 and defines the chemical nature of the cut as it will for the pure components. The characterization factor is extremely valuable and widely used in refining although the discriminatory character of the K_{UOP} is less than that obtained by more modern physical methods described in 3.2 and 3.3.

3.1.3 Characterization of a Petroleum Cut by Refractive Index, Density, and Molecular Weight (*ndM* method)

As in the case of density or specific gravity, the refractive index, *n*, for hydrocarbons varies in relation to their chemical structures. The value of *n* follows the order: *n* paraffins < *n* naphthenes < *n* aromatics and it increases with molecular weight.

With the accumulation of results obtained from various and complex analyses of narrow cuts (Waterman method), correlations have been found between refractive index, specific gravity and molecular weight on one hand, and percentages of paraffinic, naphthenic and aromatic carbon on the other.

Details are given in Table 3.3. As with all correlations, one should beware of using them if the measurements taken are outside the region of the correlation that established them. This method is commonly called *ndM* and is used mainly with vacuum distillates and lubricating oils.

With regard to the *ndM* method, one should note the following items:

- Refractive index: this is one of the most precise measurements that can be carried out on a petroleum cut. The ASTM method D 1218 indicates a reproducibility of 0.00006, which is exceptional.

n and d measured at 20°C		**n and d measured at 70°C**	
$V = 2.51 (n - 1.4750) - (d - 0.8510)$		$V = 2.42 (n - 1.4600) - (d - 0.8280)$	
$W = (d - 0.8510) - 1.11 (n - 1.4750)$		$W = (d - 0.8280) - 1.11 (n - 1.4600)$	
$V > 0$	$\% C_A = 430V + \dfrac{3660}{M}$	$V > 0$	$\% C_A = 410V + \dfrac{3660}{M}$
	$R_A = 0.44 + 0.055MV$		$R_A = 0.41 + 0.055MV$
$V < 0$	$\% C_A = 670V + \dfrac{3660}{M}$	$V < 0$	$\% C_A = 720V + \dfrac{3660}{M}$
	$R_A = 0.44 + 0.080MV$		$R_A = 0.41 + 0.080MV$
$W > 0$	$\% C_R = 820W - 3S + \dfrac{10,000}{M}$	$W > 0$	$\% C_R = 775W - 3S + \dfrac{11,500}{M}$
	$R_T = 1.33 + 0.146M (W - 0.005S)$		$R_T = 1.55 + 0.146M (W - 0.005S)$
$W < 0$	$\% C_R = 1440W - 3S + \dfrac{10,600}{M}$	$W < 0$	$\% C_R = 1440V + \dfrac{12,100}{M}$
	$R_T = 1.33 + 0.180M (W - 0.005S)$		$R_T = 1.55 + 0.180M (W - 0.005S)$

$$\% C_N = \% C_R - \% C_A$$
$$R_N = R_T - R_A$$

$\% C_A$ = weight per cent of aromatic carbon	R_A = number of aromatic rings in an average molecule
$\% C_N$ = weight per cent of naphthenic carbon	R_N = number of naphthenic rings in an average molecule
$\% C_P$ = weight per cent of paraffinic carbon	R_T = total number of rings in an average molecule
$\% C_R$ = weight per cent of cyclic carbon	S = weight per cent sulfur

Table 3.3   *The ndM method (Wuithier, 1972).*

As a consequence, other than its use in the *ndM* method, the refractive index is very often used in process operations because it can indicate small differences in product quality that would be missed by other measurements. The only restriction is that the color of the sample should be less than 5 on the ASTM D 1500 scale.

• Molecular weight: for a mixture of components such as one would encounter in petroleum cuts, the molecular weight is:

$$M = \frac{\Sigma\, n_i\, M_i}{\Sigma\, n_i}$$

where n_i = the number of molecules of component *i* having molecular weight M_i.

The measurement techniques most frequently used are derived from Raoult's and Van't Hoff's laws applied to cryometry, ebulliometry, osmometry, etc. They are not very accurate with errors on the order of ten per cent. Consequently, the molecular weight is often replaced by correlated properties. The mean average temperature or viscosity can thus replace molecular weight in methods derived from *ndM*.

3.2   Characterization of Crude Oils and Petroleum Fractions Based on Structural Analysis

As the boiling points increase, the cuts become more and more complex and the analytical means must be adapted to the degree of complexity. Tables 3.4 and 3.5 describe the most widely used petroleum product separation scheme and the analyses that are most generally applied.

Although gas chromatography can give the concentration of each component in a petroleum gas or gasoline sample, the same cannot be said for heavier cuts and one has to be satisfied with analyses by chemical family, by carbon atom distribution, or by representing the sample as a whole by an average molecule.

3.2.1 Analysis by Hydrocarbon Family

3.2.1.1   Using Mass Spectrometry for Determining Distribution by Chemical Families

In a mass spectrometer, the molecules, in the gaseous state, are ionized and fragmented. The fragments are detected as a function of their mass-to-charge ratio, *m/e*. The graphical representation of the ion intensity as a function of *m/e* makes up the mass spectrogram as illustrated in Figure 3.1.

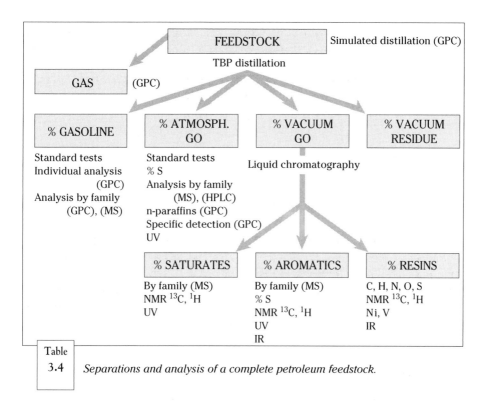

| Table 3.4 | *Separations and analysis of a complete petroleum feedstock.* |

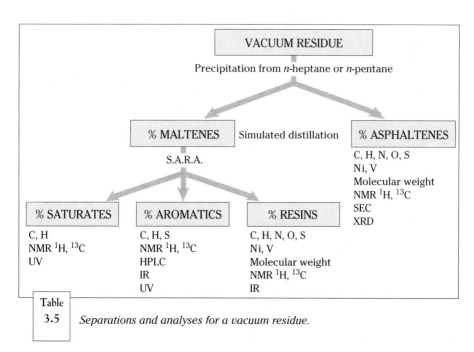

| Table 3.5 | *Separations and analyses for a vacuum residue.* |

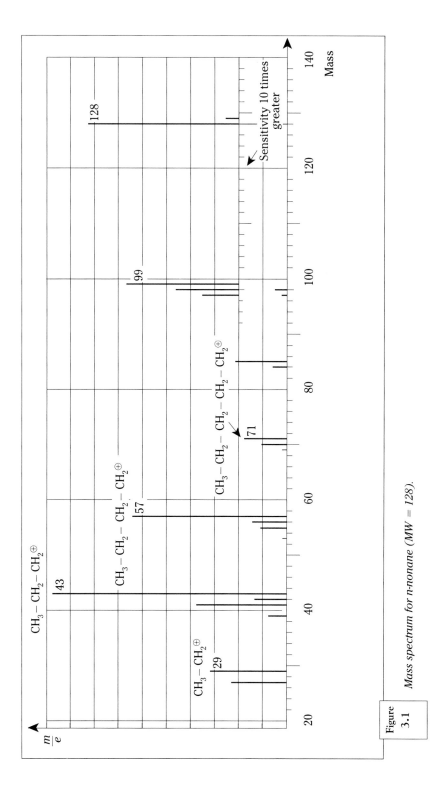

Figure 3.1 *Mass spectrum for n-nonane (MW = 128).*

Magnetic Deviation Mass Spectrometry

A schematic diagram of the apparatus is shown in Figure 3.2. The molecules are introduced under a partial vacuum of 10^{-3} torr[1] into a buffer chamber that communicates via molecular slipstream with the source itself at 10^{-5} to 10^{-6} torr in order to ensure a constant concentration in the source at all times during the analysis.

Inside the source the molecules are subjected to an electron bombardment that ionizes them with a positive charge and fragments them. In the most conventional analyses, the electrons are accelerated to 70 eV.

The source is brought to a positive potential (V) of several kilovolts and the ions are extracted by a plate at ground potential. They acquire kinetic energy and thus velocity according to their mass and charge. They enter a magnetic field whose direction is perpendicular to their trajectory. Under the effect of the field, B_0, the trajectory is curved by Lorentz forces that produce a centripetal acceleration perpendicular to both the field and the velocity.

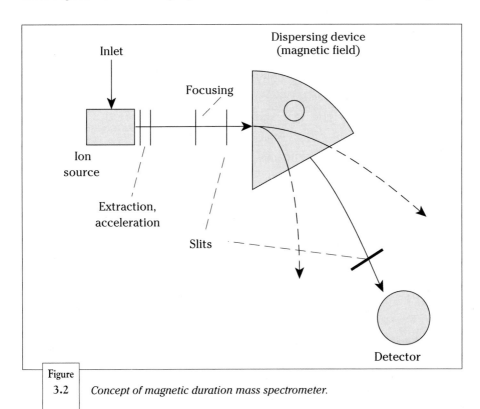

Figure 3.2   *Concept of magnetic duration mass spectrometer.*

---

[1] 1 torr = 1 mm mercury = 0.133 kPa.

The radius of trajectory is proportional to the square root of the mass-to-charge ratio, m/e, as follows:

$$R = \frac{1}{B_0} \sqrt{2V} \sqrt{\frac{m}{e}}$$

The variation of B_0 causes all ions to pass sequentially in front of the exit slit behind which is positioned the photomultiplier detector. The pressure in the apparatus is held at 10^{-7} torr in order to achieve mean free paths of ions sufficiently high that all ions emitted from the source are collected.

Qualitative Analysis by Mass Spectrometry

When subjected to an electron bombardment whose energy level is much higher than that of hydrocarbon covalent bonds (about 10 eV), a molecule of mass M loses an electron and forms the "molecular" ion, M^+; the bonds break and produce an entirely new series of ions or "fragments". Taken together, the fragments' relative intensities constitute a constant for the molecule and can serve to identify it; this is the basis of qualitative analysis.

Consider the following case as an example:

- Propane $CH_3-CH_2-CH_3$ of mass 44 gives:

    one molecular ion: mass 44, $C_3H_8^+$

- Breaking one or more carbon-hydrogen bonds results in the following:

    – ion fragments: mass 43, $C_3H_7^+$

    mass 42, $C_3H_6^+$

    mass 41, $C_3H_5^+$

    – by rupture of a $C-C$ bond:

    mass 29, $C_2H_5^+$

    mass 28, $C_2H_4^+$

    mass 27, $C_2H_3^+$

    – by rupture of a second $C-C$ bond:

    mass 15, CH_3^+

    mass 14, CH_2^+

    mass 13, CH^+

Quantitative Analysis by Mass Spectrometry

The ion intensity expressed as the amount of electricity, is proportional to the following factors:

- a sensitivity constant, k_i

- the partial pressure of the molecule in the source, and hence the concentration of the original molecule.

Example:

For a substance, A, whose partial pressure is p_a, the values (intensities) of ions 57 and 43, for example are given as:

$$h_{43} = k_1 p_a \qquad h_{57} = k_2 p_a$$

Knowing k_1 and k_2 by studying reference components, it becomes possible to calculate p_a from the measurements of ion intensities found on the mass spectrum.

In the case of mixtures, especially those of petroleum, a variety of compounds can give ions having the same mass; the mass spectrum is then the sum of the spectra of each component:

$$h_{43} = \sum k_{1i} p_i \qquad h_{57} = \sum k_{2i} p_i$$

To obtain the different values of p_i, it is only necessary to produce as many independent equations as there are components in the mixture and, if the mixture has n components, to solve a system of n equations having n unknowns. Individual analysis is now possible for mixtures having a few components but even gasoline has more than 200! It soon becomes unrealistic to have all the sensitivity coefficients necessary for analysis: in this case, 200^2.

However, a considerable amount of work was done by American petroleum companies from 1940 to 1960, who obtained very narrow range petroleum cuts and identified the chemical families present in each of them. They noted that each chemical family gave preferential fragmentation patterns allowing them to be identified. Therefore, paraffins of the general formula C_nH_{2n+2} led essentially to $C_nH_{2n+1}^+$ fragments; aromatic hydrocarbons having the formula C_nH_{2n-6} produced mainly $C_nH_{2n-6}^+$ and $C_nH_{2n-7}^+$ fragments, and so on.

The resulting mass spectrometry analysis is an analysis by chemical families of the form C_nH_{2n-z}. Using this approach, petroleum laboratories found matrices of sensitivity coefficients for each chemical family and even in certain cases by average carbon number within the chemical family. Thus an entire series of methods was created (see Table 3.6), establishing mass spectrometry as one of the foremost analytical tools for petroleum.

Note that the matrices of coefficients could be simplified if the mass spectrometer resolution were better. Consider four hydrocarbons having the same molecular weight integer of 226: the paraffin $C_{16}H_{34}(1)$, the alkylnaphthalene $C_{17}H_{22}(2)$, the indenopyrene $C_{18}H_{10}(3)$ and the alkyl dibenzothiophene $C_{15}H_{14}S(4)$, for which the exact molecular weights are, respectively: 226.266; 226.172; 226.078 and 226.081.

All have molecular weights of 226 to the nearest integer (C = 12, H = 1, S = 32), but the exact molecular weights differ slightly. A resolution of 2500 is necessary to separate molecules 1, 2 and 3 but 75,000 is required to separate molecule 4 from molecule 3 which explains why high resolution mass spectrometers are sought.

Separation of families by merely increasing the resolution evidently can not be used when the two chemical families have the same molecular formula. This is particularly true for naphthenes and olefins of the formula, C_nH_{2n}, which also happen to have very similar fragmentation patterns. Resolution of these two molecular types is one of the problems not yet solved by mass spectrometry, despite the efforts of numerous laboratories motivated by the refiner's major interest in being able to make the distinction. Olefins are in fact abundantly present in the products from conversion processes.

Analysis of such cuts by spectrometry requires a preliminary separation by chemical constituents. The separation is generally done by liquid phase chromatography described in article 3.3.5.

Petroleum Analysis by Mass Spectrometry

Mass spectrometry allows analysis by hydrocarbon family for a variety of petroleum cuts as deep as vacuum distillates since we have seen that the molecules must be vaporized. The study of vacuum residues can be conducted by a method of direct introduction which we will address only briefly because the quantitative aspects are extremely difficult to master. Table 3.6 gives some examples; the matrices used differ according to the distillation cut and the chemical content such as the presence or absence of olefins or sulfur.

Range, °C	Number of carbon atoms	Reference	Number of chemical families
IP–210	C_6–C_{12}	ASTM D 2789	6
180–350	C_{10}–C_{20}	ASTM D 2425	12 incl. 2 sulfur fam.
200–700	C_{12}–C_{55}	Fisher (1974)	24
350–550 arom. fract.	C_{16}–C_{45}	IFP (unpublished)	21 by carbon
350–550 satur. fract.	C_{16}–C_{45}	ASTM D 2786	7
350–550 arom. fract. (low resolution)	C_{16}–C_{45}	ASTM D 3239	21

Table 3.6   *Non-exhaustive summary of analytical methods using mass spectrometry.*

Three of them are given below:

(a) Analysis of IP–210°C Cuts $\left(C_6-C_{12}\right)$ (ASTM Method D 2789)

Required resolution: 1000

Results: 6 hydrocarbon families

% olefins: <3%

Reproducibility: ±5%

	Basic Formula	**Volume %**
Paraffins	C_nH_{2n+2}	41.7
Monocycloparaffins	C_nH_{2n}	29.7
Dicycloparaffins	C_nH_{2n-2}	3.3
Aromatics	C_nH_{2n-6}	21.3
Indanes + tetralins	C_nH_{2n-8}	3.1
Naphthalenes	C_nH_{2n-12}	0.9

Interest in this method has decreased since advances made in gas chromatography using high-resolution capillary columns (see article 3.3.3.) now enable complete identification by individual chemical component with equipment less expensive than mass spectrometry.

(b) Analysis of 180–350°C Cuts $\left(C_{10}-C_{20}\right)$
   (Modified ASTM Method D 2425)

Resolution required: 3000

Results: 12 hydrocarbon families, two of which are sulfur

No olefins present

Table 3.7 gives results obtained by this method for a gas oil before and after hydrotreatment.

	Basic formula	**Weight %**	
		Safanyia 250–350°C straight-run	**After hydrotreatment at 50 bar**
Paraffins	C_nH_{2n+2}	40.1	40.7
Non-condensed naphthenes	C_nH_{2n}	18.0	26.6
Condensed naphthenes	C_nH_{2n-2}	11.1	14.3
Alkylbenzenes	C_nH_{2n-6}	7.0	9.5
Indanes and tetralins	C_nH_{2n-8}	3.7	5.2
Indenes	C_nH_{2n-10}	1.6	0.9
Naphthalenes	C_nH_{2n-12}	4.7	1.5
Acenaphthenes and diphenyls	C_nH_{2n-14}	2.0	0.6
Acenaphtylenes and fluorenes	C_nH_{2n-16}	1.1	0.4
Phenanthrenes and anthracenes	C_nH_{2n-18}	1.3	0.1
Benzothiophenes	$C_nH_{2n-10}S$	6.4	0
Dibenzothiophenes	$C_nH_{2n-16}S$	3.0	0

Table 3.7   *Typical analysis of a gas oil before and after hydrotreatment.*

The variation in concentration of different chemical families readily illustrates the benefit to a refiner that such an analysis can provide as much for product quality as for the chemical reactions taking place in the process.

An important application of this type of analysis is in the determination of the calculated cetane index. The procedure is as follows: the cetane number is measured using the standard CFR engine method for a large number of gas oil samples covering a wide range of chemical compositions. It was shown that this measured number is a linear combination of chemical family concentrations as determined by the D 2425 method. An example of the correlation obtained is given in Figure 3.3.

This usefulness of the correlation is two-fold: first it provides information on the cetane indices that are not available in the literature as in the cases of polynuclear aromatics and sulfur-containing aromatics, and second it helps provide an evaluation of the cetane index based on a few milligrams of sample, instead of the liter or so required for the motor method.

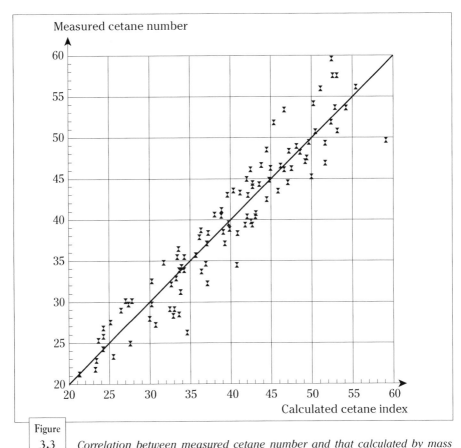

Figure 3.3 *Correlation between measured cetane number and that calculated by mass spectrometry.*

This approach is particularly useful when one wishes to characterize pilot plant effluents whose daily production can be less than a liter.

As the temperatures of the distillation cuts increase, the problems get more complicated to the point where preliminary separations are required that usually involve liquid phase chromatography (described earlier). This provides, among others, a saturated fraction and an aromatic fraction. Mass spectrometry is then used for each of these fractions.

(c) Aromatic Fractions 350–550°C $\left(C_{16}–C_{45}\right)$ (Low Ionization Voltage Method)

This method has the particularity of operating at low ionization voltage. In this case, instead of the electrons being accelerated to 70 V as in the preceding examples, they attain only 10 V, which consequently suppresses ion fragmentation. Only molecular ions are produced; the sensitivity matrices are reduced to diagonal matrices which greatly simplifies the calculations and reduces the probability of error. Nevertheless, the operation of the spectrometer's source at such a low value of ionization voltage greatly reduces the efficiency of the source and, as a result, the spectrometer's sensitivity.

Before ending this presentation on mass spectrometry, we should cite the existence of spectrometers for which the method of sorting ions coming from the source is different from the magnetic sector. These are mainly quadripolar analyzers and, to a lesser degree, analyzers measuring the ion's time of flight.

The quadripolar spectrometers whose resolution is limited to about 2000 are of simpler design than the magnetic sectors and are less costly. They are often used in conjunction with gas chromatography (see section 3.3) for purposes of identification.

3.2.1.2 Use of Ultra-Violet Spectrometry to Obtain Distribution by Hydrocarbon Families

In the electromagnetic spectrum, the ultra violet region is between that of X-rays and visible light. This corresponds to the energies $h\nu$ of one hundred to a few tens of electron-volts (wavelengths from 180 to 400 nm).

This energy region is that which corresponds to the electron energy states of the molecules, which are quantified and so can have only well-defined discrete values.

A spectrometer as shown in Figure 3.4. includes a continuous source emitting in the ultra violet spectral region, often a deuterium lamp. The polychromatic beam is analyzed by a monochromator (prism, grating). The rotation of the latter causes the different wavelengths to sweep over the sample then, after passing through slits designed to improve the resolution, the beam reaches the detector, a lead sulfide cell for example.

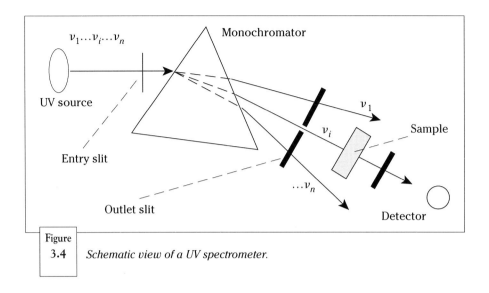

3.4 | *Schematic view of a UV spectrometer.*

The electron levels capable of being excited and therefore absorbing ultraviolet radiation are mainly those involved in π bonds and more particularly conjugated π bonds, either $\pi - \pi$, or $\pi - n$. When a photon possesses the exact amount of energy corresponding to the difference in energy of the two quantified electron levels, it causes a transition and is absorbed. A comparison of spectra obtained by passing and not passing through the sample indicates the wavelengths at which the photons are absorbed. This gives us a qualitative analytical method since the wavelength corresponds to a specific electronic configuration. Nevertheless, note, with the exception of simple cases, that ultraviolet qualitative analysis is delicate. In fact, associated with each electron level of a molecule are vibration states which are themselves associated with different rotational states that broaden the signal. With the resolution of spectrometers being insufficient, it follows that the UV spectra are more often made of very wide poorly resolved bands making the identification of individual components impossible. This method is also quantitative, the absorption being governed by the Beer Lambert theory (see Beer's law, art. 3.2.2.1).

Molecules found in petroleum products giving rise to absorption in the UV are mostly aromatics and to a lesser degree conjugated diolefins and olefins. The saturated hydrocarbons, alkanes or naphthenes, give no signal for wavelengths greater that 180 nm. This particularity might seem restrictive but in fact is an advantage because knowing the aromatic content is very often necessary in refining. UV absorption is of interest also because for aromatics, ring condensation causes absorption towards the higher wavelengths, as well as large variations in the sensitivity coefficients (called extinction or absorption coefficients). Table 3.8 gives the average molar absorptivities for different wavelengths as a function of ring condensation for an optical path of 1 cm and a concentration of 1 mol/l.

	197 nm	**220 nm**	**260 nm**
Monoaromatics	50,000	6,000	500
Diaromatics	10,000	100,000	5,000
Polynuclear aromatics	5,000	10,000	20,000

Table 3.8	*Average absorptivities for aromatic compounds.*

This characteristic is used to analyze aromatics in gas oil cuts; an example of a UV absorption spectrum is shown in Figure 3.5.

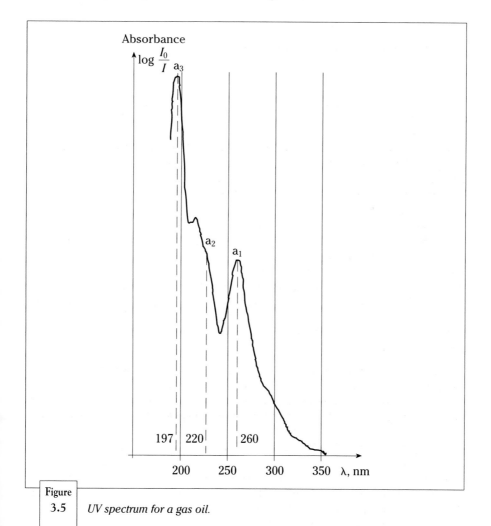

Figure 3.5	*UV spectrum for a gas oil.*

Using this concept, Burdett developed a method in 1955 to obtain the concentrations in mono-, di- and polynuclear aromatics in gas oils from the absorbances measured at 197, 220 and 260 nm, with the condition that sulfur content be less than 1%. Knowledge of the average molecular weight enables the calculation of weight per cent from mole per cent. As with all methods based on statistical sampling from a population, this method is applicable only in the region used in the study; extrapolation is not advised and usually leads to erroneous results.

The specific nature of UV absorption for certain structures when combined with the high sensitivity of the method enables trace quantities (≈ 1 ppm) of molecules in a matrix transparent to UV beams to be analyzed. Benzene in cyclohexane is an example.

3.2.2 **Characterization of a Petroleum Fraction by Carbon Atom Distribution**

One has seen that the number of individual components in a hydrocarbon cut increases rapidly with its boiling point. It is thereby out of the question to resolve such a cut to its individual components; instead of the analysis by family given by mass spectrometry, one may prefer a distribution by type of carbon. This can be done by infrared absorption spectrometry which also has other applications in the petroleum industry. Another distribution is possible which describes a cut in terms of a set of structural patterns using nuclear magnetic resonance of hydrogen (or carbon); this can thus describe the average molecule in the fraction under study.

3.2.2.1 **Using Infrared Spectrometry to Characterize Petroleum Fractions according to the Nature of the Carbon Atoms**

Principle of Infrared Spectrometry

One knows that molecular structures have vibrational states, associated with electron levels in the UV region, which are quantified: that is, which can give only discrete values. If a molecular oscillator such as the carbonyl group $C = O$ in a ketone is subjected to an electromagnetic beam having an energy of $E = h\nu$ exactly equal to the resonant frequency, the energy will be absorbed and the oscillator passes into an excited state of vibration. The corresponding energy region is that of the infrared. These vibrations depend on the nature of the atoms making up the molecule and the strength of the bonds between them. For example, if two atoms of mass m_1 and m_2 are held together by a bond whose force constant is k, the vibration frequency ν will be as follows:

$$\nu = \frac{1}{2\pi} \sqrt{\frac{k}{\mu}}$$

where μ is the reduced mass equal to: $\dfrac{m_1 \cdot m_2}{m_1 + m_2}$

Although a diatomic molecule can produce only one vibration, this number increases with the number of atoms making up the molecule. For a molecule of N atoms, 3N-6 vibrations are possible. That corresponds to 3N degrees of freedom from which are subtracted 3 translational movements and 3 rotational movements for the overall molecule for which the energy is not quantified and corresponds to thermal energy. In reality, this number is most often reduced because of symmetry. Additionally, for a vibration to be active in the infrared, it must be accompanied by a variation in the molecule's dipole moment.

For the two modes of vibration for CO_2:

$$O = C = O \qquad O = C = O$$
$$\blacktriangleleft \quad 1 \quad \blacktriangleright \qquad \blacktriangleleft \quad 2 \quad \blacktriangleleft$$

only mode 2 (non-symmetrical) allows absorption to take place. Mode 1 is said to be inactive.

Nevertheless, a molecule possesses sufficient vibrations so that all its frequencies taken together can be used to characterize it. In this sense the infrared spectrum is generally considered to be a molecule's "fingerprint."

Infrared Absorption Spectrometers

The first requirement is a source of infrared radiation that emits all frequencies of the spectral range being studied. This polychromatic beam is analyzed by a monochromator, formerly a system of prisms, today diffraction gratings. The movement of the monochromator causes the spectrum from the source to scan across an exit slit onto the detector. This kind of spectrometer in which the range of wavelengths is swept as a function of time and monochromator movement is called the dispersive type.

If the sample is placed in the path of the infrared beam, usually between the source and the monochromator, it will absorb a part of the photon energy having the same frequency as the vibrations of the sample molecule's atoms. The comparison of the source's emission spectrum with that obtained by transmission through the sample is the sample's transmittance spectrum.

To obtain this spectrum directly, double beam instruments such as shown in Figure 3.6 are used most often. The beam is divided into two optically equivalent parts, that is, equivalent optical paths and the same number of reflections on identical mirrors. One of the beams traverses the sample while the second is the reference beam and can be made to travel through an empty cell identical to the sample cell, or through a cell filled with the solvent used to dilute the sample, should that be the case. Hence, any differences between the two beams can come only from the sample itself. By means of a rotating sector, the two beams are recombined and, after passing through the monochromator, reach the detector, usually a thermocouple, where an electric signal is produced. For one-fourth of the cycle time, the detector receives the reference signal; for the next quarter cycle, it receives the signal after absorption by the sample. Since the detector is synchronized with the movement of the rotating sector it sees only the variations due to absorption by the sample.

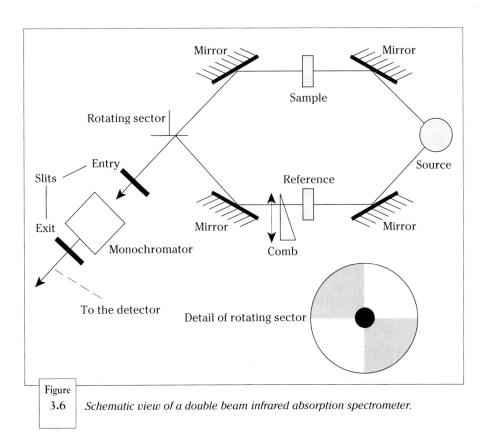

Figure 3.6 *Schematic view of a double beam infrared absorption spectrometer.*

Today another kind of spectrometer is replacing the dispersive spectrometers. They are called interferometers and are based on the principle of Michelson's interferometer. These instruments allow measurement of the variation of energy as a function of the displacement of a moving mirror, thus with time. By applying a Fourier transform to the interferogram, one obtains the frequency domain equivalent, that is, one observes the variation of energy as a function of frequency which makes up the infrared spectrum as in Figure 3.8. This technique has been known for a long time, but only since the explosion of computer power and the creation of fast Fourier transform algorithms has the development of interferometry been possible.

Interferometers have several advantages:

- The high luminosity of the instrument with no slits to limit the size of the beam. This is the Jacquinot's advantage also called "etendue."
- An improved signal/noise ratio because all signals are seen simultaneously along with the instrument's own noise (called the multiplex or Fellgett advantage).
- The rapidity with which information is received: one second instead of an average 10 minutes for dispersive spectrometry.

In current practice, rationalized units are not used in IR; the absorption bands have long been identified in terms of wavelengths, i. e., in micrometers. The general trend now is to express energy by a scale proportional to the frequency: the wave number designated by $\bar{\nu}$ is defined as:

$$\lambda = \frac{c}{\nu} = \frac{10^4}{\bar{\nu}}$$

where:

$\bar{\nu}$ = wave number in cm^{-1}

λ = wavelength in μm

ν = frequency in Hertz (cycles per second)

c = speed of light in a vacuum in $\mu m/s$

(the frequency in Hz would be hard to work with because the IR region is around $30 \ 10^{12}$ Hz). Thus the middle infrared, of interest to us here, ranges from 4000 to 400 cm^{-1} (corresponding to wavelengths from 2.5 to 25 μm). With these units, the molecular vibration of CO at 2350 cm^{-1} enables the calculation of a force constant for the oscillator to be made:

$$k = 18 \times 10^{-8} \, N/m.$$

Qualitative Analysis by Infrared Absorption Spectrometry

Strictly speaking, a group of atoms cannot be isolated from the rest of the molecule. However, with a few exceptions, the same chemical group always absorbs energy in the same IR spectral region. For instance the vibrations from the $C-H$ bonds of a methyl group cause absorption of infrared radiation at 2960, 1450, 1390 cm^{-1}.

The OH vibrator absorbs at 3600 and 1620 cm^{-1}, the benzene ring at 1600, 1500 cm^{-1} and, depending on the position of its substitution groups, between 1200 and 1000 and between 900 and 600 cm^{-1}, etc.

These characteristic absorption regions called group frequencies allow the analyst to detect the different elemental patterns and from them to reconstruct the molecule either by deduction or by comparison with library reference spectra. The libraries contain several hundred thousand spectra.

Quantitative Analysis Using Infrared Spectrometry

Using $I_0(\bar{\nu})$, the incident radiation energy for wave number $\bar{\nu}$ and $I(\bar{\nu})$, the radiation energy after passing through the sample at the same wave number, the following are defined:

- Transmittance: $T_{\bar{\nu}} = \dfrac{I(\bar{\nu})}{I_0(\bar{\nu})}$ expressed as %

Transmittance is measured as shown in Figure 3.7.

- Absorbance: $A_{\bar{\nu}} = -\log T_{\bar{\nu}} = \log \dfrac{I_0(\bar{\nu})}{I(\bar{\nu})}$

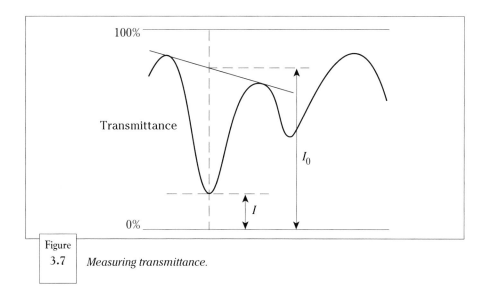

Figure
3.7 | *Measuring transmittance.*

Absorbance is linked to the concentration by the Beer-Lambert relation:

$$A_{\bar{\nu}} = a(\bar{\nu}) \cdot c \cdot l$$

with

$a(\bar{\nu})$ = sample absorptivity for the wave number $\bar{\nu}$, in l/(g·cm)

c = concentration in g/l

l = cell length in centimeters

This formula shows that if quantitative analysis in the infrared is to be possible, it is necessary to know the coefficients $a(\bar{\nu})$, therefore, either to have the pure substance, or to be able to obtain them from the literature

This same principle, as indicated earlier, is used in atomic absorption spectroscopy and UV absorption.

Analysis of Petroleum Products by Infrared Spectrometry

Aside from its obvious qualitative functions, infrared spectrometry is applied to several other kinds studies of petroleum products.

(a) Distribution by Carbon Type (method of Brandes, 1958)

Correlations have been found between certain absorption patterns in the infrared and the concentrations of aromatic and paraffinic carbons given by the *ndM* method (see article 3.1.3.). The absorptions at 1600 cm^{-1} due to vibrations of valence electrons in carbon-carbon bonds in aromatic rings and at 720 cm^{-1} (see the spectrum in Figure 3.8) due to paraffinic chain deformations are directly related to the aromatic and paraffinic carbon concentrations, respectively.

$$C_A(\%) = 0.98\, A\left(1600\ \text{cm}^{-1}\right) + 1.2$$
$$C_P(\%) = 0.66\, A\left(720\ \text{cm}^{-1}\right) + 29.9$$

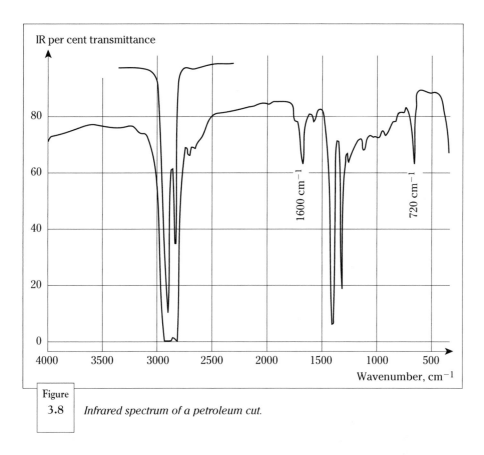

Figure 3.8   *Infrared spectrum of a petroleum cut.*

The naphthenic carbon percentage is the difference between 100 and the sum of the above concentrations.

This method is applicable for mineral oil fractions whose molecular weight is between 290 and 500 and for $C_A < 60\%$ and $40\% < C_P < 70\%$. The analysis is fast, approximately 10 minutes, and the correlation with other methods is satisfactory.

Adaptations of this method have been proposed in order to take into account the band displacement due either to substitution on the aromatic ring, or to chains of different lengths. The variations consist, instead of measuring the absorbance at maximum absorption, of an integration of the absorbance curve over a specified range (Oelert's method, 1971). More exact, this method is used less often mainly because the Brandes method is simpler to use.

Starting from these methods, as we will see further on, nuclear magnetic resonance (NMR) of carbon has provided an absolute percentage of aromatic, paraffinic, and naphthenic carbons.

Other correlations between NMR and infrared have been studied because the latter technique is less cumbersome than NMR. Correlations are obtained not just on the two absorption bands but on the whole of the IR spectrum after reduction of the spectrum into its principal components.

(b) Monitoring the Oxidation of a Motor Lubricating Oil during Operation

Oxidation of a motor lubricating oil is manifested by the presence of oxygenated functional groups: in particular, esters or acids. Formation of compounds containing these groups can be followed by infrared spectrometry by observing the characteristic absorption measurements for carbonyl groups $C = O$, between 1700 and 1800 cm^{-1}, precisely where non-oxidized hydrocarbons do not show absorption. Following the trend is usually the best that can be done because measuring the exact concentration is difficult. In fact, an average absorptivity $a(\bar{\nu})$ for all the esters and/or acids cannot be evaluated because they can be quite different from one molecule to another. Exact quantitative analysis is not possible.

(c) Identification and Monitoring of Additive
Concentrations in Lubricating Oils

This is an analysis frequently conducted on oil lubricants. Generally, the additive is known and its concentration can be followed by direct comparison of the oil with additive and the base stock. For example, concentrations of a few ppm of dithiophosphates or phenols are obtained with an interferometer. However, additive oils today contain a large number of products; their identification or their analysis by IR spectrometry most often requires preliminary separation, either by dialysis or by liquid phase chromatography.

3.2.2.2 Determining the Parameters of a Petroleum Fraction by Nuclear Magnetic Resonance

Principles of Nuclear Magnetic Resonance

Nuclear magnetic resonance involves the transitions between energy levels of the fourth quantum number, the spin quantum number, and only certain nuclei whose spin is not zero can be studied by this technique. Atoms having both an even number of protons and neutrons have a zero spin: for example, carbon 12, oxygen 16 and silicon 28.

In the following pages, only hydrogen ^1H and the carbon 13 isotope, both of which have a nuclear spin of 1/2 will be discussed.

In a magnetic field B_0 (see Figure 3.9.), nuclei whose spin is 1/2 can only align in two directions said to be parallel or anti-parallel to the direction of the magnetic field (see Figure 3.9). Each of these directions corresponds to one energy level. For a given field B_0, the energy difference between the two states is characteristic of the nucleus.

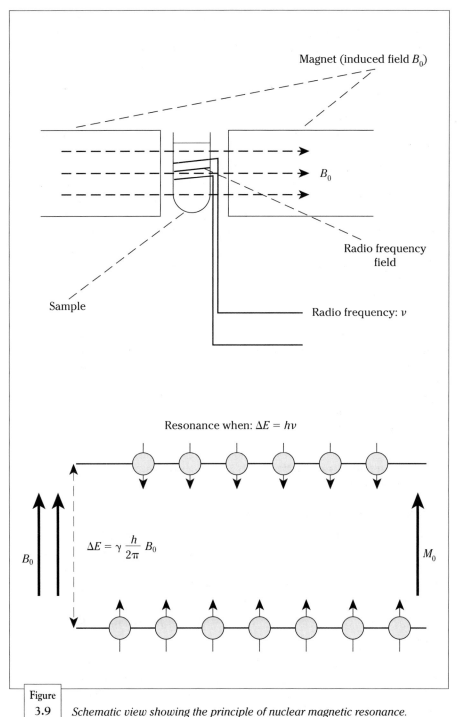

Figure

3.9   *Schematic view showing the principle of nuclear magnetic resonance.*

Measurement of the energy difference is achieved by a resonance method. The population of nuclei in a given state is governed by the Boltzman distribution that leads to an excess of nuclei in the state of lowest energy and confers to the lattice a magnetization, M_0, parallel to B_0. The distribution (hence the magnetization) can be changed by applying to the sample a radiation whose frequency is in resonance with the spin system.

In an NMR spectrometer, the sample placed in an homogeneous magnetic field is surrounded by a coil carrying an alternating electric current of variable frequency ν of a few hundred MHz and therefore emitting photons having an energy $h\nu$. When the frequency ν is such that this energy is equal to that separating the nuclear energy states, the magnetization is changed bringing about a variation of current in the coil. This variation is used to detect the resonance and thus characterize the nuclei species. Table 3.9 gives the resonance frequencies in a 9.4 tesla magnetic field.

Nucleus	Resonance frequency
^{1}H	400.13
^{19}F	376
^{31}P	162
^{13}C	100.6
^{29}Si	79.46

Table 3.9   *Resonance frequencies in MHz in a 9.4 tesla field.*

This could provide method to distinguish between the different types of nuclei but it is of minor interest to the physical chemist. On the other hand, if the experiment is conducted using spectrometers with strong, very uniform magnetic fields, and if the spectrometer is equipped with a very precise frequency synthesizer, it is possible to detect frequency differences for a given isotope. These differences are on the order of 10^{-7}, 10^{-8} with respect to the nominal frequency and are due to the chemical environment surrounding the nucleus. They are called chemical shifts and are expressed in parts per million, ppm, of the base frequency given by the resonance of a reference material. There are two ways to find the resonance, by exciting in succession the different types of nuclei; *continuous wave NMR*, by exciting all the nuclei of an isotope at the same time and extracting the resonances of each chemical species from the signal obtained; by *pulsed NMR*, often called "Fourier transform NMR".

Continuous Wave NMR

The sample is placed in a constant magnetic field, B_0, and the variation in frequency throughout the domain being explored excites one by one the different resonances. The scan lasts a few minutes. Inversely, one can maintain a constant frequency and cause the magnetic field to vary.

Impulse NMR

The sample is again subjected to a constant magnetic field but all the nuclei are excited by a very short radio frequency pulse. The frequency ν (e.g., 400 MHz for a proton at 9.4 tesla) is applied over a period of several microseconds. Immediately after the pulse, the excited system's return to equilibrium is observed by recording the variation of magnetic field with time. The Fourier transform of this signal gives the variation as a function of the frequency which constitutes the NMR spectrum. The advantage of this second method is in the speed of data collection; a few microseconds for excitation, a few seconds, usually less than five, for acquisition. The time factor is very important in NMR because it allows accumulation of a signal which, except in the case of proton NMR, is generally weak. Note that adding n identical spectra multiplies the signal-to-noise ratio by $\sqrt{n}$.

As in the case of infrared, progress in computing and the development of powerful algorithms for Fourier transforms has made the development of pulse NMR possible.

Although continuous wave NMR is sufficient for naturally abundant nuclei with strong magnetic moments such as hydrogen, fluorine and phosphorous, the study of low abundance nuclei and/or weak magnetic moments such as carbon 13 or silicon 29 requires pulse NMR.

Hydrogen NMR

With respect to a reference compound, tetra-methyl silane (TMS), to which the chemical displacement of 0 ppm is attributed arbitrarily, the chemical shifts obtained for hydrogen cover a region of about 15 ppm; going from an alkane CH_3 (0.9 ppm) to acidic hydrogen ($H^+ \sim 15$ ppm), passing through olefinic CH (~ 5 ppm) and the aromatic CH (about 7–8 ppm).

NMR is principally a structural analysis method, the chemical displacement enabling the various protons of a molecule to be identified. Moreover, the signals can be finely structured so that the different groups identified can be classed and the structure can be easily described. This supplemental information is given by the influence, on the nucleus being examined, of the magnetic fields of neighboring protons constituting so many magnetic dipoles. Thus the signal for the methyl group on ethyl benzene:

will be split into three because of the different distributions that the two protons of neighboring CH_2 can have. The methylene group signal itself will be split into four. The multiplicity of the fine structure, due to magnetic moment coupling called spin-spin coupling provides information as to the number of neighboring protons. The distance (expressed in Hz and called the coupling constant) separating the spectral lines of this structure, gives information on the spatial geometry of the molecule.

In straight-run petroleum products which do not usually contain olefins, four types of hydrogen can be easily differentiated as seen in the spectrum in Figure 3.10:

(a) Hydrogens attached to carbons on aromatic rings (H_a).

(b) Hydrogens attached to carbons in the α position on the rings (H_α).

(c) Hydrogens attached to methyl groups (CH_3) located in the γ position and beyond relative to the aromatic rings (H_γ).

(d) All other hydrogens, i.e., the CH, CH_2 of chains in the β position and beyond, and the CH_3 hydrogens in the β position (H_β).

An important advantage of NMR over other spectral analysis methods comes from the signal's area being directly proportional to the number of protons. Therefore, in a spectrum, the area percentages for the different signals are related to the percentages of atoms. From that, it is easy to obtain the percentage of hydrogen for each of the above species. Furthermore, using the hypothesis that the average number of hydrogen atoms attached to the carbon atoms is two and using the results from elemental analysis of carbon and hydrogen, it is possible to deduce the following parameters:

• The aromaticity factor (ratio of the number of aromatic carbons to the total number of carbons), identical to that given by the *ndM* method or the Brandes method in the infrared region.

• The parameter giving the ratio of the number of effectively substituted aromatic carbon atoms to the number of substitutable carbons giving a

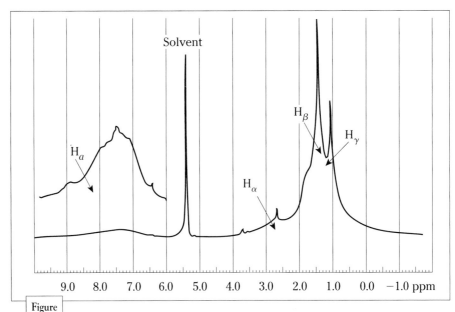

| Figure 3.10 | *Hydrogen NMR spectrum for a Boscan resin.* |

measurement of the degree of substitution of an "average" molecule representative of the sample.

- The parameter giving the ratio of the number of substitutable aromatic carbon atoms to the total number of aromatic carbons which gives a measure of the average condensation of aromatic rings.

- The average length of the chains attached to aromatic rings (Brown and Ladner, 1960).

One can see that the hydrogen NMR spectrum contains implicitly the molecular structure of all hydrocarbons contained in a petroleum cut. This same structure being the essential origin of the properties of cuts, correlations based on ^{1}H NMR structures can be studied. Correlations have been established allowing cetane indices to be calculated for gas oils. Once having determined the cetane number by the motor method for about one hundred samples, one can produce an equation to calculate a cetane index based on percentages of different types of hydrogen atoms measured using the NMR spectrum. Figure 3.11 is an example of one such correlation. The advantage of the NMR method is its rapidity, taking only a few minutes, and the small sample volume (a few ml) required; this is compatible with a micro-pilot unit's production.

Carbon NMR

Carbon 12, the most abundant naturally occurring isotope, has zero spin and thus cannot be studied by NMR. On the other hand, its isotope carbon 13 has an extra neutron and can be; its low natural occurrence (1.1%) nevertheless makes the task somewhat difficult. Only pulsed NMR can be utilized.

^{13}C NMR has several advantages over that of hydrogen:

- The range of chemical displacement is much wider (200 ppm as opposed to 15).

- For hydrocarbon studies, analyses can be made without prior assumptions, since the carbons not carrying protons can be excited directly, this of course not being the case for hydrogen (e.g., quaternary carbons in alkanes, substituted carbons in aromatic rings).

- Additionally, more sophisticated pulse sequences (the procedure is called spectral editing) enable one to obtain spectra, after addition or subtraction, where only the following are present (see, for example Bouquet, 1986):
  - only CH_3 and CH
  - only CH_2
  - only quaternary carbons.

The changes of three types of carbon atoms can be monitored during processing with this method:

(a) Aromatic CH groups.

(b) Aromatic carbons substituted by an alkyl chain.

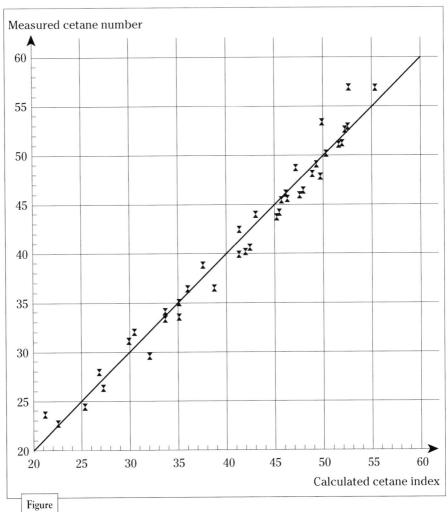

Figure
3.11 *Correlation between measured cetane numbers and those calculated by hydrogen NMR.*

(c) "Condensed" aromatic carbons belonging to two attached rings:

Types of aromatic carbons differentiated by carbon NMR (a, b, c are defined in the text and are reported in Figure 3.12).

Comparing the overall concentrations of these different carbons designated generally as "structural patterns", measured before and after a process such as FCC or hydrocracking (see Chapter 10), enables the conversion to be monitored; the simple knowledge of the percentage of condensed aromatic carbon of a feedstock gives an indication of its tendency to form coke.

Figure 3.12 shows the spectrum of carbon 13 obtained from a distillation residue and Table 3.10 gives average parameters for two FCC feedstocks as measured by NMR.

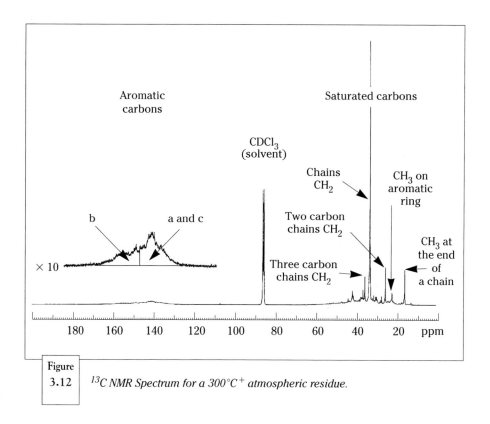

Figure 3.12 ^{13}C *NMR Spectrum for a $300°C^{+}$ atmospheric residue.*

It must be emphasized that NMR is first and foremost a tool for structural analysis and, in addition to the petroleum analyses described above, the technique (phosphorus NMR and sometimes nitrogen NMR) is abundantly used in all petrochemical synthesis operations.

Carbon NMR is also applied using the same principles as those for hydrogen to measure the cetane indices for diesel fuels.

Sample	Feed 1	Feed 2
Aromatic carbons		
Condensed quaternary C	5.2	3.2
Branched quaternary C	10.1	8.6
CH	9.5	9.7
Total aromatic C	24.8	21.6
C/H of aromatic C	2.611	2.222
Ring condensation index	2.1	1.6
Ring substitution index	0.515	0.471
Saturated carbons		
Quaternary C	4	0.7
CH	8.8	12.9
CH_2	49.6	54.3
CH_3	12.8	10.5
Total saturated carbon	75.2	78.4
C/H of saturated C	0.514	0.513

Table
3.10   *Structural patterns by ^{13}C NMR.*

3.3   **Characterization of Petroleum Fractions by Chromatographic Techniques**

Chromatographic techniques, particularly gas phase chromatography, are used throughout all areas of the petroleum industry: research centers, quality control laboratories and refining units. The applications covered are very diverse and include gas composition, search and analysis of contaminants, monitoring production units, feed and product analysis. We will show but a few examples in this section to give the reader an idea of the potential, and limits, of chromatographic techniques.

3.3.1   **Analysis of Permanent Gases and Noncondensable Hydrocarbons by Gas Phase Chromatography**

The most commonly encountered permanent gases are CO_2, H_2S, and N_2 associated with gas or crude oil production and N_2, CO, H_2, and O_2 found in refining and petrochemicals.

The noncondensable hydrocarbons comprise the hydrocarbons having less than five carbon atoms: methane, ethane, propane and butanes encountered in production; refining will add the olefins and diolefins:

ethylene, propylene, butenes, propadiene and butadienes, some of which are important petrochemical intermediates.

We will give two examples of analysis of these components, bearing in mind that the distinction between condensable and noncondensable hydrocarbons rarely holds in actual refining streams, most of them producing both classes of hydrocarbons simultaneously.

In this case, a preliminary separation will have taken place either in the plant by stabilization, or by the chromatograph which will have had a prefractionating column. This column will isolate the components having boiling points higher than pentane, allowing only the noncondensable hydrocarbons and a fraction of the pentanes to pass through to the analytical column.

3.3.1.1 **Natural Gas Analysis**

This type of analysis requires several chromatographic columns and detectors. Hydrocarbons are measured with the aid of a flame ionization detector FID, while the other gases are analyzed using a katharometer. A large number of combinations of columns is possible considering the commutations between columns and, potentially, backflushing of the carrier gas. As an example, the hydrocarbons can be separated by a column packed with silicone or alumina while O_2, N_2 and CO will require a molecular sieve column. H_2S is a special case because this gas is fixed irreversibly on a number of chromatographic supports. Its separation can be achieved on certain kinds of supports such as Porapak® which are styrene-divinylbenzene copolymers. This type of phase is also used to analyze CO_2 and water.

Natural gas analysis has considerable economic importance. In fact, commercial contracts increasingly specify not just volume but the calorific or heating value as well. Today the calorific value of a natural gas calculated from its composition obtained by chromatography is recognized as valid. There is therefore a large research effort devoted to increasing the precision of this analysis.

3.3.1.2 **Refinery Gas Analysis by Gas Phase Chromatography**

It is clear that these gases have widely varying compositions according to the processes used, but refinery gas is distinguished from natural gases by the presence of hydrogen, mono- and diolefins, and even acetylenes.

As in the preceding case, the analysis should be done in two steps: hydrogen will be separated first in a column and analyzed by katharometry. If the hydrogen concentration is small, there should be a large difference in thermal conductivity between the carrier gas and hydrogen in order to reduce the threshold of detection. Argon or nitrogen are among the choices.

The hydrocarbons are separated in another column and analyzed by a flame ionization detector, FID. As an example, Figure 3.13 shows the separation obtained for a propane analyzed according to the ISO 7941 standard. Note that certain separations are incomplete as in the case of ethane-ethylene. A better separation could be obtained using an alumina capillary column, for instance.

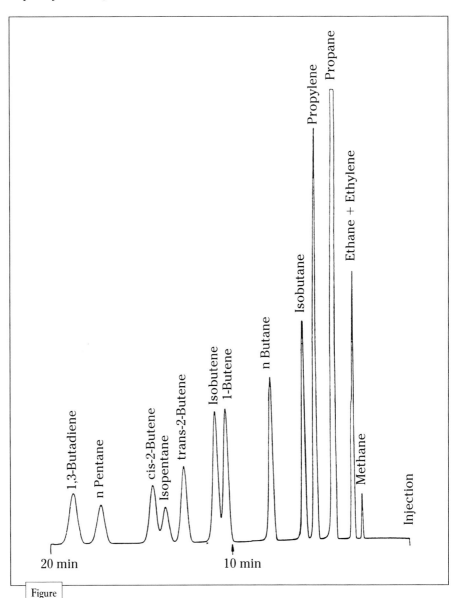

Figure 3.13   *Typical chromatogram obtained with a sebaconitrile column (mixture of reference containing LPG components).*

3.3.2 **Analysis of Hydrocarbons Contained in a Gasoline by Gas Phase Chromatography**

The resolution of capillary columns enables the separation of all principal components of a straight-run gasoline. The most frequently used stationary phases are silicone-based, giving an order of hydrocarbon elution times close to the order of increasing boiling point.

Gas chromatography is not an identification method; the components must be identified after their separation by capillary column. This is done by coupling to the column a mass spectrometer by which the components can be identified with the aid of spectra libraries. However the analysis takes a long time (a gasoline contains about two hundred components) so it is not practical to repeat it regularly. Furthermore, analysts have developed techniques for identifying components based on indices relating their retention times to those of normal paraffins (Kovats (1959) or Van den Dool (1963) indices). Using these indices and owing to the stability and reproducibility achieved in chromatographic ovens, automatic identification of chromatograms is now possible.

The analyst now has available the complete details of the chemical composition of a gasoline; all components are identified and quantified. From these analyses, the sample's physical properties can be calculated by using linear or non-linear models: density, vapor pressure, calorific value, octane numbers, carbon and hydrogen content.

It should be noted there is a cloud on this bright horizon; the best capillary columns, in spite of having the equivalent of hundreds of thousands of theoretical plates, lack sufficient resolving power to separate olefins having more than eight carbon atoms. Beyond that, the number of possible olefin isomers increases rapidly owing to combinations of double bond position, branching, and cis-trans isomers. This results in an unresolved chromatographic region encountered mainly in heavy gasolines from catalytic cracking and thermal conversion such as coking and visbreaking.

3.3.3 **Specific Analysis for Normal Paraffins by Gas Phase Chromatography**

This analysis is necessary because of the particular temperature behavior of these components. Normal paraffins are the first to crystallize as the temperature is reduced.

Knowledge of their quantity and their distribution by number of carbon atoms is indispensable for the evaluation of low temperature behavior of diesel motor fuels as well as the production and transport characteristics of paraffinic crudes.

Identification of normal paraffins by chromatography presents no special problems; with the exception of biodegraded crudes, they are clearly distinguished. The problem encountered is to quantify, as shown in Figure 3.14, the normal paraffin peaks that are superimposed on a background representing other hydrocarbons.

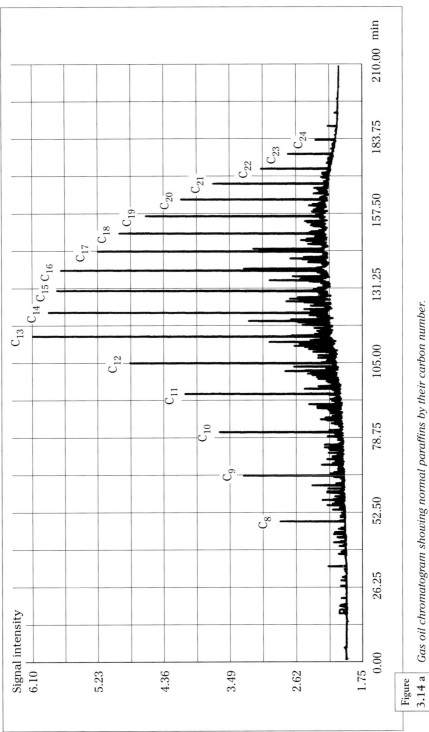

Figure
3.14 a   *Gas oil chromatogram showing normal paraffins by their carbon number.*

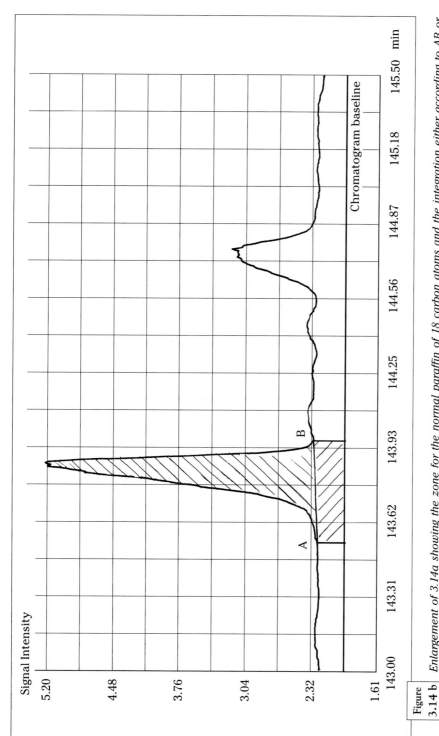

Figure 3.14 b *Enlargement of 3.14a showing the zone for the normal paraffin of 18 carbon atoms and the integration either according to AB or according to the base line.*

We have seen that it is the peak area which represents the concentration of a component. It is clear that this area will be overstated with respect to the base line or understated if the integration is calculated using the line AB, known as valley-to-valley integration. The limitation of this technique becomes apparent and it is difficult to overcome even with the use of very high resolution columns. In fact, integration limited to the line AB provides an acceptable result. This has been verified after having selectively extracted the normal paraffins by passage through 5 Å molecular sieves followed by a chromatographic analysis of this extract.

3.3.4 **Specific Detectors in Gas Phase Chromatography**

It is not possible to present all special detectors used in gas phase chromatography, but instead we will mention some recent applications.

By specific detectors (as compared to universal detectors such as FIDs or mass or infra red spectrometers) it is meant those that respond only to the presence of a particular atom in molecule. In the petroleum industry, the atoms said to be heteroatoms are essentially sulfur, nitrogen, oxygen, and to a lesser extent, nickel and vanadium. As seen previously, knowledge of the groups carrying these atoms is essential in the petroleum industry which seeks to eliminate them in order to improve product quality and to avoid catalyst poisoning.

3.3.4.1 **Specific Detection for Sulfur**

SCDs (Sulfur Chemiluminescence Detectors) have recently joined the photometric and electrochemical detectors. Upon combustion, sulfur compounds form sulfur monoxide, SO, which reacts with ozone to form a molecule of SO_2 in the excited state. The return to the normal molecular state is accompanied by an emission between 300 and 450 nm. It is this emission that is measured and thus allows detection and a selective quantification of sulfur molecules. The detector is very sensitive; in theory it responds to a few parts per billion of sulfur. Figure 3.15 illustrates one of the possible uses: a chromatogram shown in Figure 3.15a for a gas oil feedstock to a hydrotreating unit and, in Figure 3.15b, for this same gas oil after hydrotreatment. The efficiency of hydrodesulfurization is thus controlled showing the molecules most resistant to desulfurization and allowing modelling studies of the process reactions to be conducted.

3.3.4.2 **Specific Detection for Nitrogen**

Among the various detectors specific for nitrogen, the NPD (Nitrogen Phosphorus Thermionic Detector) we will consider, is based on the following concept: the eluted components enter a conventional FID burner whose air and hydrogen flows are controlled to eliminate the response for hydrocarbons.

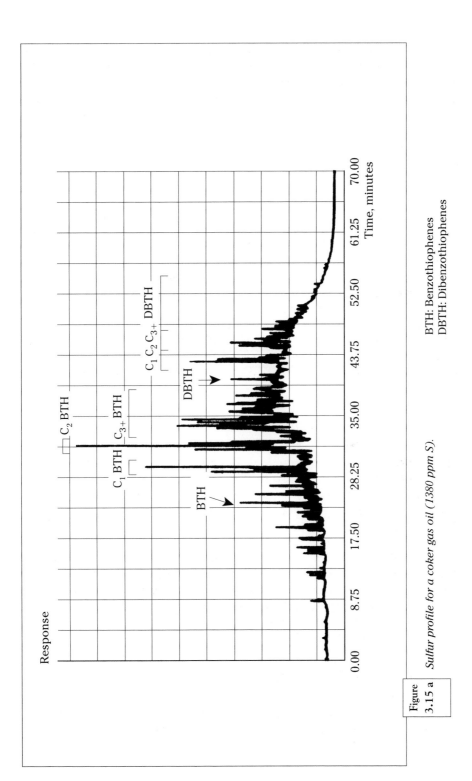

Figure 3.15 a *Sulfur profile for a coker gas oil (1380 ppm S).*

BTH: Benzothiophenes
DBTH: Dibenzothiophenes

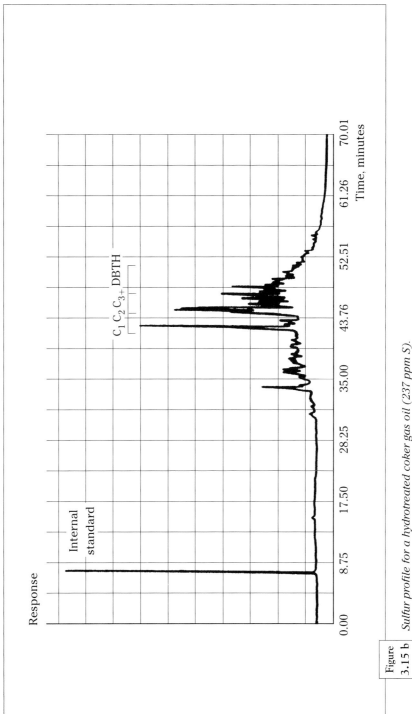

Figure
3.15 b   *Sulfur profile for a hydrotreated coker gas oil (237 ppm S).*

The collector contains an electrically-heated rubidium salt used as the thermionic source. During the elution of a molecule of a nitrogen compound, the nitrogen is ionized and the collection of these ions produces the signal. The detector is very sensitive but its efficiency is variable subject to the type of nitrogen molecule, making quantification somewhat delicate.

Figure 3.16 illustrates the chromatogram with NPD detection of basic nitrogen compounds extracted from a catalytically-cracked gas oil (LCO). The interest of this example is identical to that of the preceding one: interpretation of catalytic cracking reactions; study of the behavior of different species during a subsequent denitrification hydrotreatment; correlations between the content of certain nitrogen components and certain characteristics of LCO such as color and stability during storage.

3.3.4.3 **Specific Detection for Oxygen**

The need for this detector arose when gasolines containing oxygenates such as alcohols, and especially ethers, were introduced for which composition and content were subject to regulations. Following separation in a chromatographic column, the molecules pass through a cracker where the hydrocarbons are retained and where the molecules containing oxygen give CO. This gas is then sent to a methanizer where it is converted to CH_4 and detected thence by an FID. When well adjusted, the type of detector is very selective and sensitive enough to measure oxygen contents as low as a few dozen ppm.

3.3.4.4 **Atomic Emission Detectors**

On leaving the column, the components enter a plasma functioning in identical manner to that described in article 2.2.6.3. All the atoms present emit their own specific emission spectra and can be detected and quantified. The sensitivity is excellent for carbon, hydrogen, and sulfur and average for nitrogen and oxygen.

The sensitivity is very good for nickel and vanadium but for these metals for which distribution data would be of great value, the chromatographic process is the limiting factor, heavy molecules are not eluted from the column with the exception of some porphyrins. This detector can be used to supply H/C and S/C profiles for hydrocarbon cuts with the chromatograph operating in the "simulated distillation" mode.

3.3.5 **Analysis by Fluorescent Indicator in Liquid Chromatography**

This analysis, abbreviated as FIA for Fluorescent Indicator Adsorption, is standardized as ASTM D 1319 and AFNOR M 07-024. It is limited to fractions whose final boiling points are lower than 315°C, i.e., applicable to gasolines and kerosenes. We mention it here because it is still the generally accepted method for the determination of olefins.

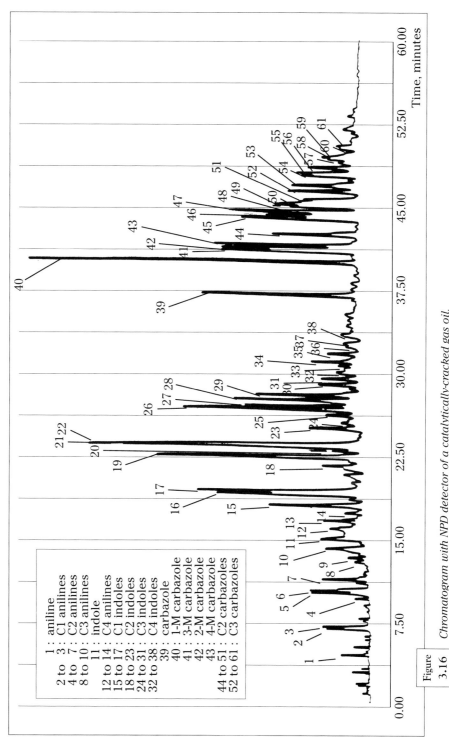

Figure 3.16   *Chromatogram with NPD detector of a catalytically-cracked gas oil.*

As stated earlier, these hydrocarbons are difficult to quantify with accuracy. The FIA method, which is a chromatographic adsorption on silica, gives volume percentages of saturated hydrocarbons, olefins and aromatics.

Some attempts, not yet very convincing, have been made to extend the field of application to gas oils by heating the column.

3.3.6 **Analysis of Aromatics in Diesel Motor Fuels by Liquid Chromatography**

Knowledge of the overall aromatics content and their distribution in mono-, di-, and polynuclear aromatics in diesel fuels has become necessary because of environmentally-related problems. These components are suspected of being at least partly responsible for diesel engine emissions. The need exists for a reliable and easy method (these considerations exclude spectrometry and NMR) to analyze aromatics. The method accepted by a commission of the European Committee of Standardization is liquid chromatography. The fixed phase is a silica modified by NH_2 groups, the eluent is normal heptane, and detection is by refractometry although this type of detector has some disadvantages described in section 2.1.2.4. Figure 3.17 shows a chromatogram of a diesel motor fuel. Monoaromatics are quantified by comparing their response to that of an orthoxylene standard, the diaromatics to a standard of α-methylnaphthalene and the polynuclear aromatics to a phenanthrene standard. The accuracy and reproducibility obtained during inter-laboratory round-robin tests are satisfactory and mean this method may be adopted in the near future as a European standard.

We would add that this method is applicable to other gas oil cuts by ultimately changing if necessary the standards for refractometric response.

3.3.7 **SARA Analysis of Heavy Fractions by Preparative Liquid Chromatography**

SARA (Saturates, Aromatics, Resins, Asphaltenes) analysis is widely practiced on heavy fractions such as vacuum and atmospheric residues and vacuum distillates for two purposes:

- knowledge of the weighted quantities of these four classes of components which is in itself extremely valuable to the refiner
- preparation of fractions that could be subject to further analysis: mass spectrometry for the vaporizable fractions, proton or carbon NMR for all fractions.

The SARA separation is conducted in several stages.

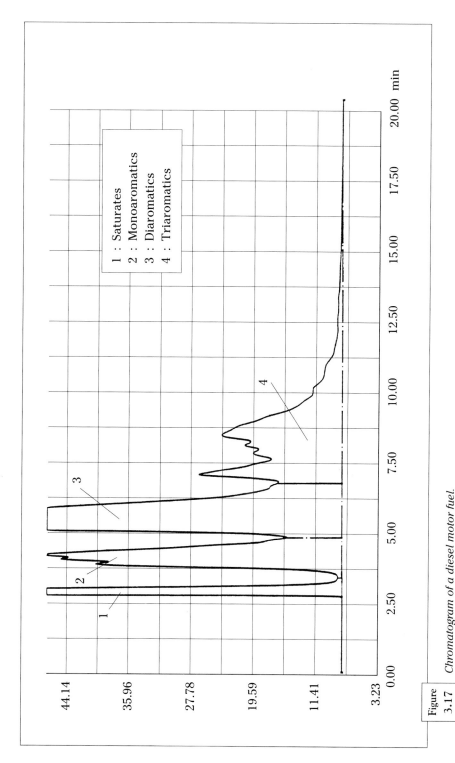

Figure
3.17

Chromatogram of a diesel motor fuel.

Liquid chromatography is preceded by a precipitation of the asphaltenes, then the maltenes are subjected to chromatography. Although the separation between saturated hydrocarbons and aromatics presents very few problems, this is not the case with the separation between aromatics and resins. In fact, resins themselves are very aromatic and are distinguished more by their high heteroatom content (this justifies the terms, "polar compounds" or "N, S, O compounds", also used to designate resins).

Chromatograms do not give a sharp differentiation between aromatics and resins.

The yield of each of these fractions will depend on their retention volume which in turn will depend on the adsorbent selected and the eluting force of the solvents.

Because this separation is not subject to precise standards today, the resulting wide variations make comparisons between laboratories risky.

Nevertheless, within the same work group, once the chromatographic procedures are established, SARA analyses are very often performed to characterize heavy feedstocks or to follow their conversion.

Without going into details of the chromatographic method, a SAR separation (asphaltenes having been eliminated) can be performed in a mixed column of silica followed by alumina. The saturated hydrocarbons are eluted by heptane, the aromatics by a 2:1 volume mixture of heptane and toluene, and the resins by a 1:1:1 mixture of dichloromethane, toluene and methanol.

3.4 Characterization of Petroleum Fractions Based on Chemical Reactions

Despite numerous efforts using various techniques, precise knowledge of olefin content remains an unresolved problem. That is why it is worthwhile to mention two methods commonly employed which provide an answer to the problem.

3.4.1 Bromine Number

This method follows the ASTM D 1159 and D 2710 procedures and the AFNOR M 07-017 standard. It exploits the capacity of the double olefinic bond to attach two bromine atoms by the addition reaction. Expressed as grams of fixed bromine per hundred grams of sample, the bromine number, BrN, enables the calculation of olefinic hydrocarbons to be made if the average molecular weight of a sufficiently narrow cut is known.

The reactivity of polyolefins to bromine addition may not be complete or it can depend on the relative positions of the double bonds and the branching within the molecules provoking steric hindrance to addition.

3.4.2 **Maleic Anhydride Value**

Represented by its abbreviation, MAV, the Maleic Anhydride Value is based on the fact that olefinic conjugated double bonds can be added to maleic anhydride by the reaction below:

$$
\begin{array}{ccc}
\underset{HC}{\overset{CH_2}{\diagup}} \quad + \quad \underset{CH-CO}{\overset{CH-CO}{|\!|}}\diagdown O & \longrightarrow & \underset{HC}{\overset{CH_2}{\diagup}}\underset{CH-CO}{\overset{CH-CO}{|\!|}}\diagdown O \\
\underset{HC}{\diagdown}\underset{CH_2}{\diagup} & & \underset{HC}{\diagdown}\underset{CH_2}{\diagup}
\end{array}
$$

The reaction takes place with an excess of reactant which is analyzed later in accordance with a method described by the Amoco company.

MAV is expressed in mg of anhydride per gram of sample. It is still widely used to evaluate the quantity of conjugated olefins in a fraction. This type of molecule is highly undesirable in a large number of end products because of its propensity to polymerize spontaneously and to form gums.

Note that styrenes are believed to react incompletely with the anhydride.

Methods for the Calculation
of Hydrocarbon Physical Properties

Henri Paradowski

Knowledge of physical properties of fluids is essential to the process engineer because it enables him to specify, size or verify the operation of equipment in a production unit. The objective of this chapter is to present a collection of methods used in the calculation of physical properties of: mixtures encountered in the petroleum industry, different kinds of hydrocarbon components, and some pure compounds.

The methods that will be described are widely used and, for the most part, are integrated into commercial simulation softwares such as PRO2, ASPEN[+] or HYSIM. They constitute the *de facto* standards and one forgets too often that they do have limited accuracy and range of application. During their integration into software programs, these methods sometimes are subjected to questionable modifications and generalizations.

In spite of considerable development of thermodynamics and molecular theory, most of the methods used today are empirical and their operation requires knowledge of experimental values. However, the rate of accumulation of experimental data seems to be slowing down even though the need for precise values is on the rise. It is then necessary to rely on methods said to be predictive and which are only estimates.

The generalized use of computers makes seemingly complex calculations quite easy to perform; however, curves and tables are still invaluable when one needs to obtain approximate values or to take into account the sensitivity of a property to operating conditions or to a mixture's characteristics.

4.1 Characterization Required for Calculating Physical Properties

The current calculation methods are based on the hypothesis that each mixture whose properties are sought can be characterized by a set of pure components and petroleum fractions of a narrow boiling point range and by a composition expressed in mass fractions.

The set should have a finite dimension and it is customary to restrict its number to 50. In exceptional cases, it is necessary to use up to 100 constituents.

Because of the existence of numerous isomers, hydrocarbon mixtures having a large number of carbon atoms can not be easily analyzed in detail. It is common practice either to group the constituents around key components that have large concentrations and whose properties are representative, or to use the concept of petroleum fractions. It is obvious that the grouping around a component or in a fraction can only be done if their chemical natures are similar. It should be kept in mind that the accuracy will be diminished when estimating certain properties particularly sensitive to molecular structure such as octane number or crystallization point.

In the rest of this chapter, we will review the important items that are necessary or useful for the calculations.

4.1.1 Characteristics of Pure Hydrocarbons

Characteristics are the experimental data necessary for calculating the physical properties of pure components and their mixtures. We shall distinguish several categories:

- principal characteristics
- critical constants required for applying the principle of corresponding states
- special coefficients for calculating certain properties.

a. Principal Characteristics

We will limit the principal characteristics to the following items:

- name
- chemical formula
- normal boiling point
- standard specific gravity or density at a given temperature
- liquid viscosity at two temperatures
- molecular weight.

b. Critical Constants

- critical temperature
- critical pressure
- critical compressibility factor.

c. Special Coefficients

These are necessary for precise determination of certain physical properties; they include following items:

- acentric factor
- coefficients for calculating the enthalpy, entropy, and specific heat of an ideal gas
- latent heat of vaporization at the normal boiling point
- melting point temperature
- latent heat of fusion
- liquid conductivity at two temperatures.

d. Other Special Coefficients

- coefficient *"m"* used in Soave's method
- parachor
- interfacial tension at 20°C
- solubility coefficient at 25°C.

The above coefficients are not absolutely required because they can be estimated from other characteristics.

4.1.1.1 **Available Data for Pure Hydrocarbons**

In Appendix 1, the reader will find the data required to calculate the properties of the most common hydrocarbons as well as those components that most frequently accompany them in refinery process streams. The data are grouped in seven categories:

- non-hydrocarbons $(H_2, N_2, CO, O_2, CO_2, COS, H_2S, NH_3, CH_3SH,$ $C_2H_5SH, H_2O)$
- light hydrocarbons $(C_1$ to $C_4)$
- n-paraffins $(C_5$ to $C_{20})$
- iso-paraffins $(C_5$ to $C_9)$
- naphthenes $(C_5$ to $C_9)$
- olefins $(C_5$ to $C_9)$
- aromatics $(C_6$ to $C_{10}).$

This data bank is obviously not complete.

Supplementary information can be found in the following references and programs:

(a) Reid and Sherwood *"The Properties of Gases and Liquids"*. The original edition was in 1958. It is preferable to refer to the third edition of 1977, authored by Reid, Prauznitz and Sherwood or the fourth edition published in 1987, under the signature of Reid, Prausnitz and Poling. These works are published by Mac Graw-Hill.

(b) API Technical Data Book of Petroleum Refining – Volumes 1, 2 and 3, published by the American Petroleum Institute.

(c) DIPPR (Design Institute for Physical Property Data).

   The DIPPR is a research organization sponsored by the AIChE (American Institute of Chemical Engineers). Its objective is to develop a thermophysical data bank for the components most frequently encountered in the chemical industry.

   Data for pure components are available in software edited by TDS (Technical Data Services):

   • The DIPPR Pure Component Data Compilation, Numerica (TM) Version 9.2 (1994).

4.1.1.2 **Methods of Contributing Groups for Pure Hydrocarbons**

When data for a particular component can not be obtained from data banks, the engineer has two choices:

   • Group the component in a petroleum fraction, which is possible if the normal boiling temperature and the standard specific gravity are known. This method gives correct results when the chemical structure is simple as in the case of a paraffin or naphthene.

   • Generate the data from the method of contributing groups which requires knowledge of the chemical structure and some careful attention.

a. *Contributing Groups for Determining Critical Constants of a Pure Component*

The critical constants are obtained by applying Lydersen's method (1955):

$$T_c = \frac{T_b}{0.567 + \Sigma\, dT_{c_i} - \left(\Sigma\, dT_{c_i}\right)^2} \qquad [4.1]$$

where

T_c = critical temperature   [K]

T_b = normal boiling point   [K]

i = group index

dT_{c_i} = increments; see Table 4.1

$$P_c = \frac{1.013\,M}{\left(0.34 + \Sigma\ dP_{c_i}\right)^2} \qquad\qquad [4.2]$$

where

P_c = critical pressure   [bar]

M = molecular weight   [kg/kmol]

dP_{c_i} = increments; see Table 4.1

The average error is about 2% for the critical temperatures and pressures. The error increases with molecular weight and can reach 5%.

Aliphatic groups	dT_{c_i}	dP_{c_i}
– CH$_3$	0.020	0.227
– CH$_2$ –	0.020	0.227
> CH –	0.012	0.210
> C <	0.000	0.210
= CH$_2$	0.018	0.198
= CH –	0.018	0.198
= C <	0.000	0.198
= C =	0.000	0.198
≡ CH	0.005	0.153
≡ C –	0.005	0.153
Cyclic groups		
– CH$_2$ –	0.013	0.184
> CH –	0.012	0.192
> C <	−0.007	0.154
= CH –	0.011	0.154
= C <	0.011	0.154
= C =	0.011	0.154

Table 4.1

Increments for Lydersen's method (1955) used for calculating critical constants.

The acentric factor is calculated using Edmister's equation (1948):

$$\omega = \frac{3}{7}\left(\frac{x}{1-x}\right)\left(\log P_c - 1.0057\right) \qquad\qquad [4.3]$$

where $x = \dfrac{T_b}{T_c}$

and

T_b = normal boiling point   [K]

T_c = critical temperature   [K]

log = common logarithm (base 10)

P_c = critical pressure   [bar]

The critical compressibility factor is estimated using the Lee and Kesler equation (1975):

$$z_c = 0.291 \div 0.08\omega \qquad [4.4]$$

where

z_c = critical compressibility factor

ω = acentric factor

b. Contributing Groups for Determining Enthalpy, Entropy, and Specific Heat of a Component in the Ideal Gas State

The thermal properties of an ideal gas, enthalpy, entropy and specific heat, can be estimated using the method published by Rihani and Doraiswamy in 1965:

$$H_{gp} = AT + \frac{B}{2} 10^{-2} T^2 + \frac{C}{3} 10^{-4} T^3 + \frac{D}{4} 10^{-6} T^4 \qquad [4.5]$$

$$A = \sum a_i \qquad B = \sum b_i \qquad C = \sum c_i \qquad D = \sum d_i$$

where

T = temperature   [K]

H_{gp} = enthalpy of the gas in the ideal state   [kJ/kmol]

a_i = group contributions: see Table 4.2

b_i = group contributions: see Table 4.2

c_i = group contributions: see Table 4.2

d_i = group contributions: see Table 4.2

This method is also applicable to other organic compounds for which one should refer to the original publication. The average error is around 2%.

c. Contributing Groups for Determining the Parachor of a Pure Component

The Parachor is a parameter used to determine the interfacial tension. It can be estimated by a simple method proposed by Quayle in 1953:

$$Pa = \sum pa_i \qquad [4.6]$$

where

Pa = parachor

pa_i = group contributions given in Table 4.3

d. Contributing Groups for Determination of the Liquid Viscosity of a Pure Component

Van Velzen's method provides an estimation of hydrocarbon viscosities and their variation with temperature:

$$\log \mu = B\left(\frac{1}{T} - \frac{1}{T_0}\right) \qquad [4.7]$$

Aliphatic groups	a_i	b_i	c_i	d_i
$- CH_3$	2.54741	8.96971	−0.35656	0.04750
$- CH_2 -$	1.65098	8.94042	−0.50094	0.01086
$= CH_2$	2.20382	7.68240	−0.39925	0.00816
$> CH -$	−14.74459	14.29512	−1.17850	0.03354
$> C <$	−24.40148	18.64041	−1.76105	0.05286
$- CH = CH_2$	1.16050	14.47173	−0.80268	0.01728
$> C = CH_2$	−1.74640	16.26165	−1.16469	0.03082
$cis - CH = CH -$	3.92427	12.51482	−0.73196	0.01640
$trans - CH = CH -$	−13.06138	15.92811	−0.98724	0.02303
$> C = C <$	1.98202	14.72409	−1.31828	0.03852

Aromatic groups	a_i	b_i	c_i	d_i
$- CH -$	−6.09838	8.01302	−0.51601	0.01249
$a - C <$	−5.81004	6.34404	−0.44738	0.01113
$- C <$	0.51015	5.09314	−0.35782	0.00888

Corrections for rings	a_i	b_i	c_i	d_i
Cyclopentane	−51.41273	7.78787	−0.43398	0.00898
Cyclopentene	−28.79824	3.27183	−0.14438	0.00247
Cyclohexane	−56.04678	8.95255	−0.17954	−0.00781
Cyclohexene	−33.57960	9.30702	−0.80143	0.02290

Table 4.2   *Coefficients of Rihani's and Doraiswamy's method (1965) for calculating enthalpy, entropy, and the C_p for an ideal gas.*
Note: (a) signifies aliphatic.

Group	pa_i
$- H$	15.5
$- CH_3$	55.5
$- CH_2 -$	40.0
$> CH -$	24.5
$> C <$	9.0
$= CH_2$	59.1
$= CH -$	42.2
$= C <$	26.7
$= C =$	44.4
$\equiv CH$	65.1
$\equiv C -$	49.6
Phenyl	189.6

Corrections for rings	pa_i
C_5 rings	3.0
C_6 rings	0.8

Table 4.3   *Group contributions for estimating the parachor by Quayle's method (1953).*

$$B = Ba + \sum dB_i \qquad\qquad N = N_c + \sum dN_i$$

when the number of carbon atoms, $N_c \leq 20$:

$$T_0 = 28.86 + 37.439\,N - 1.3547\,N^2 + 0.0207\,N^3$$
$$Ba = 24.79 + 66.885\,N - 1.3173\,N^2 + 0.00377\,N^3$$

and when $N_c > 20$:

$$T_0 = 238.59 + 8.164\,N \quad Ba = 530.59 + 13.74\,N$$

where

N_c = number of carbon atoms

dN_i = group contributions for calculating N, see Table 4.4

dB_i = group contributions for calculating B, see Table 4.4

T = temperature   [K]

μ = viscosity   [mPa·s]

The average error with this method is about 20%.

e. Contributing Groups for Calculating other Properties of Pure Components

Contributing group methods have been developed and published for calculating numerous other properties. However, for our purposes, it is not necessary to employ them. Moreover, there are some properties for which the method is not recommended.

Group		dN_i	dB_i
Isoparaffins		$1.389 - 0.238\,N_c$	15.5100
Linear olefins		$-0.152 - 0.042\,N_c$	$-44.9400 + 5.41\,N$
Linear diolefins		$-0.304 - 0.084\,N_c$	$-44.94 + 5.41\,N$
Non linear olefins		$1.237 - 0.280\,N_c$	$-36.0100 + 5.41\,N$
Non linear diolefins		$1.085 - 0.322\,N_c$	$-36.0100 + 5.41\,N$
Alkyl cyclopentanes	$N_c < 16$	$0.205 + 0.069\,N_c$	$-45.9600 + 2.224\,N$
Alkyl cyclopentanes	$N_c > 16$	$3.971 - 0.172\,N_c$	$-339.6700 + 23.135\,N$
Alkyl cyclohexanes	$N_c < 16$	1.480	$-272.8500 + 25.041\,N$
Alkyl cyclohexanes	$N_c > 16$	$6.571 - 0.311\,N_c$	$-272.8500 + 25.041\,N$
Alkyl benzenes	$N_c < 16$	0.600	$-140.0400 + 13.869\,N$
Alkyl benzenes	$N_c > 16$	$3.055 - 0.161\,N_c$	$140.0400 + 13.869\,N$
Ortho-methyl group		0.510	54.8400
Meta-methyl group		0.110	27.2500
Para-methyl group		-0.040	-17.8700
Polyphenyls		$-5.340 + 0.815\,N_c$	$-188.4000 - 9.558\,N$

Table 4.4   *Group contributions for calculating the liquid viscosity.*

Superfluous Usage

Using the principle of corresponding states for the following characteristics avoids the use of the contributing groups' method:

- latent heat of vaporization
- vapor pressure.

These properties are calculated directly from the critical constants.

Usage not Recommended

The method of contributing groups does not apply with sufficient accuracy for the following calculations:

- latent heat of fusion
- melting point.

4.1.2 **Characterization of Petroleum Fractions**

We will use the term **petroleum fraction** to designate a mixture of hydrocarbons whose boiling points fall within a narrow temperature range, typically as follows:

- 10°C for light fractions with boiling points less than 200°C
- 15°C for fractions with boiling points between 200 and 400°C
- 20°C for fractions with boiling points between 400 and 600°C
- 30°C for fractions with boiling points beyond 600°C.

In a general manner, the following expression should be obeyed:

$$0.020 < \frac{dT_b}{T_b} < 0.035$$

where

dT_b = normal boiling point interval   [K]

T_b = average normal boiling point   [K]

4.1.2.1 **Normal Boiling Point of Petroleum Fractions, T_b**

This is the average boiling temperature at atmospheric pressure (1.013 bar abs). This characteristic is obtained by direct laboratory measurement and is expressed in K or °C.

When the boiling point is measured at a pressure other than normal atmospheric, the normal boiling point can be calculated by a method described in article 4.1.3.4.

If the boiling temperature is not known, it is somewhat risky to estimate it. One could, if the Watson characterization factor is known, use the following relation:

$$T_b = \frac{\left(K_W \cdot S\right)^3}{1.8} \qquad [4.8]$$

where

T_b = normal boiling point   [K]

K_W = Watson characterization factor (see article 4.1.2.5)

S = standard specific gravity (see article 4.1.2.2)

4.1.2.2 Standard Specific Gravity of Petroleum Fractions, S

The standard specific gravity is the ratio of the density of a hydrocarbon at 15.55°C (60°F) to that of water at the same temperature. It differs from the specific gravity d_4^{15} which is the ratio of the density of a hydrocarbon at 15°C to that of water at 4°C.

The standard specific gravity can be estimated from d_4^{15} using the following relation:

$$S = 1.001 \, d_4^{15}$$ [4.9]

Gravity is also expressed in degrees API:

$$A = \frac{141.5}{S} - 131.5$$ [4.10]

where

A = gravity in degrees API

S = standard specific gravity

This characteristic is obtained by laboratory measurement

It is common that a mixture of hydrocarbons whose boiling points are far enough apart **(petroleum cut)** is characterized by a distillation curve and an average standard specific gravity. It is then necessary to calculate the standard specific gravity of each fraction composing the cut by using the relation below [4.8]:

$$S_i = \frac{\left(1.8 \, T_{b_i}\right)^{1/3}}{K_W}$$

where

S_i = standard specific gravity of the petroleum
        cut under consideration

T_{b_i} = normal boiling point of the fraction   [K]

K_W = Watson characterization factor (see article 4.1.2.5)

This factor is presumed identical for all the petroleum fractions of the cut under consideration.

4.1.2.3 Liquid Viscosity of Petroleum Fractions
at Two Temperatures

The absolute or dynamic viscosity is defined as the ratio of shear resistance to the shear velocity gradient. This ratio is constant for Newtonian fluids.

The viscosity is expressed in Pa·s. The commonly used unit is mPa·s, formerly called centipoise, cP.

The kinematic viscosity is defined as the ratio between the absolute viscosity and the density. It is expressed in m^2/s. The most commonly used unit is mm^2/s formerly called centistoke, cSt.

The liquid dynamic viscosities at 100°F and 210°F are used to characterize petroleum fractions, notably the heavy fractions.

The temperatures 100°F and 210°F (37.8°C, 98.9°C) have been selected because they were initially used in the ASTM procedure for calculating the viscosity index of petroleum cuts (ASTM D 2270).

In 1991 they were replaced by the temperatures, 40°C and 100°C, in the definition of viscosity index.

When the viscosities are not known, they can be estimated by the relations of Abbott et al. (1971):

$$\log v_{100} = 4.39371 - 1.94733 K_W + 0.12769 K_W^2$$

$$+ 3.2629 \cdot 10^{-4} A^2 - 1.18246 \cdot 10^{-2} K_W A$$

$$+ \frac{\left(0.171617 K_W^2 + 10.9943 A + 9.50663 \cdot 10^{-2} A^2 - 0.860218 K_W A\right)}{\left(A + 50.3642 - 4.78231 K_W\right)}$$

$$[4.11]$$

$$\log v_{210} = -0.463634 - 0.166532 A + 5.13447 \cdot 10^{-4} A^2 - 8.48995 \cdot 10^{-3} K_W A$$

$$+ \frac{\left(8.0325 \cdot 10^{-2} K_W + 1.24899 A + 0.19768 A^2\right)}{\left(A + 26.786 - 2.6296 K_W\right)}$$

$$[4.12]$$

where

K_W = Watson characterization factor (see. 4.1.2.5)

A = gravity in degrees API

v_{210} = viscosity at 210°F $[mm^2/s]$

v_{100} = viscosity at 100°F $[mm^2/s]$

log = common logarithm (base 10)

These relations should not be used if $K_W < 10$ and $A < 0$.

Their use is recommended for within the following range:

$$0.5 < v_{100} < 20 \ mm^2/s \quad \text{and} \quad 0.3 < v_{210} < 40 \ mm^2/s$$

Average error is on the order of 20%.

4.1.2.4 **Molecular Weight of Petroleum Fractions**

When direct measurement is not available, the molecular weight can be estimated by two different means:

- from the normal boiling point and the standard specific gravity
- from the viscosities at 210°F and 100°F and the standard specific gravity.

API recommends a formula established by Riazi in 1986:

$$M = 42.965 \left[\exp\left(2.097 \cdot 10^{-4} T_b - 7.78712 S + 2.08476 \cdot 10^{-3} T_b S \right) \right]$$
$$\left(T_b^{1.26007} S^{4.98308} \right) \qquad [4.13]$$

For heavy fractions whose boiling temperatures exceed 600 K, it is better to use the method published by Lee and Kesler in 1975:

$$M = -12272.6 + 9486.4 S + T_b (8.3741 - 5.9917 S)$$

$$+ \frac{10^7}{T_b} \left(1 - 0.77084 S - 0.02058 S^2 \right) \left(0.7465 - \frac{222.466}{T_b} \right)$$

$$+ \frac{10^{12}}{T_b^{3}} \left(1 - 0.80882 S + 0.02226 S^2 \right) \left(0.32284 - \frac{17.3354}{T_b} \right) \qquad [4.14]$$

where

M = molecular weight [kg/kmol]

T_b = normal boiling point [K]

S = standard specific gravity

Riazi's method applies to fractions whose specific gravities are less than 0.97 and whose boiling points are less than 840 K. The Lee and Kesler method is applicable for fractions having molecular weights between 60 and 650.

The average error for both methods is around 5%.

The molecular weight can be also estimated for petroleum fractions whose boiling point is not known precisely starting with a relation using the viscosities at 100 and 210°F:

$$M = 223.56 \, \nu_{100}^{(1.1228 S - 1.2435)} \, \nu_{210}^{(3.4758 - 3.038 S)} \, S^{-0.6665} \qquad [4.15]$$

where

M = molecular weight [kg/kmol]

ν_{100} = kinematic viscosity at 100°F (37.8°C) [mm^2/s]

ν_{210} = kinematic viscosity at 210°F (98.9°C) [mm^2/s]

S = standard specific gravity

The average error is about 10%.

4.1.2.5 **The Watson Characterization Factor for Petroleum Fractions**

This factor is expressed by the following relationship:

$$K_W = \frac{\left(1.8\,T_b\right)^{1/3}}{S} \tag{4.8}$$

where

T_b = normal boiling point [K]
S = standard specific gravity
K_W = Watson characterization factor

This factor is the intermediate parameter employed in numerous calculational methods. For petroleum cuts obtained by distillation from the same crude oil, the Watson factor K_W is generally constant when the boiling points are above 200°C.

The K_W values for hydrocarbons of different chemical families (see Chapter 3) are as follows:

- paraffins show a K_W of about 13
- naphthenes have a K_W of about 12
- aromatics show a K_W of about 10.

4.1.2.6 **Pseudo-Critical Constants and Acentric Factors for Petroleum Fractions**

Using the principle of corresponding states requires knowledge of pseudo-critical constants of petroleum fractions; these should be estimated starting from characteristic properties which are the normal boiling temperature and the standard specific gravity.

The estimation of the three parameters — pseudo-critical temperature, pseudo-critical pressure, and the acentric factor — should be done using the same method because these constants should be coherent.

We will use the method established by Lee and Kesler in 1975 because it is related to the calculation of thermal properties method we have selected and will discuss later.

a. *Pseudo-Critical Temperature*

$$T_c = 189.8 + 450.6\,S + T_b\,(0.4244 + 0.1174\,S) + \frac{(14{,}410 - 100{,}688\,S)}{T_b} \tag{4.16}$$

where

T_c = pseudo-critical temperature [K]
S = standard specific gravity
T_b = normal boiling point [K]

The average error is about 10 K.

b. Pseudo-Critical Pressure

$$\ln P_c = 5.68925 - \frac{0.0566}{S} - 10^{-3} T_b \left(0.436392 + \frac{4.12164}{S} + \frac{0.213426}{S^2} \right)$$

$$+ 10^{-7} T_b^2 \left(4.75794 + \frac{11.819}{S} + \frac{1.53015}{S^2} \right) - 10^{-10} T_b^3 \left(2.45055 + \frac{9.901}{S^2} \right) \quad [4.17]$$

where

P_c = pseudo-critical pressure   [bar]

ln = Napierian logarithm

The average error is about 5%.

c. Acentric Factor

When the reduced boiling point is greater than 0.8, it is not recommended to use the conventional formula [4.3]. The acentric factor should be estimated by the following relation:

$$\omega = -7.904 + 0.1352 K_W - 0.007465 K_W^2 + 8.359 T_{br} + \frac{\left(1.408 - 0.01063 K_W \right)}{T_{br}} \quad [4.18]$$

$$T_{br} = \frac{T_b}{T_c}$$

where

ω = acentric factor

T_{br} = reduced boiling point temperature

K_W = Watson characterization factor

4.1.3 Characterization of Mixtures of Pure Hydrocarbons and Petroleum Fractions (Petroleum Cuts)

As seen in Chapter 2, mixtures of hydrocarbons and petroleum fractions are analyzed in the laboratory using precise standards published by ASTM (American Society for Testing and Materials) and incorporated for the most part into international (ISO), European (EN) and national (NF) collections. We will recall below the methods utilizing a classification by boiling point:

- D 2892 Petroleum distillation method employing a 15 theoretical plate column, called TBP (True Boiling Point).
- D 2887 Method for determining the distribution of boiling points of petroleum cuts by gas chromatography called SD (Simulated Distillation).
- D 3710 Method for determining the distribution of boiling points of light gasolines by gas phase chromatography.

- D 86   Distillation method for light petroleum products.
- D 1160 Method for reduced pressure distillation of high-boiling petroleum products.
- D 1078 Distillation method for volatile organic liquids.

Non-standard distillation equipment having up to 100 plates and operating at high reflux rates is also used. The fractionation is very efficient and gives a precise distribution of boiling points.

Tests employing the less-efficient distillations, D 86, D 1160, and D 1078 are generally conducted on refined products while those giving a detailed analysis, D 2887 and D 2892, are concerned mostly with crude oils and feeds to and effluents from conversion units.

From the analytical results, it is possible to generate a model of the mixture consisting of an N_c number of constituents that are either pure components or petroleum fractions, according to the schematic in Figure 4.1. The real or simulated results of the atmospheric TBP are an obligatory path between the experimental results and the generation of bases for calculation of thermodynamic and thermophysical properties for different cuts.

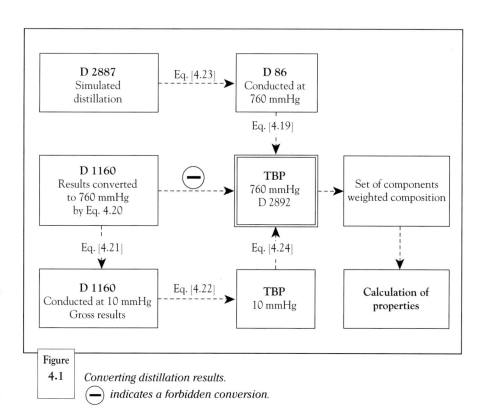

Figure
4.1   *Converting distillation results.*
⊖ *indicates a forbidden conversion.*

Transformation equations are as follows:

- Transformation of ASTM D 86 results into an atmospheric TBP, equation [4.19].
- Transformation of TBP results at 10 mmHg into an atmospheric TBP, equation [4.24].
- Transformation of simulated distillations, D 3710 and D 2887 results into an ASTM D 86, equation [4.23].
- Transformation of ASTM D 1160 results into a TBP at 10 mmHg, equation [4.22].
- Transformation of an ASTM D 1160 at 760 mmHg data into ASTM D 1160 results at 10 mmHg, equation [4.21].

4.1.3.1 **ASTM D 86 Distillation for Light Petroleum Cuts**

This is the most common method. It is used for gasolines, kerosenes, gas oils and similar products. The test is conducted at atmospheric pressure and is not recommended for gasolines having high dissolved gas contents or solvents whose cut points are close together.

The result is a distillation curve showing the temperature as a function of the per cent volume distilled (initial point, 5, 10, 20, 30, 40, 50, 60, 70, 80, 90 and 95% distilled volume, and final boiling point).

The results can be converted into an atmospheric TBP by using an equation equivalent to that proposed by Riazi and published by the API:

$$T' = a \cdot T^b \tag{4.19}$$

where

T' = temperature of the simulated TBP test   [K]

T = D 86 test temperature   [K]

a, b = constants depending on the fraction distilled

Table 4.5 gives the coefficients a and b as well as the conversion of D 86 results into an atmospheric TBP.

The accuracy of the conversion depends on the smoothness of the D 86 curve. Errors affect essentially the points in the low % distilled ranges. Average error is on the order of 5°C for conversion of a smooth curve.

Daubert has recently published a new method (Hydrocarbon Processing, September 1994, page 75) to convert D 86 data to TBP results using the following equations:

$$T'_{50} = A_4 \cdot (T_{50})^{B4} \tag{4.19a}$$

$$\begin{aligned}
T'_{30} &= T'_{50} - \Delta T'_3 & \Delta T_3 &= T_{50} - T_{30} \\
T'_{10} &= T'_{30} - \Delta T'_2 & \Delta T_2 &= T_{30} - T_{10} \\
T'_i &= T'_{10} - \Delta T'_1 & \Delta T_1 &= T_{10} - T_i \\
T'_{70} &= T'_{50} + \Delta T'_5 & \Delta T_5 &= T_{70} - T_{50} \\
T'_{90} &= T'_{70} + \Delta T'_6 & \Delta T_6 &= T_{90} - T_{70}
\end{aligned}$$

$$T'_{95} = T'_{90} + \Delta T'_7 \qquad \Delta T_7 = T_f - T_{90}$$
$$\Delta T'_i = A_i \cdot (\Delta T_i)^{Bi}$$

where

T'_i = TBP temperature at initial point   [°F]

T'_f = TBP temperature at final point   [°F]

T'_n = TBP temperature at n % volume distilled   [°F]

T_n = D 86 temperature at n % volume distilled   [°F]

A_i and B_i are coefficients given in Table 4.5

Table 4.5 shows the results for an example. These results differ significantly from those obtained by the method of Riazi for the initial and 10% distilled points. The reported average error for this method is about 3°C, except for the initial point where it reaches 12°C.

% distilled volume	Coefficient *a*	Coefficient *b*	Temperature D 86, °C	Temperature TBP, °C
0	0.9177	1.0019	36.5	14
10	0.5564	1.0900	54.0	33
30	0.7617	1.0425	77.0	69
50	0.9013	1.0176	101.5	102
70	0.8821	1.0226	131.0	135
90	0.9552	1.0110	171.0	181
95	0.8177	1.0355	186.5	194
				example

Table 4.5a *Conversion of D 86 Test results into an atmospheric TBP (Riazi's method).*

% distilled volume	Coefficient *a*	Coefficient *b*	Temperature D 86, °C	Temperature TBP, °C	Index *i*
0	7.4012	0.6024	36.5	−5	1
10	4.9004	0.7164	54.0	28	2
30	3.0305	0.8008	77.0	67	3
50	0.8718	1.0258	101.5	102	4
70	2.5282	0.8200	131.0	138	5
90	3.0419	0.7550	171.0	181	6
95	0.1180	1.6606	186.5	197	7

Table 4.5b *Conversion of D 86 test results into an atmospheric TBP (Daubert's method).*

4.1.3.2 **ASTM D 1160 Distillation for Heavy Petroleum Cuts**

This method is reserved for heavy fractions. The distillation takes place at low pressure: from 1 to 50 mmHg. The results are most often converted into equivalent atmospheric temperatures by using a standard relation that neglects the chemical nature of the components:

$$T' = \frac{748.1\,A}{\dfrac{1}{T} + 0.3861\,A - 5.1606 \cdot 10^{-4}} \qquad [4.20]$$

$$A = \frac{5.9991972 - 0.9774472\log P}{2663.129 - 95.76\log P}$$

where

T' = temperature equivalent at atmospheric pressure   [K]

T = experimental temperature taken at pressure P [K]

P = pressure   [mmHg]

log = common logarithm (base 10)

The results of the D 1160 distillation converted to 760 mmHg are not to be converted directly to an atmospheric TBP. The D 1160 results converted to 760 mmHg must be converted into the D 1160 equivalent at 10 mmHg, then these D 1160 results at 10 mmHg have to be transformed into a TBP at 10 mmHg and finally the TBP results at 10 mmHg are transformed into a TBP at 760 mmHg.

a. To transform the temperatures converted at 760 mmHg to temperatures at 10 mmHg, the formula [4.20] is used which is written as follows:

$$T'' = \frac{0.683398\,T'}{1 - 1.6343 \cdot 10^{-4}\,T'} \qquad [4.21]$$

where

T'' = temperature of the D 1160 at 10 mmHg   [K]

T' = temperature of the D 1160 at 760 mmHg   [K]

b. The D 1160 data at 10 mmHg can be transformed to TBP at 10 mmHg by using the following equations:

$$\left.\begin{aligned}
T_{50} &= T''_{50} \\
T_{30} &= T_{50} - f_1\left(T''_{50} - T''_{30}\right) \\
T_{10} &= T_{30} - f_1\left(T''_{30} - T''_{10}\right) \\
T_i &= T_{10} - f_2\left(T''_{10} - T''_i\right)
\end{aligned}\right\} \qquad [4.22]$$

where

T_{50} = temperature of the TBP at 50% volume distilled   [°C]

T_{30} = temperature of the TBP at 30% volume distilled   [°C]

T_{10} = temperature of the TBP at 10% volume distilled   [°C]

T_i = temperature of the TBP initial point   [°C]

T''_{50} = temperature of the ASTM D 1160 at 50% volume distilled   [°C]

T''_{30} = temperature of the ASTM D 1160 at 30% volume distilled   [°C]

T''_{10} = temperature of the ASTM D 1160 at 10% volume distilled   [°C]

T''_i = temperature of the ASTM D 1160 initial point   [°C]

The functions, f_1 (dT) and f_2 (dT) are obtained by interpolating the values given in Table 4.6.

For the fractions distilled at higher than 50%, the TBP and D 1160 curves are identical.

	dT, °C	f_2 (dT), °C	f_1 (dT), °C
Table	0	0.0	0.0
4.6	10	20.0	13.0
	20	35.5	24.0
Values of $f(dT)$ *functions for*	30	47.5	34.5
transforming an ASTM D 1160 at	40	57.0	44.0
10 mmHg into a TBP at 10 mmHg	50	64.0	53.5
dT = *temperature interval on*	60	70.0	63.0
the ASTM D 1160	70	75.0	72.0
f_1 (dT) = *temperature interval on*	80	82.5	81.5
the TBP	90	91.0	90.5
f_2 (dT) = *temperature interval on*	100	100.0	100.0
the TBP			

c. **The TBP at 0.0133 bar (10 mmHg) is transformed into an atmospheric TBP by the Maxwell and Bonnel formulas given in article 4.1.3.4 [Equation 4.24].**

4.1.3.3 Simulated Distillation for Petroleum Cuts (ASTM D 2887)

Distillation simulated by gas chromatography is a reproducible method for analyzing a petroleum cut; it is applicable for mixtures whose end point is less than 500°C and the boiling range is greater than 50°C. The results of this test are presented in the form of a curve showing temperature as a function of the weight per cent distilled equivalent to an atmospheric TBP.

For paraffinic materials, the results are close to those of a TBP, the equipment being calibrated using n-paraffins. For aromatics, the differences are larger.

The API has recommended the use of a method to convert the D 2887 results into those of an ASTM D 86, developed by Riazi using the following equation:

$$T' = a\,T^b\,F^c \qquad [4.23]$$

$$F = 0.01411\,(T_{10})^{0.05434}\,(T_{50})^{0.6147}$$

where

T' = ASTM D 86 test temperature for a volume % distilled equal to the wt % from the D 2887 test   [K]

T = temperature result from the ASTM D 2887 test   [K]

a, b, c = are conversion coefficients (refer to Table 4.7)

T_{10} = temperature from the 10 wt % distilled by D 2887   [K]

T_{50} = temperature from the 50 wt % distilled by D 2887   [K]

The table gives the coefficients a, b, c, as well as an example of the conversion.

% distilled*	Coefficient a	Coefficient b	Coefficient c	Temperature D 2887, °C	Temperature D 86, °C
0	5.1766	0.7445	0.2879	11.1	50
10	3.7451	0.7944	0.2671	58.3	79
30	4.2748	0.7719	0.3450	102.2	107
50	18.4448	0.5425	0.7132	133.9	128
70	1.0750	0.9867	0.0486	155.6	147
90	1.0850	0.9834	0.0354	178.3	165
100	1.7992	0.9007	0.0625	204.4	185
					example

Table 4.7a   *Coefficients for converting a D 2887 curve into an ASTM D 86 curve [Equation 4.23] (Riazi's method). Factor F for the example: 0.7774.*

% distilled*	Coefficient E	Coefficient F	Temperature D 86, °C	Temperature TBP, °C	Index i
0	0.3047	1.1259	11.1	56	1
10	0.0607	1.5176	58.3	81	2
30	0.0798	1.5386	102.2	107	3
50	0.7760	1.0395	133.9	129	4
70	0.1486	1.4287	155.6	145	5
90	0.3079	1.2341	178.3	161	6
100	2.6029	0.6596	204.4	180	7

Table 4.7b   *Coefficients for converting a D 2887 curve into an ASTM D 86 curve [Equation 4.23] (Daubert's method).*

---

* Pertains to the weight per cent for the D 2887 simulated distillation results and the volume per cent for the D 86 simulated distillation.

Daubert (1994) recommends a new method based on the following equations:

$$T'_{50} = E_4 \cdot (T_{50})^{F4} \qquad\qquad\qquad\qquad\qquad\qquad [4.19b]$$

$$T'_{30} = T'_{50} - \Delta T'_3 \qquad\qquad \Delta T_3 = T_{50} - T_{30}$$

$$T'_{10} = T'_{30} - \Delta T'_2 \qquad\qquad \Delta T_2 = T_{30} - T_{10}$$

$$T'_i = T'_{10} - \Delta T'_1 \qquad\qquad \Delta T_1 = T_{10} - T_i$$

$$T'_{70} = T'_{50} + \Delta T'_5 \qquad\qquad \Delta T_5 = T_{70} - T_{50}$$

$$T'_{90} = T'_{70} + \Delta T'_6 \qquad\qquad \Delta T_6 = T_{90} - T_{70}$$

$$T'_f = T'_{90} + \Delta T'_7 \qquad\qquad \Delta T_7 = T_f - T_{90}$$

$$\Delta T'_i = E_i \cdot (\Delta T_i)^{Fi}$$

where

T'_i = D 86 temperature at initial point [°F]

T'_f = D 86 temperature at final point [°F]

T'_n = D 86 temperature at n % volume distilled [°F]

T_n = D 2887 temperature at n % weight distilled [°F]

E_i and F_i are coefficients given in Table 4.7

The average error is about 4°C except for the initial and final points where it attains 12°C.

4.1.3.4 Conversion of the Low Pressure Distillation Results into Equivalent Results for Atmospheric Pressure

To convert low pressure distillation results into those of atmospheric pressure, the Maxwell and Bonnel (1955) equations are used.

These relations are given below:

$$T' = \frac{748.1 \times T}{1 + T(0.3861 x - 0.00051606)} \qquad\qquad [4.24]$$

$$x = \frac{5.994296 - 0.972546 \log P}{2\,663.129 - 95.76 \log P}$$

$$T'' = T' + 1.389\, f \left(K_W - 12 \right) (\log P - 2.8808)$$

$$f = 0 \quad \text{if} \quad T'' < 366\,K$$

$$f = 1 \quad \text{if} \quad T'' > 477\,K$$

$$f = \frac{T'' - 366}{111} \quad \text{if} \quad 366\,K < T'' < 477\,K$$

where

T = temperature observed at pressure P \qquad [K]

P = pressure at which the distillation took place \qquad [mmHg]

T' = temperature calculated for a K_W equal to 12   [K]

T'' = temperature calculated for a K_W different from 12   [K]

K_W = Watson characterization factor

log = common logarithm (base 10)

These equations differ from those of [4.20] in that the Watson factor is taken into account.

4.1.3.5 **Special Case for Crude Oil**

Crude oil is generally characterized by a TBP analysis whose results are expressed as temperatures equivalent to atmospheric pressure as a function of the fraction of volume and weight distilled

Each petroleum cut obtained by mixing the TBP distilled fractions (and thus characterized by the TBP cut points) is described by a collection of properties including the viscosity at two temperatures.

The C_1 to C_5 light components are analyzed by gas chromatography and the results are presented as weight per cents of the crude oil.

Cutting petroleum into fractions is done by the method illustrated in Figure 4.2. The petroleum fractions should correspond to the characteristics described in article 4.1.2. If certain characteristics are estimated, it is mandatory to compare the calculational results with the properties of the cuts and to readjust the estimated characteristics.

The methods for the evaluation of crude oils are examined in Chapter 8.

4.2 **Basic Calculations of Physical Properties**

In the absence of a single accurate theory representing the physical reality of liquids and gases and, consequently, all their physical properties, a property can be calculated in various ways.

A panoply of methods whose results can be widely scattered are available to the process engineer; not knowing the pitfalls attached to this activity, he would like to have a unique method or an exact guideline for applying these methods.

Using computer programs complicates the problem because the calculation accuracy is never given for commercial reasons. Furthermore, the ways in which the methods are executed are not explicit and the data banks are often considered secret and inaccessible.

The nature of the calculational bases, that is, the components and conditions of temperature and pressure, are too varied for it to be possible to make absolute recommendations as to the choice of methods.

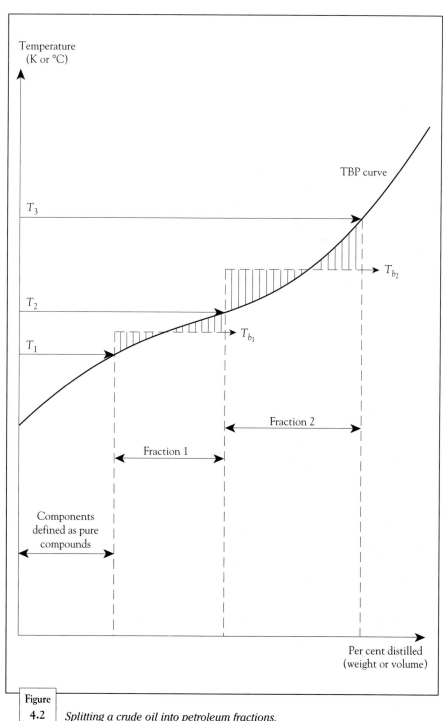

Figure
4.2 *Splitting a crude oil into petroleum fractions.*

However, the rules governing method selection can be given:

(a) Exclude the methods which do not use the characteristic properties of the components.

(b) Prefer the methods based on experimental results specific to the mixture or the components to be studied.

(c) Use general methods only with appropriate adjustment factors.

(d) Use "predictive methods" only as a last resort.

Incidentally, it is advisable to compare the results of several methods to uncover possible anomalies and to analyze the results very carefully.

4.2.1 **Properties of Pure Hydrocarbons and Petroleum Fractions**

A certain number of properties depend only on temperature; these have to do with properties of ideal gases and saturated liquids.

The properties of real gases and liquids under pressure are calculated by adding a pressure correction to the properties determined for the ideal gas or the saturated liquid.

The correction for pressure is often determined by applying the principle of corresponding states.

4.2.1.1 **Influence of Temperature**

When the properties depend only on temperature, they can be expressed in two ways:

- either as empirical functions of temperature
- or as functions of reduced temperature.

The empirical function of temperature is used for the following properties:

- enthalpy of the ideal gas
- specific heat of the saturated liquid
- viscosity of the saturated liquid
- conductivity of the saturated liquid.

The expression in terms of reduced temperature provides a way to calculate the following properties:

- viscosity of the ideal gas
- conductivity of the ideal gas
- density of the saturated liquid
- vapor pressure of the liquid.

4.2.1.2 **Influence of Pressure**

This influence must be taken into consideration when calculating real gas and compressible liquid properties.

Pressure correction can be done in four ways:

- directly as a function of the difference in pressure of saturation and the pressure applied to the fluid under consideration as in the case for liquid density
- by solving an equation of state; this method is mainly used to determine the densities, enthalpies, specific heats and fugacities of gases and compressible liquids
- by using the reduced density which is the case for the calculation of viscosities and conductivities of gases and liquids under pressure
- by using another property, itself dependent on pressure, which is used to find the diffusivity.

4.2.2 **Properties of Mixtures**

4.2.2.1 **Basis of Calculation**

The calculation of the properties of mixtures by modern methods requires that the composition be known and that the component parameters have been determined previously.

Next the properties of each component must be determined at the temperature being considered in the ideal gas state and, if possible, in the saturated liquid state.

The amounts of each phase and their compositions are calculated by resolving the equations of phase equilibrium and material balance for each component. For this, the partial fugacities of each constituent are determined:

- in the gas phase, by solving the equation of state for this phase
- in the liquid phase, either as in the gas phase, or from the component properties in the saturated liquid state.

It is then possible to calculate all the properties of each phase.

4.2.2.2 **Concept of Weighting**

Generally the properties of mixtures in the ideal gas state and saturated liquids are calculated by weighting the properties of components at the same temperature and in the same state. Weighting in these cases is most often linear with respect to composition:

$$f\left(\psi_m^0\right) = \sum \left[x_i f\left(\psi_i^0\right)\right]$$

where

f = a function of the property ψ

ψ^0_m = property of a mixture as an ideal gas or as saturated liquid at temperature T

ψ^0_i = property of the component i in the ideal gas state or as the saturated liquid at temperature T

x_i = mole fraction of component i

4.2.2.3 **Correction for Pressure**

Properties of mixtures as a real gas or as a liquid under pressure are determined starting from the properties of mixtures in the ideal gas state or saturated liquid after applying a pressure correction determined as a function of a property or a variable depending on pressure:

$$g\left(\psi_m\right) = g\left(\psi^0_m\right) + h\left(\xi_r\right)$$

where

g = function of the property ψ

ψ^0_m = property of a mixture as an ideal gas or as saturated liquid at temperature T

ψ_m = property of the mixture as a real gas or as liquid under pressure at temperature T and pressure P

$h(\xi_r)$ = pressure correction

ξ_r = property or variable dependent on pressure, for example reduced volume. Refer to 4.2.3.1.

4.2.3 **Principle of Corresponding States**

4.2.3.1 **Concept of Corresponding States**

The principle of corresponding states has a double origin:

- from theoretical thermodynamics which, starting from the concept of inter-molecular forces, demonstrate that the thermodynamic properties are all functions of the two parameters of the expression of the potential (Vidal, 1973)

- from empirical observation, which, admitting the existence of variables and reduced properties, invented the acentric factor and gave the needed accuracy to the calculation of properties using the principle of corresponding states for it to be universally adopted.

According to this concept, a reduced property is expressed as a function of two variables, T_r and V_r and of the acentric factor, ω:

$$\psi_r = \frac{\psi}{\chi_c} = f\left(T_r, V_r, \omega\right)$$ [4.25]

$$T_r = \frac{T}{T_c}$$

$$V_r = \frac{VP_c}{RT_c} \qquad\qquad [4.26]$$

where

T = temperature   [K]

T_c = critical temperature   [K]

P_c = critical pressure   [bar]

R = ideal gas constant $[0.08314 \text{ m}^3{\cdot}\text{bar}/(\text{K}{\cdot}\text{kmol})]$

T_r = reduced temperature

V_r = reduced volume

V = molar volume at the given conditions $[\text{m}^3/\text{kmol}]$

ω = acentric factor

ψ = property

ψ_r = reduced property

X_C = reduction group having the same dimensions as the property

For non-polar components like hydrocarbons, the results are very satisfactory for calculations of vapor pressure, density, enthalpy, and specific heat and reasonably close for viscosity and conductivity provided that V_r is greater than 0.10.

This concept can be extended to mixtures if the pseudo-critical constants of the mixture and a mixture reduction group are defined. This gives the following:

$$\psi_{r_m} = \frac{\psi_m}{X_{C_m}} = f\left(T_{r_m}, V_{r_m}, \omega_m\right)$$

$$T_{r_m} = \frac{T}{T_{C_m}} \qquad V_{r_m} = \frac{VP_{C_m}}{RT_{C_m}}$$

where

T_{C_m} = pseudo-critical temperature of the mixture   [K]

P_{C_m} = pseudo-critical pressure of the mixture   [bar]

X_{C_m} = reduction group for the property ψ,
        for example $\left(R{\cdot}T_{C_m}\right)$ for enthalpy

T_{r_m} = pseudo-reduced temperature of the mixture

V_{r_m} = pseudo-reduced volume of the mixture

4.2.3.2 **Pseudocritical Constants for Mixtures**

These should form a uniform group. Among several that have been proposed, we will use only those of Lee and Kesler (1975).

a. *Calculation of the Pseudocritical Molar Volume of a Mixture*

The pseudocritical molar volume of a mixture is obtained by weighting the pseudocritical volumes of each component:

$$V_{C_m} = \frac{1}{4} \Sigma \left(x_i V_{C_i} \right) + \frac{3}{4} \Sigma \left(x_i V_{C_i}^{2/3} \right) \cdot \Sigma \left(x_i V_{C_i}^{1/3} \right) \qquad [4.27]$$

where

V_{C_m} = pseudocritical molar volume of the mixture $\qquad$ [m^3/kmol]
V_{C_i} = critical molar volume of the component i $\qquad$ [m^3/kmol]
x_i $\quad$ = mole fraction of component i

The critical molar volume is defined using the acentric factor by the following relations:

$$V_{C_i} = \frac{z_{C_i} R T_{C_i}}{P_{C_i}} \qquad [4.28]$$

$$z_{C_i} = 0.291 - 0.08 \, \omega_i \qquad [4.29]$$

where

ω_i $\:$ = acentric factor of the component i
T_{C_i} = critical temperature of the component i $\qquad$ [K]
P_{C_i} = critical pressure of the component i $\qquad$ [bar]
z_{C_i} = critical compressibility factor of the component i

b. *Calculation of the Pseudocritical Temperature of a Mixture*

The pseudocritical temperature of a mixture is obtained by weighting the pseudocritical temperatures and volumes for each component:

$$T_{C_m} = \frac{\left(S_1 + 3 S_2 S_3 \right)}{4 V_{C_m}} \qquad [4.30]$$

$$S_1 = \Sigma \left(x_i V_{C_i} T_{C_i} \right)$$

$$S_2 = \Sigma \left(x_i V_{C_i}^{2/3} T_{C_i}^{1/2} \right)$$

$$S_3 = \Sigma \left(x_i V_{C_i}^{1/3} T_{C_i}^{1/2} \right)$$

where

V_{C_m} = pseudocritical molar volume of the mixture $\qquad$ [m^3/kmol]
V_{C_i} = critical molar volume of the component i $\qquad$ [m^3/kmol]

x_i = mole fraction of component i

T_{c_m} = pseudocritical temperature of the mixture $\qquad$ [K]

T_{c_i} = critical temperature of the component i $\qquad$ [K]

c. Calculation of the Acentric Factor of a Mixture

The acentric factor of a mixture is calculated as follows:

$$\omega_m = \sum \left(x_i\, \omega_i \right) \qquad [4.31]$$

where

ω_m = acentric factor of the mixture

ω_i = acentric factor of component i

x_i = mole fraction of component i

d. Calculation of the Pseudocritical Compressibility Factor of a Mixture

The pseudocritical compressibility factor is obtained directly from the acentric factor using the expression:

$$z_{c_m} = 0.291 - 0.08\, \omega_m \qquad [4.29]$$

e. Calculation of the Pseudocritical Pressure of a Mixture

The pseudocritical pressure is calculated as a function of other constants:

$$P_{c_m} = \frac{z_{c_m} R T_{c_m}}{V_{c_m}} \qquad [4.32]$$

where

P_{c_m} = pseudocritical pressure of the mixture $\qquad$ [bar]

R = ideal gas constant $\qquad$ $\left[0.08314\ \mathrm{m^3 \cdot bar/(K \cdot kmol)} \right]$

4.2.3.3 **Comments on the Meaning of the Acentric Factor**

The factor enabling interpolation of reduced properties of a pure compound or mixture between two reduced properties calculated on two reference fluids merits attention in order to understand its meaning.

Hexane, for example, is a component whose properties are well known and follow the principle of corresponding states very closely. The acentric factor recommended by the DIPPR is 0.3046 and is considered by convention not to vary with temperature.

The acentric factor can be determined as a function of temperature by finding the exact properties supplied by the DIPPR.

If the vapor pressure is of interest, the acentric factor is calculated by the Lee and Kesler formula or by the Soave method, which are given in article 4.5.2.

To calculate the heat of vaporization, the Lee and Kesler method in article 4.3.1.3 is used.

The values obtained for the acentric factor differ significantly from one another. As shown in Figure 4.3, this factor depends on the temperature, the physical property being considered, and the method used.

Nevertheless, an average value between 0.30 and 0.31 is acceptable and the calculated error is reasonable.

Hexane is an easy example. The variations in acentric factors are much more pronounced for heavy polar or polarizable components. It comes as no surprise that the values reported from different sources are not identical.

The acentric factor is also dependent on the critical coordinates being used.

To avoid confusion, the only acentric factor that we will use is that employed to find the boiling point by the Lee and Kesler method.

4.3 **Properties of Liquids**

4.3.1 **Thermodynamic Properties of Liquids**

4.3.1.1 **Liquid Densities**

Liquid densities can be calculated according two types of methods, both based on the principle of corresponding states.

In the first type of method, the density at saturation pressure is calculated, then this density is corrected for pressure. The COSTALD and Rackett methods belong to this category. Correction for pressure is done using Thompson's method. These methods are applicable only if the reduced temperature is less than 0.98.

For reduced temperatures higher than 0.98, a second type of method must be used that is based on an equation of state such as that of Lee and Kesler.

For our needs, the saturation pressure of a mixture will be defined as the vapor pressure of a pure component that has the same critical constants as the mixture:

$$P_s = P_{c_m} \exp\left[f\left(T_{r_m}, \omega_m\right)\right]$$

where

P_s = saturation pressure for the mixture   [bar]

P_{c_m} = pseudocritical pressure for the mixture   [bar]

$f\left(T_{r_m}, \omega_m\right)$ = Lee and Kesler function used for
           the calculation of vapor pressure

ω_m = acentric factor for the mixture

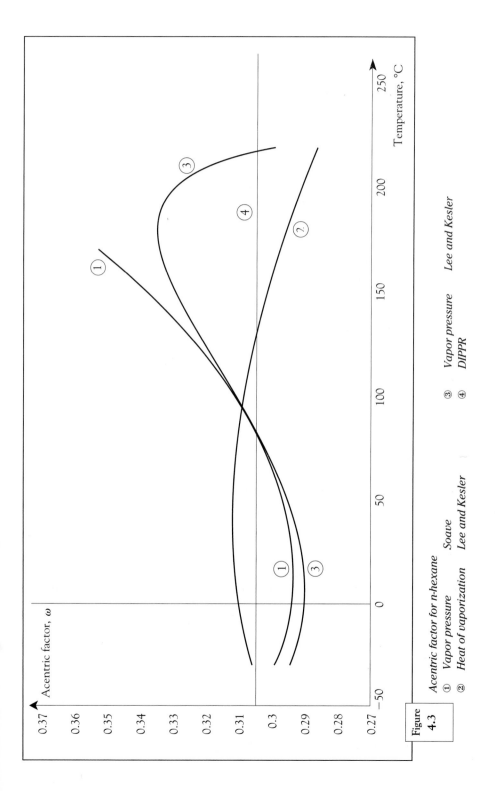

Figure 4.3

Acentric factor for n-hexane

① Vapor pressure   Soave   ③ Vapor pressure   Lee and Kesler
② Heat of vaporization Lee and Kesler ④ DIPPR

$$f\left(T_{r_m}, \omega_m\right) = \ln P_{r_0} + \omega_m \ln P_{r_1} \qquad [4.33]$$

$$\ln P_{r_0} = 5.92714 - \frac{6.09648}{T_{r_m}} - 1.28862 \ln T_{r_m} + 0.169347\, T_{r_m}^6$$

$$\ln P_{r_1} = 15.2518 - \frac{15.6875}{T_{r_m}} - 13.4721 \ln T_{r_m} + 0.43577\, T_{r_m}^6$$

The saturation pressure, P_s, is different from the bubble point pressure (see. Vidal, 1973) and has no physical reality; it merely serves as an intermediate calculation.

a. Calculation of the Density at Saturation Pressure Using the Rackett Method

The density at saturation pressure is expressed as a function of reduced temperature:

$$\frac{1}{\rho_s} = \left(\frac{R\, T_{c_m}}{P_{c_m}\, M}\right) z_{Ra_m}^{\left[1 + \left(1 - T_{r_m}\right)^{2/7}\right]} \qquad [4.34]$$

$$z_{Ra_m} = \sum \left(x_i\, z_{Ra_i}\right)$$

$$T_{r_m} = \frac{T}{T_{c_m}}$$

where

T_{r_m}	= reduced temperature of the mixture	
T	= temperature	[K]
T_{c_m}	= pseudocritical temperature of the mixture	[K]
P_{c_m}	= pseudocritical pressure of the mixture	[bar]
R	= ideal gas constant	$\left[0.08314\ \text{bar·m}^3/(\text{kmol·K})\right]$
ρ_s	= density at saturation pressure	$[\text{kg/m}^3]$
z_{Ra_i}	= parameter of Rackett's method, determined for each component, i	
z_{Ra_m}	= Rackett's parameter for the mixture	
M_m	= molecular weight for the mixture	[kg/kmol]

The error is about 3% when the z_{Ra_i} are known with accuracy.

For petroleum fractions, the z_{Ra_f} values should be calculated starting with the standard specific gravity according to the relation:

$$z_{Ra_f} = \left[\frac{P_{c_f} M_f}{\rho_{H_2O}^4 S_f R T_{c_f}} \right]^{E_f} \tag{4.35}$$

$$E_f = \frac{1}{1 + \left(1 - \dfrac{288.7}{T_{c_f}}\right)^{2/7}}$$

where

T_{c_f}	= pseudocritical temperature of the fraction f	[K]
P_{c_f}	= pseudocritical pressure of the fraction f	[bar]
S_f	= standard specific gravity of the fraction f	
M_f	= molecular weight of the fraction f	[kg/kmol]
$\rho_{H_2O}^4$	= specific gravity of water at 4°C	$\left[1000 \text{ kg/m}^3\right]$

b. Calculation of the Density at Saturation Pressure Using the COSTALD Method

The COSTALD (Corresponding states liquid density) method was originally developed for calculating the densities of liquefied gases; its use has become generally widespread.

The method was published in 1979 by Hankinson and Thompson. The relations are the following:

$$\frac{1}{\rho_S} = \frac{V_{S_m} \cdot V_{r_m} \left[1 - \omega_m f\left(T_{r_m}\right)\right]}{M_m} \tag{4.36}$$

$$V_{r_m} = 1 + \sum_{k=1}^{4}\left(a_k \, \theta^k\right)$$

$$\theta = \left(1 - T_{r_m}\right)^{1/3}$$

$$f\left(T_{r_m}\right) = \sum_{k=1}^{4}\left(\frac{b_k \, T_{r_m}^{k-1}}{\left(T_{r_m} - 1\right)}\right)$$

$a_1 = -1.52816$		$a_3 = -0.81446$
$a_2 = 1.43907$		$a_4 = 0.90454$
$b_1 = -0.296123$		$b_3 = 0.386914$
$b_2 = -0.0427258$		$b_4 = -0.0480645$

$$V_{S_m} = \frac{1}{4}\sum\left(x_i V_{S_i}\right) + \frac{3}{4}\sum\left(x_i V_{S_i}^{1/3}\right) \cdot \sum\left(x_i V_{S_i}^{2/3}\right)$$

where

T_{r_m} = reduced temperature of the mixture

V_{s_i} = molar volume characteristic for the component i [m³/kmol]

V_{s_m} = molar volume characteristic for the mixture   [m³/kmol]

M_m = molecular weight of the mixture   [kg/kmol]

ρ_s = density of the mixture at the saturation pressure   [kg/m³]

ω_m = acentric factor of the mixture

The average error for this method is about 2%.

The molar volume characteristics, V_{s_i}, for petroleum fractions and hydrocarbons can be obtained from the known density at temperature T_1:

$$V_{s_i} = \frac{M_i}{\rho_{ls_i}^1 V_{r_i}^1 \left(1 - \omega_i f\left(T_{r_i}^1\right)\right)} \qquad [4.37]$$

where

V_{s_i} = molar volume characteristic of the component i [m³/kmol]

M_i = molecular weight of the component i [kg/kmol]

$\rho_{ls_i}^1$ = density of component i at temperature T_1
and saturation pressure   [kg/m³]

ω_i = acentric factor of the component i

$V_{r_i}^1$ = value of the function V_r for component i at temperature T_1

$f\left(T_{r_i}^1\right)$ = value of the $f\left(T_r\right)$ function for component i à la température T_1

c. Pressure Correction for Density

The density of a liquid depends on the pressure; this effect is particularly sensitive for light liquids at reduced temperatures greater than 0.8. For pressures higher than saturation pressure, the density is calculated by the relation published by Thompson et al. in 1979:

$$\frac{1}{\rho} = \frac{1 - C\ln\dfrac{B+P}{B+P_s}}{\rho_s} \qquad [4.38]$$

$$C = 0.0861488 + 0.0344483\,\omega_m$$

$$B = P_{c_m} \sum_{k=1}^{4} \left(a_k\,\theta^k - 1\right)$$

$$\theta = \left(1 - T_{r_m}\right)^{1/3}$$

$a_1 = -9.070217 \qquad a_2 = 62.45326 \qquad a_3 = -135.1102$

$a_4 = \exp\left(4.79594 + 0.250047\,\omega_m + 1.14188\,\omega_m^2\right)$

where

P_{c_m} = pseudocritical pressure for the mixture   [bar]

P_s = saturation pressure for the mixture at T [bar]

P = pressure   [bar]

ω_m = acentric factor for the mixture

T_{r_m} = reduced temperature for the mixture

ρ_s = density at the saturation pressure $[kg/m^3]$

ρ = density at T and P $[kg/m^3]$

ln = Napierian logarithm

The average error for this method is about 2%.

d. Calculation of Density by the Lee and Kesler Method

This method utilizes essentially the concept developed by Pitzer in 1955. According to the principle of three-parameter corresponding states, the compressibility factor z, for a fluid of acentric factor ω, is obtained by interpolating between the compressibilities z_1 and z_2 for the two fluids having acentric factors ω_1 and ω_2:

$$z = z_1 + (z_2 - z_1)\left(\frac{\omega - \omega_1}{\omega_2 - \omega_1}\right) \tag{4.39}$$

The two reference fluids are methane $(\omega_1 = 0)$ and n-octane $(\omega_2 = 0.3978)$.

Each fluid is described by a BWR equation of state whose coefficients are adjusted to obtain simultaneously: the vapor pressure, enthalpies of liquid and gas as well as the compressibilities. The compressibility z of any fluid is calculated using the equation below:

$$z = z_1 + 2.5138 \,\omega \,(z_2 - z_1) \tag{4.40}$$

The compressibility factors z_i are obtained after solving the following equations in V_{r_i} for $i = 1$ and $i = 2$:

$$z_i = \frac{P_r V_{r_i}}{T_r} = 1 + \frac{B_i}{V_{r_i}} + \frac{C_i}{V_{r_i}^2} + \frac{D_i}{V_{r_i}^5} + \frac{c_{4i}\left(\beta_i + \dfrac{\gamma_i}{V_{r_i}^2}\right)\exp\left(\dfrac{-\gamma_i}{V_{r_i}^2}\right)}{T_r^3 V_{r_i}^2} \tag{4.41}$$

$$B_i = b_{1i} - \frac{b_{2i}}{T_r} - \frac{b_{3i}}{T_r^2} - \frac{b_{4i}}{T_r^3} \qquad C_i = c_{1i} - \frac{c_{2i}}{T_r} + \frac{c_{3i}}{T_r^3} \qquad D_i = d_{1i} + \frac{d_{2i}}{T_r}$$

$b_{11} = 0.1181193$ $b_{21} = 0.265728$ $b_{31} = 0.154790$

$b_{41} = 0.030323$ $c_{11} = 0.0236744$ $c_{21} = 0.0186984$

$c_{31} = 0$ $c_{41} = 0.042724$ $d_{11} = 1.55488 \times 10^{-5}$

$$d_{21} = 6.23689 \times 10^{-5} \qquad \beta_1 = 0.65392 \qquad \gamma_1 = 0.060167$$
$$b_{12} = 0.2026579 \qquad b_{22} = 0.331511 \qquad b_{32} = 0.027655$$
$$b_{42} = 0.203488 \qquad c_{12} = 0.0313385 \qquad c_{22} = 0.0503618$$
$$c_{32} = 0.016901 \qquad c_{42} = 0.041577 \qquad d_{12} = 4.8736 \times 10^{-5}$$
$$d_{32} = 0.740336 \times 10^{-5} \qquad \beta_2 = 1.226 \qquad \gamma_2 = 0.03754$$

Equation [4.41] has, for $T_r < 1$, at least two solutions for V_{r_i}; the smallest value corresponds to the liquid and the largest to the gas.

When the liquid compressibility, z_l, has been obtained, the density is calculated as follows:

$$\rho_l = \frac{PM}{z_l RT} \qquad\qquad [4.42]$$

where

P = pressure   [bar]

M = molecular weight   [kg/kmol]

R = ideal gas constant $\left[0.08314 \text{ bar·m}^3/(\text{K·kmol})\right]$

T = temperature   [K]

ρ_l = liquid density $\left[\text{kg/m}^3\right]$

This equation has the following limits:

$$0.3 < T_r < 4 \quad \text{and} \quad P_r < 10$$

The accuracy is around 10% for densities of liquids.

4.3.1.2 **Specific Heat of Liquids**

The specific heat for a liquid at constant pressure, written as C_{pl} and expressed in kJ/(kg·K), can be calculated for mixtures in two ways:

a. *Concept of Weighting the Specific Heats of the Components*

The simplest way is to weight the C_{pl} of each component according to the following equation:

$$C_{pl_m} = \Sigma \left(x_{w_i} C_{pl_i} \right) \qquad\qquad [4.43]$$

where

C_{pl_m} = specific heat of the mixture in the liquid state   [kJ/(kg·K)]

C_{pl_i} = specific heat of the component i in the liquid state   [kJ/(kg·K)]

x_{w_i} = mass fraction of the component i

The effect of pressure is neglected. The limits of this model are easy to understand: each component must exist in the liquid state for the C_{pl} to be known; equally important is that the effect of pressure must be negligible which is the case for $T_r < 0.8$ and $P_r < 1$.

Specific Heats for Liquid Petroleum Fractions

The isobaric specific heat for a petroleum fraction is estimated by a correlation attributed to Watson and Nelson in 1933, which was used again by Johnson and Grayson in 1961 as well as by Lee and Kesler in 1975. This relation is valid at low pressures:

$$C_{pl} = 4.185 \left(0.35 + 0.055\, K_W\right) \left(0.3065 - 0.16734\, S \right. $$
$$\left. + T \left(1.467 \cdot 10^{-3} - 5.508 \cdot 10^{-4}\, S\right)\right) \tag{4.44}$$

where

K_W = Watson characterization factor
S = standard specific gravity
T = temperature   [K]
C_{pl} = isobaric mass specific heat   [kJ/(kg·K)]

This relation should not be applied for temperatures less than 0°C. Its average accuracy is on the order of 5%. For a Watson factor of 11.8, the C_{pl} can be obtained from the curve shown in Figure 4.4. For different K_W values, the following correction is used:

$$C_{pl} = C'_{pl} \left(0.35 + 0.055\, K_W\right)$$

where

C_{pl} = specific heat of the liquid   [kJ/(kg·K)]
C'_{pl} = specific heat of the liquid whose K_W is 11.8   [kJ/(kg·K)]

b. *Principle of Corresponding States for Calculation of C_{pl}*

The other method is to employ the principle of corresponding states and calculate the C_{pl} of the mixture in the liquid phase starting from the mixture in the ideal gas state and applying an appropriate correction:

$$C_{pl_m} = C_{pgp_m} + \frac{R}{M_m}\, dC_{pr} \tag{4.45}$$

where

C_{pl_m} = C_{pl} of the mixture in the liquid state at T and P [kJ/(kg·K)]
C_{pgp_m} = C_{pg} of the mixture in the ideal gas state at T [kJ/(kg·K)]
M_m = molecular weight of the mixture   [kg/kmol]
dC_{pr} = reduced correction for C_{pl}, function of T_{r_m}, P_{r_m} and ω_m
R = ideal gas constant   [8.314 kJ/(kmol·K)]

The term dC_{pr} is calculated by the following expressions:

$$dC_{pr} = dC_{pr_1} + 2.5138\, \omega_m \left(dC_{pr_2} - dC_{pr_1}\right) \tag{4.46}$$

where

dC_{pr_1} = C_p reduced correction calculated for reference fluid 1 (methane)
dC_{pr_2} = C_p reduced correction calculated for reference fluid 2 (n-octane)

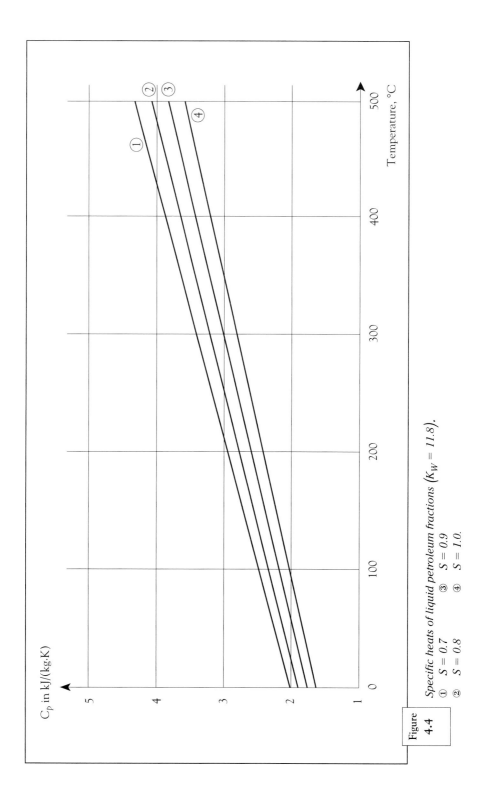

Figure 4.4 *Specific heats of liquid petroleum fractions* ($K_W = 11.8$).

① $S = 0.7$ ③ $S = 0.9$
② $S = 0.8$ ④ $S = 1.0$.

This expression is also written as:

$$dC_{pr} = dC_{pr_1} + \omega \, dC_{pr_b} \quad \text{and} \quad dC_{pr_b} = 2.5138 \left(dC_{pr_2} - dC_{pr_1} \right)$$

When the reduced temperature is less than 0.85, dC_{pr} depends very little on pressure. Table 4.8 gives values for C_{pl} and for enthalpy correction factors calculated by the Lee and Kesler method.

In practice, dC_{pr} can be calculated by finite difference starting from reduced corrections for enthalpies by a relation of the following type:

$$dC_{pr} \left(T_{r_m}, P_{r_m} \right) = 1000 \left(dH'_r \left(T'_{r_m}, P_{r_m} \right) - dH_r \left(T_{r_m}, P_{r_m} \right) \right) \qquad [4.47]$$

$$T'_{r_m} = 0.999 \, T_{r_m}$$

where

dH_r = reduced correction for enthalpy calculated at T_{r_m} and P_{r_m}

dH'_r = reduced correction for enthalpy at T'_{r_m} and P_{r_m}

T_{r_m} = reduced temperature of the mixture

P_{r_m} = reduced pressure of the mixture

T_{r_m}	dH_{r_1}	dH_{r_b}	dC_{pr_1}	dC_{pr_b}
0.300	6.046	11.099	2.798	8.481
0.350	5.908	10.656	2.807	9.780
0.400	5.764	10.120	2.931	11.482
0.450	5.616	9.515	2.989	12.653
0.500	5.466	8.868	3.005	13.113
0.550	5.316	8.209	3.004	13.040
0.600	5.166	7.567	3.012	12.683
0.650	5.015	6.945	3.058	12.153
0.700	4.860	6.353	3.177	11.550
0.750	4.695	5.790	3.418	10.939
0.800	4.515	5.254	3.854	10.414
0.850	4.302	4.748	4.851	9.753

Table 4.8 — *Reduced corrections for C_{pl} and enthalpy H_l at zero pressure for $0.3 < T_{r_m} < 0.85$.*

4.3.1.3 Liquid Enthalpy

The mass specific enthalpy for a liquid is written as H_l and is expressed in kJ/kg; as in the case of the C_{pl} it can be calculated in two ways:

a. Concept of Weighting the Liquid Specific Enthalpies of the Components

$$H_{l_m} = \sum \left(x_{w_i} H_{l_i} \right) \qquad [4.48]$$

where

H_{l_m} = liquid specific enthalpy of the mixture   [kJ/kg]

H_{l_i} = liquid specific enthalpy of the component i [kJ/kg]

x_{w_i} = weight fraction of the component i

The limits of this relation are the same as those given for C_{pl}.

Enthalpy of Petroleum Fractions

The enthalpy of a petroleum fraction is obtained by integration of the Watson and Nelson relations:

$$H_l = \int_{T_{ref}}^{T} \left(C_{pl}\, dT \right) + H_{l_{ref}} \qquad [4.49]$$

If the reference temperature is taken to be 0 K, the following relation is obtained:

$$H_l = H_{l_{ref}} + 4.185 \left(0.35 + 0.055\, K_W \right) \left(T \left(0.3065 - 0.16734\, S \right) \right.$$
$$\left. + T^2 \left(7.355\ 10^{-4} + 2.754\ 10^{-4}\, S \right) \right)$$

where

H_l = mass specific enthalpy of the liquid   [kJ/kg]

$H_{l_{ref}}$ = reference enthalpy at the reference temperature   [kJ/kg]

S = standard specific gravity

T_{ref} = reference temperature for the enthalpy origin   [K]

This relation is used only for temperatures greater than 0°C. The average error is about 5 kJ/kg. Figure 4.5 gives the enthalpy for petroleum fractions whose K_W is 11.8 as a function of temperature. For K_W factors different from 11.8, a correction identical to that used for C_{pl} is used:

$$H_l = H'_l \left(0.35 + 0.055\, K_W \right)$$

where

H_l = specific enthalpy for the liquid   [kJ/kg]

H'_l = specific enthalpy for the liquid whose K_W is 11,8   [kJ/kg]

b. Principle of Corresponding States Applied to the Calculation of Liquid Enthalpy

The principle of corresponding states enables the enthalpy of a liquid mixture to be expressed starting from that of an ideal gas mixture and a reduced correction for enthalpy:

$$H_{l_m} = H_{g p_m} + \frac{R\, T_{c_m}}{M_m}\, dH_r \qquad [4.50]$$

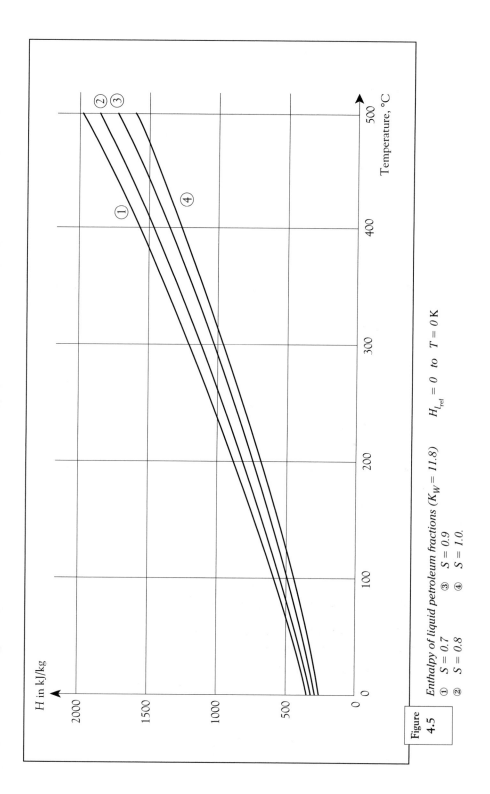

Figure 4.5 *Enthalpy of liquid petroleum fractions ($K_W = 11.8$) $H_{l_{ref}} = 0$ to $T = 0\,K$*

① $S = 0.7$ ③ $S = 0.9$
② $S = 0.8$ ④ $S = 1.0$.

where

H_{l_m} = liquid specific enthalpy of the mixture   [kJ/kg]

H_{gp_m} = enthalpy of the mixture in the ideal gas state   [kJ/kg]

T_{c_m} = pseudocritical temperature of the mixture   [K]

dH_r = reduced correction for enthalpy

M_m = molecular weight of the mixture   [kg/kmol]

The reduced correction for enthalpy employed in the preceding equation is obtained by the Lee Kesler model:

$$dH_r = dH_{r_1} + 2.5138\, \omega_m \left(dH_{r_2} - dH_{r_1}\right) \qquad [4.51]$$

where

dH_r = reduced enthalpy correction at T_{r_m}, P_{r_m} and ω_m

dH_{r_1} = reduced enthalpy correction calculated for methane at T_{r_m}, P_{r_m}

dH_{r_2} = reduced enthalpy correction calculated for n-octane à T_{r_m}, P_{r_m}

ω_m = acentric factor of the mixture

The terms dH_{r_1} and dH_{r_2} are calculated after having calculated the reduced volumes, V_{r_1} and V_{r_2}, and the compressibilities, z_1 and z_2, of fluids 1 and 2 at T_{r_m} and P_{r_m}. For each fluid, the dH_{r_i} is expressed in terms of the following relation:

$$dH_{r_i} = -T_{r_m}\left(z_i - 1 - B_{b_i} - C_{c_i} + D_{d_i} + 3E_i\right) \qquad [4.52]$$

$$B_{b_i} = \frac{\left(b_{2i} + \dfrac{2b_{3i}}{T_{r_m}} + \dfrac{3b_{4i}}{T_{r_m}^2}\right)}{T_{r_m} V_{r_i}} \qquad C_{c_i} = \frac{\left(c_{2i} - \dfrac{3c_{3i}}{T_{r_m}^2}\right)}{2\, T_{r_m} V_{r_i}^2} \qquad D_{d_i} = \frac{\left(d_{2i}\right)}{5\, T_{r_m} V_{r_i}^5}$$

$$E_i = \frac{c_{4i}}{2\, T_{r_m}^3 \gamma_i}\left[\beta_i + 1 - \left(\beta_i + 1 + \frac{\gamma_i}{V_{r_i}^2}\right)\exp\left(\frac{-\gamma_i}{V_{r_i}^2}\right)\right]$$

The average error is about 7 kJ/kg. This model is valid only for:

$$0.3 < T_{r_m} < 4 \quad \text{and} \quad P_{r_m} < 10$$

4.3.2 Thermophysical Properties of Liquids

4.3.2.1 Viscosity and Viscosity Index

Liquid viscosity is one of the most difficult properties to calculate with accuracy, yet it has an important role in the calculation of heat transfer coefficients and pressure drop. No single method is satisfactory for all temperature and viscosity ranges. We will distinguish three cases for pure hydrocarbons and petroleum fractions:

- liquids of low viscosity, low density, and close to the critical temperature
- liquids at saturation pressure, of medium viscosity
- highly viscous liquids at low pressure.

We will distinguish two cases for the mixtures:

- low-viscosity mixtures near their pseudocritical temperatures
- medium and high viscosity liquid mixtures.

a. Pure Hydrocarbons of Low Viscosity

As a liquid approaches its critical conditions, its density decreases and consequently the distance between molecules increases resulting in a rapid decrease in viscosity.

Viscosity can be expressed as a function of reduced density to which the viscosity of the ideal gas must by added. We will use the formulation proposed by Dean and Stiel in 1965:

$$\mu_l = \mu_{gp} + \mu_{sc} f(\rho_r) \qquad [4.53]$$

$$f(\rho_r) = 1.07 \ 10^{-4} \left[\exp\left(1.439 \ \rho_r\right) - \exp\left(- \ 1.11 \ \rho_r^{1.858}\right) \right]$$

where

μ_l = viscosity of the liquid at T and P [mPa·s]

μ_{gp} = viscosity of the ideal gas at T [mPa·s]

ρ_r = reduced density

μ_{sc} = viscosity reduction group   [mPa·s]

The reduced density ρ_r is expressed by the following relation:

$$\rho_r = \frac{V_c}{V} \qquad [4.54]$$

where

V_c = critical molar volume $[\text{m}^3/\text{kmol}]$

V = molar volume $[\text{m}^3/\text{kmol}]$

The coefficient μ_{sc} can be modified to include an experimental viscosity at a reduced temperature between 0.85 and 0.95. This method applies only if the reduced density is less than 2.5 and the reduced temperature is greater than 0.85. Its average accuracy is about 30%.

The coefficient μ_{sc} can also be estimated starting with the critical constants by the following formula:

$$\mu_{sc} = M^{1/2} P_c^{2/3} T_c^{-1/6} \qquad [4.55]$$

where

M = molecular weight   [kg/kmol]

P_c = critical pressure   [bar]

T_c = critical temperature   [K]

b. Pure Hydrocarbons of Medium Viscosity

At the saturation pressure, the viscosity variation with temperature follows a law analogous to that of Clapeyron for the vapor pressure:

$$\log \mu_l = B\left(\frac{1}{T} - \frac{1}{T_0}\right)$$

where

μ_l = viscosity of the liquid   [mPa·s]
T = temperature   [K]
B = viscosity coefficient   [K]
T_0 = viscosity coefficient   [K]

The coefficients B and T_0 are obtained either by smoothing the experimental viscosity data, or by Van Velzen's method of contributing groups (see article 4.1.1.2).

They can also be calculated from the kinematic viscosities at 100 and 210°F. After having calculated the densities, ρ_1 and ρ_2, one obtains:

$$\mu_1 = \frac{\nu_1 \rho_1}{1000} \quad \text{and} \quad \mu_2 = \frac{\nu_2 \rho_2}{1000}$$

where

μ_1, μ_2 = absolute viscosities   [mPa·s]
ν_1, ν_2 = kinematic viscosities $\left[mm^2/s\right]$
ρ_1, ρ_2 = densities $\left[kg/m^3\right]$

Coefficients B and T_0 are obtained by the following relations:

$$B = 1892.9 \log \frac{\mu_1}{\mu_2}$$

[4.56]

$$T_0 = \frac{1}{3.2161 \cdot 10^{-3} - \log \dfrac{\mu_1}{B}}$$

This relation is applicable to hydrocarbons whose viscosities vary between 0.1 and 10 mPa·s.

The average error is around 10%.

c. Highly Viscous Hydrocarbons

At atmospheric pressure, hydrocarbon viscosities can be estimated by two methods: the ASTM method and that of Mehrotra (1990).

In Mehrotra's method, the dynamic viscosity is expressed as a function of temperature according to the relation:

$$\log\left(\mu_l + 0.8\right) = a\,T^b$$

[4.57]

where

T = temperature   [K]

μ_l = dynamic viscosity of the liquid   [mPa·s]

a = viscosity correction factor

b = correction factor for the variation of viscosity with temperature

For a large number of hydrocarbons a can be estimated as a function of b:

$$a = a_1 \, a_2^b \qquad\qquad a_1 = 100 \qquad\qquad a_2 = 0.01 \qquad [4.58]$$

The coefficients b have been given by the author for a number of components.

Coefficients a and b can also be determined from the kinematic viscosities at 100 and 210°F.

ASTM proposes representing the kinematic viscosity of hydrocarbons by a straight line on graph paper, called viscometric, for which the scales are such that:

$$\log\,(\log v + f(v)) = a \log T + b \qquad\qquad [4.59]$$

$f(v) = 0.7$ for $v \geq 1.5$ mm²/s

$f(v) = 0.7 + 0.085\,(v - 1.5)^2$ for $v < 1.5$ mm²/s

where

T = temperature   [K]

a, b = straight line coefficients

v = kinematic viscosity   [mm²/s]

log = common logarithm (base 10)

The straight line and consequently its coefficients are determined starting from two known viscosity values.

The Mehrotra and ASTM methods apply with acceptable accuracy only for viscosities between 1 and 1000 mPa·s. The average error is about 20%. The largest spreads are obtained at low and very high viscosities.

d. Hydrocarbon Mixtures of Low Gravity

When the reduced density is low, the Dean and Stiel (1965) relationship extended to mixtures can be used to determine the viscosity:

$$\mu_m = \mu_{gp_m} + \mu_{sc_m} f\left(\rho_r\right) \qquad\qquad [4.60]$$

$$\mu_{sc_m} = \mu_{a_m} M_m^{1/2} P_{c_m}^{2/3} T_{c_m}^{-1/6}$$

$$\mu_{a_i} = \frac{\mu_{sc_i}}{M_i^{1/2} P_{c_i}^{2/3} T_{c_i}^{-1/6}}$$

$$\mu_{a_m} = \sum_i \left(x_i \, \mu_{a_i}\right)$$

where

M_i = molecular weight of the component i [kg/kmol]

P_{c_i} = critical pressure of the component i [bar]

T_{c_i} = critical temperature of the component i [K]

μ_{sc_i} = viscosity scaling coefficient for component i [mPa·s]

μ_{a_i} = viscosity coefficient for component i

M_m = molecular weight of the mixture   [kg/kmol]

P_{c_m} = pseudocritical pressure of the mixture   [bar]

T_{c_m} = pseudocritical temperature of the mixture   [K]

x_i = mole fraction of component i

μ_m = viscosity of the liquid at T and P [mPa·s]

μ_{gp_m} = viscosity of the mixture in the ideal gas state at T [mPa·s]

ρ_r = reduced density of the mixture

$f(\rho_r)$ = function given in article 4.3.2.1.a

The average error is about 30%. The relation can be used only if the reduced density is less than 2.5 and the reduced temperature of the mixture is greater than 0.80.

e. Mixtures of Viscous Hydrocarbons

Usually it is not easy to predict the viscosity of a mixture of viscous components. Certain binary systems, such as methanol and water, have viscosities much greater than either compound.

For hydrocarbons, three weighting methods are known to give satisfactory results:

- conventional weighting:

$$\mu_m = \left[\sum_i \left(x_i \mu_i^{1/3} \right) \right]^3 \qquad [4.61]$$

- Mehrotra method:

$$\log(\mu_m + 0.8) = \frac{\sum_i \left[x_i M_i^{1/2} \log(\mu_i + 0.8) \right]}{M_m^{1/2}} \qquad [4.62]$$

- ASTM method:

$$\log\left(\log\left(\nu_m + f\left(\nu_m\right)\right)\right) = \sum_i \left[x_{w_i} \log\left(\log\left(\nu_i + f\left(\nu_i\right)\right)\right) \right] \qquad [4.63]$$

where

x_i = mole fraction of component i

x_{w_i} = weight fraction of component i

M_i = molecular weight of the component i [kg/kmol]

M_m = molecular weight of the mixture   [kg/kmol]

μ_m = absolute viscosity of the mixture   [mPa·s]

μ_i = viscosity of component i [mPa·s]

ν_i = kinematic viscosity of the component i [mm²/s]

ν_m = kinematic viscosity of the mixture   [mm²/s]

log = common logarithm (base 10)

It is difficult to judge the accuracy of these methods because data are scarce. Table 4.9 compares the values obtained by different weighting methods with experimental values for a mixture of n-hexane–n-hexadecane at 25°C. The ASTM method shows results very close to those obtained experimentally.

Mole fraction of hexane	Experimental viscosity	Calculated viscosity method 1	Calculated viscosity method 2	Calculated viscosity method 3	Molecular weight	Density
	mPa·s	mPa·s	mPa·s	mPa·s	kg/kmol	kg/m³
0.0	3.078	3.078	3.078	3.078	226	769.7
0.2	2.240	2.183	2.424	2.270	198	758.5
0.4	1.510	1.481	1.823	1.592	170	744.6
0.6	0.991	0.949	1.270	1.042	142	725.9
0.8	0.584	0.562	0.763	0.603	114	698.9
1.0	0.298	0.298	0.298	0.298	86	657.2

Table 4.9
Comparison of weighting methods for liquid phase viscosities. Mixture of n-hexane - n-hexadecane at 298 K.

Method 1   conventional weighting.
Method 2   weighting by Mehrotra's method.
Method 3   weighting by the ASTM method.

f. Influence of Pressure on the Viscosity of Liquids

The viscosity of a liquid increases with pressure. This behavior is relatively small for liquids of high molecular weight and low compressibility; the effect can be estimated by Kouzel's method (1965):

$$\log \frac{\mu}{\mu_s} = (P - P_s)\left(5.829 \cdot 10^{-4} \mu_s^{0.181} - 1.479 \cdot 10^{-4}\right) \qquad [4.64]$$

where

P = pressure   [bar]

P_s = saturation pressure   [bar]

μ = viscosity at pressure P [mPa·s]

μ_s = viscosity at saturation pressure   [mPa·s]

The average error of this method is about 10%. The method is applicable only for pressures less than 1000 bar.

g. Viscosity Index

The viscosity index is an empirical number, determined from the kinematic viscosities at 40 and 100°C; it indicates the variation in viscosity with temperature.

The exact calculation of the index is given in the ASTM D 2270 standard. The kinematic viscosity at 40°C (U) of an oil whose viscosity index (VI) is being calculated is compared with those of two reference oils for which the viscosity indices are 0 and 100 respectively, and which have at 100°C the same kinematic viscosity as that of the oil being examined:

$$VI = 100 \frac{(L - U)}{(L - H)}$$

where

VI = viscosity index

U = kinematic viscosity at 40°C of the oil being tested $[\text{mm}^2/\text{s}]$

L = kinematic viscosity at 40°C of the reference oil of index 0 $[\text{mm}^2/\text{s}]$

H = kinematic viscosity at 40°C of the reference oil of index 100

$[\text{mm}^2/\text{s}]$

4.3.2.2 **Thermal Conductivities of Liquids**

Thermal conductivity is expressed in W/(m·K) and measures the ease in which heat is transmitted through a thin layer of material. Conductivity of liquids, written as λ_l, decreases in an essentially linear manner between the triple point and the boiling point temperatures. Beyond a reduced temperature of 0.8, the relationship is not at all linear. For estimation of conductivity we will distinguish two cases:

- liquids of low density at $T_r > 0.8$
- liquids of low compressibility at $T_r < 0.8$.

The conductivity of petroleum fractions whose nature is not known is estimated using the following relation:

$$\lambda_l = 0.17 - 1.418 \ 10^{-4} \ T \qquad [4.65]$$

where

λ_l = thermal conductivity of the liquid $[\text{W}/(\text{m·K})]$

T = temperature $[\text{K}]$

Although widely used, this relation can be very inaccurate, notably in the case of aromatics and iso-paraffins, as shown in Table 4.10.

a. Conductivity of Pure Hydrocarbon Liquids of Low Density $\left(\rho_r < 2.8 \right)$

When the reduced temperature is greater than 0.8, the conductivity of pure hydrocarbons can be calculated by the method of Stiel and Thodos (1963):

$$\lambda_l = \lambda_{gp} + \lambda_{sc} \ f \left(\rho_r \right) \qquad [4.66]$$

Composant	Temperature, °C	Experimental conductivity, W/(m·K)	Calculated conductivity, W/(m·K)	Difference, %
Propane	50	0.078	0.124	59.2
n-Pentane	20	0.114	0.128	12.7
n-Hexane	20	0.122	0.128	5.3
n-Heptane	20	0.126	0.128	1.9
n-Octane	20	0.129	0.128	−0.4
n-Nonane	20	0.131	0.128	−2.0
n-Decane	41	0.127	0.125	−1.2
2-Methylpentane	32	0.108	0.127	17.3
2,2,4-Trimethylpentane	38	0.097	0.126	29.8
Benzene	20	0.148	0.128	−13.2

Table 4.10   *Comparison of calculated and experimental thermal conductivities.*

$$\lambda_{sc} = \lambda_a P_c^{2/3} T_c^{-1/6} M^{-1/2} z_c^{-5}$$

when $\rho_r \leq 2$ $f(\rho_r) = 5.48 \cdot 10^{-5} [\exp(0.67\, \rho_r) - 1.069]$

when $\rho_r > 2$ $f(\rho_r) = 1.245 \cdot 10^{-5} [\exp(1.155\, \rho_r) + 2.016]$

where

λ = thermal conductivity of the liquid at T and P [W/(m·K)]

λ_{sc} = reduction group for the conductivity   [W/(m·K)]

λ_{gp} = conductivity of the ideal gas at T [W/(m·K)]

λ_a = conductivity scaling coefficient

ρ_r = reduced density

M = molecular weight   [kg/kmol]

T_c = critical temperature   [K]

P_c = critical pressure   [bar]

z_c = critical compressibility

The coefficient λ_a can be calculated starting from a liquid conductivity at T_r near 0.8. It is generally close to 1.

The average error of this method is around 10%. The method is not applicable for reduced densities greater than 2.8.

For mixtures of hydrocarbons and petroleum fractions, the same formula applies:

$$\lambda_{l_m} = \lambda_{gp_m} + \lambda_{sc_m}\, f(\rho_r) \qquad [4.66]$$

$$\lambda_{sc_m} = \lambda_{a_m} P_{c_m}^{2/3} T_{c_m}^{-1/6} M_m^{-1/2} z_{c_m}^{-5}$$

$$\lambda_{a_m} = \Sigma \left(x_i \lambda_{a_i} \right)$$

where

λ_{l_m} = conductivity of the liquid mixture at T and P [W/(m·K)]

λ_{gp_m} = conductivity of the mixture of the ideal gas at T [W/(m·K)]

ρ_r = reduced density of the mixture

P_{c_m} = pseudocritical pressure of the mixture   [bar]

T_{c_m} = pseudocritical temperature of the mixture   [K]

M_m = molecular weight of the mixture   [kg/kmol]

z_{c_m} = pseudocritical compressibility of the mixture

$f(\rho_r)$ = function identical to that of the pure hydrocarbon

b. Conductivity of Liquid at Low Temperature

At low temperature and pressure, the conductivity of a pure hydrocarbon is obtained by linear interpolation between two known conductivities:

$$\lambda_l = \lambda_{l_1} + \left(\lambda_{l_2} - \lambda_{l_1}\right)\frac{\left(T - T_1\right)}{\left(T_2 - T_1\right)} \qquad [4.67]$$

where

λ_l = thermal conductivity of the liquid at temperature T [W/(m·K)]

λ_{l_1} = thermal conductivity of the liquid at temperature T_1 [W/(m·K)]

λ_{l_2} = thermal conductivity of the liquid at temperature T_2 [W/(m·K)]

T, T_1, T_2 = temperatures   [K]

Usually, T_1 is taken as the triple point and T_2 as the normal boiling point. The average error is about 5%.

When only one conductivity λ_{l_1} is known at temperature T_1, the conductivity can be estimated using the following relation:

$$\lambda_l = \lambda_{l_1}\frac{\left(1 - 0.7\,T_r\right)}{\left(1 - 0.7\,T_{r_1}\right)} \qquad [4.68]$$

$$T_{r_1} = \frac{T_1}{T_c} \quad \text{and} \quad T_r = \frac{T}{T_c}$$

where

T_r, T_{r_1} = reduced temperatures

T = temperature   [K]

T_c = critical temperature   [K]

T_1 = temperature at which the conductivity is known   [K]

λ_{l_1} = known conductivity at T_1 [W/(m·K)]

The conductivity of hydrocarbons and petroleum fractions at 20°C, can be estimated by the relations:

- for n-paraffins: $\lambda_{l_1} = 0.18\,S$ [4.69]
- for cyclic hydrocarbons: $\lambda_{l_1} = 0.4\,M^{-0.2}\,S$

where

λ_{l_1} = conductivity of the liquid at 20°C [W/(m·K)]

M = molecular weight [kg/kmol]

S = standard specific gravity

The formula for cyclic hydrocarbons has been established from data on hydrocarbons whose molecular weights are lower than 140 and must not be used for higher molecular weights.

The average error for these methods is about 10% at 20°C and about 15% at other temperatures.

c. Conductivities of Mixtures at Low Temperatures

When the conductivities of each component have been determined, the conductivity of the mixture is obtained by the rules set forth by Li in 1976:

$$\lambda_{l_m} = \sum_{i=1}^{n}\left[\sum_{j=1}^{n}\left(x_{v_i}\,x_{v_j}\,\lambda_{i_j}\right)\right]$$ [4.70]

$$\lambda_{i_j} = \frac{2}{\dfrac{1}{\lambda_{l_i}} + \dfrac{1}{\lambda_{l_j}}} \qquad x_{v_i} = \frac{x_i\,V_{l_i}}{\sum\left(x_i\,V_{l_i}\right)}$$

where

V_{l_i} = molar volume of component i [m³/kmol]

x_{v_i} = volume fraction of component i

x_i = mole fraction of component i

λ_{l_i} = conductivity of the component i in the liquid state [W/(m·K)]

λ_{l_m} = conductivity of the mixture in the liquid state [W/(m·K)]

n = number of components in the mixture

The method has an average error of 5% for all mixtures of hydrocarbons whose conductivities of its components are known.

d. Influence of Pressure on Liquid Conductivity

Liquid conductivities increase with pressure. The conductivity at pressure P_2 can be calculated from that at pressure P_1 by the Lenoir method proposed in 1957:

$$\lambda_{l_2} = \lambda_{l_1}\frac{C_2}{C_1}$$ [4.71]

$$C_i = 17.77 + 0.065\, P_{r_i} - 7.764\, T_r - \frac{2.054\, T_r^2}{\exp\!\left(0.2\, P_{r_i}\right)}$$

where

λ_{l_2} = conductivity of a liquid at pressure P_2 [W/(m·K)]

λ_{l_1} = conductivity of a liquid at pressure P_1 [W/(m·K)]

T_r = reduced temperature

P_{r_i} = reduced pressures

The effects of pressure are especially sensitive at high temperatures. The analytical expression [4.71] given by the API is limited to reduced temperatures less than 0.8. Its average error is about 5%.

4.3.2.3 **Estimation of Diffusivity**

Diffusivity measures the tendency for a concentration gradient to dissipate to form a molar flux. The proportionality constant between the flux and the potential is called the diffusivity and is expressed in m^2/s. If a binary mixture of components A and B is considered, the molar flux of component A with respect to a reference plane through which the exchange is equimolar, is expressed as a function of the diffusivity and of the concentration gradient with respect to an axis Ox perpendicular to the reference plane by the following relation:

$$J_A = - D_{AB} \frac{dC_A}{dx}$$

where

J_A = the molar flux of component A $[kmol/(m^2 \cdot s)]$

D_{AB} = diffusivity $[m^2/s]$

C_A = concentration of component A $[kmol/m^3]$

$\dfrac{dC_A}{dx}$ = concentration gradient of component A with respect to the Ox axis $[kmol/m^4]$

The value of coefficient D_{AB} depends on the composition. As the mole fraction of component A approaches 0, D_{AB}, approaches D_{AB}^0 the diffusion coefficient of component A in the solvent B at infinite dilution. The coefficient D_{AB}^0 can be estimated by the Wilke and Chang (1955) method:

$$D_{AB}^0 = \frac{1.17 \cdot 10^{-9} \sqrt{\phi_B M_B} \cdot T}{\mu_B\, V_A^{0.6}} \qquad [4.73]$$

where

D_{AB}^0 = diffusivity of A in solvent B at infinite dilution $[m^2/s]$

T = temperature   [K]

M_B = molecular weight of solvent B   [kg/kmol]

μ_B = dynamic viscosity of solvent B   [mPa·s]

V_A = molar volume of solute A $[m^3/kmol]$

ϕ_B = association coefficient of solvent B

The authors recommend the following values for association coefficients:

- water   2.6
- methanol   1.9
- ethanol   1.5.

For hydrocarbon solvents the association coefficient is equal to 1.

This method should not be used when water is the solute.

4.4 **Properties of Gases**

4.4.1 **Thermodynamic Properties of Gases**

4.4.1.1 **Gas Density**

a. *Density of Ideal Gases*

The ideal gas theory has given us the following relation:

$$PV = RT$$

where

P = pressure   [bar]

V = molar volume $[m^3/kmol]$

T = temperature   [K]

R = ideal gas constant $[0.08314 \ m^3 \cdot bar/(kmol \cdot K)]$

This relation is easily transformed to express the ideal gas density:

$$\rho_{gp} = \frac{MP}{RT} = 12.03 \frac{MP}{T} \qquad [4.74]$$

where

ρ_{gp} = density of the ideal gas $[kg/m^3]$

M = molecular weight   [kg/kmol]

b. *Density of Real Gases*

For real gases, density is expressed by the following relationship:

$$\rho_g = \frac{MP}{RTz} = 12.03 \frac{MP}{Tz} \qquad [4.75]$$

where

z = compressibility factor

ρ_g = gas density $[kg/m^3]$

There have been several equations of state proposed to express the compressibility factor. Remarkable accuracy has been obtained when specific equations for certain components are used; however, the multitude of their coefficients makes their extension to mixtures complicated.

Hydrocarbon mixtures are most often modeled by the equations of state of Soave, Peng Robinson, or Lee and Kesler.

The average accuracy of the Lee and Kesler model is much better than that of all cubic equations for pressures higher than 40 bar, as well as those around the critical point.

The Lee and Kesler method for calculating densities is given in article 4.3.1.1; its average accuracy is about 1%, when the pressure is less than 100 bar.

4.4.1.2 **Specific Heat of Gases**

The specific heat of gases at constant pressure C_{pg} is calculated using the principle of corresponding states. The C_{pg} for a mixture in the gaseous state is equal to the sum of the C_{pg} of the ideal gas and a pressure correction term:

$$C_{pg_m} = C_{pgp_m} + \frac{R}{M_m} dC_{pr} \qquad [4.76]$$

where

C_{pg_m} = C_p of the gas mixture at T et P [kJ/(kg·K)]
C_{pgp_m} = C_p of the mixture in the ideal gas state at T [kJ/(kg·K)]
R = ideal gas constant   [8.314 kJ/(kmol·K)]
dC_{pr} = reduced correction for C_p, a function of
         T_{r_m} and P_{r_m} and of the acentric factor ω_m
M_m = molecular weight of the mixture   [kg/kmol]

a. *Specific Heats of Pure Hydrocarbons in the Ideal Gas State*

The C_{pg} is expressed as the derivative of the enthalpy with respect to temperature at constant pressure. For an ideal gas it is a total derivative:

$$C_{pgp} = \frac{dH_{gp}}{dT}$$

The enthalpy of pure hydrocarbons in the ideal gas state has been fitted to a fifth order polynomial equation of temperature. The corresponding C_p is a polynomial of the fourth order:

$$H_{gp} = 2.325 \left(A + BT + CT^2 + DT^3 + ET^4 + FT^5 \right) \qquad [4.77]$$

$$C_{pgp} = 4.185 \left(B + 2CT + 3DT^2 + 4ET^3 + 5FT^4 \right) \qquad [4.78]$$

$$T = 1.8 \, T'$$

where

H_{gp} = specific enthalpy of the ideal gas [kJ/kg]

C_{pgp} = specific heat of the ideal gas [kJ/(kg·K)]

A, B, C, D, E, F = coefficients of the polynomial expression

T' = temperature [K]

The coefficients A, B, C, D, E and F have been tabulated in the API technical data book for a great number of hydrocarbons. See Appendix 1.

b. Specific Heat of Petroleum Fractions in the Ideal Gas State

The coefficients B, C, and D from equations 4.77 and 4.78 can be estimated from a relation of Lee and Kesler, cited in the API Technical Data Book; the terms E and F are neglected:

$$B = -0.35644 + 0.02972 K_W + \alpha \left(0.29502 - \frac{0.2846}{S} \right) \qquad [4.79]$$

$$C = \frac{-10^{-4}}{2} \left(2.9247 - 1.5524 K_W + 0.05543 K_W^2 + C' \right)$$

$$C' = \alpha \left(6.0283 - \frac{5.0694}{S} \right)$$

$$D = \frac{-10^{-7}}{3} \left(1.6946 + 0.0844\, \alpha \right)$$

if $10 < K_W < 12.8$ and $0.7 < S < 0.885$

then $\alpha = \left[\left(\frac{12.8}{K_W} - 1 \right) \left(1 - \frac{10}{K_W} \right) (S - 0.885)(S - 0.7) \cdot 10^4 \right]^2$

if not, then $\alpha = 0$,

where

K_W = Watson characterization factor

S = standard specific gravity

This relation should be used preferentially when the reduced temperature for the fraction under consideration is greater than 0.8.

When the reduced temperature is less than 0.8, it is better to estimate the C_{pgp} starting from the C_{pl} of the fraction in the liquid state by the following relationship:

$$C_{pgp} = C_{pl} - \frac{R}{M} dC_{pr} \qquad [4.80]$$

$$dC_{pr} = dC_{pr_1} + 2.5138\, \omega \left(dC_{pr_2} - dC_{pr_1} \right)$$

The C_p corrections have been calculated for the reduced saturation pressure P_{r_s}

$$P_{r_s} = \frac{P_s}{P_c}.$$

where

C_{pgp} = specific heat of the ideal gas   [kJ/(kg·K)]

C_{pl} = specific heat of the liquid   [kJ/(kg·K)]

R = ideal gas constant   [8.314 kJ/(kmol·K)]

M = molecular weight   [kg/kmol]

dC_{pr_1} = reduced correction for C_p calculated at T_r and P_{r_s} for liquid methane

dC_{pr_2} = reduced correction for C_p calculated at T_r and P_{r_s} for liquid n-octane

dC_{pr} = reduced correction for C_p of the Lee Kesler model for the liquid

P_s = saturation pressure for the petroleum fraction at T [bar]

P_c = pseudocritical pressure for the petroleum fraction   [bar]

c. Specific Heat for Mixtures in the Ideal Gas State

For mixtures, the C_{pg} of the ideal gas is equal to:

$$C_{pgp_m} = \Sigma \left(x_{w_i} C_{pgp_i} \right) \qquad\qquad [4.81]$$

where

C_{pgp_m} = C_p of the mixture in the ideal gas state   [kJ/(kg·K)]

C_{pgp_i} = C_p of the component i in the ideal gas state   [kJ/(kg·K)]

x_{w_i} = weight fraction of component i

d. Specific Heat of a Real Gas

The C_{pg} of real gas is calculated using the equation derived from the Lee and Kesler model:

$$C_{pg} = C_{pgp} + \frac{R}{M} dC_{pr} \qquad\qquad [4.76]$$

where

C_{pg} = specific heat of the gas at T and P [kJ/(kg·K)]

C_{pgp} = specific heat of the ideal gas at T [kJ/(kg·K)]

R = ideal gas constant   [8.314 kJ/(kmol·K)]

M = molecular weight   [kg/kmol]

dC_{pr} = reduced C_p correction calculated from the Lee and Kesler model

From a practical point of view, as for liquids, it is possible to calculate dC_{pr} using finite differences by applying the relation [4.47].

4.4.1.3 **Specific Enthalpy of a Gas**

The specific enthalpy of a gas is calculated using the principle of corresponding states. The enthalpy of a gas mixture is equal to the sum of the ideal gas enthalpy and a correction term:

$$H_{g_m} = H_{g_{P_m}} + \frac{R\,T_{c_m}}{M_m}\,dH_r$$
[4.82]

where

H_{g_m} = specific enthalpy of the gas mixture at T and P [kJ/kg]

$H_{g_{P_m}}$ = specific enthalpy of the mixture in the ideal gas
    state at T [kJ/kg]

T_{c_m} = pseudocritical temperature of the mixture   [K]

M_m = molecular weight of the mixture   [kg/kmol]

dH_r = reduced enthalpy correction, a function of T_{r_m},
    P_{r_m} and of the mixture's acentric factor ω_m

a. *Enthalpy of an Ideal Gas*

The enthalpy of a pure compound or petroleum fraction is obtained by relations [4.77] and [4.79] described in articles 4.4.1.2 a and b.

For petroleum fractions, there is a problem of coherence between the expression for liquid enthalpy and that of an ideal gas. When the reduced temperature is greater than 0.8, the liquid enthalpy is calculated starting with the enthalpy of the ideal gas. On the contrary, when the reduced temperature is less than 0.8, it is preferable to calculate the enthalpy of the ideal gas starting with the enthalpy of the liquid:

at $T_r < 0.8$ $H_{gp} = \left(H_l - H_{l_{ref}}\right) + dH_{lgp} + k$ at $T_r > 0.8$ $H_l = H_{gp} - dH_{lgp}$

The constant k is such that the these two expressions are identical at $T_r = 0.8$; this results in the following relation:

$$H_{gp} = H'_{gp} + H_l - H'_l + dH_{lgp} - dH'_{lgp}$$
[4.83]

where

H_{gp} = enthalpy of the ideal gas at $T_r < 0.8$ [kJ/kg]

H'_{gp} = enthalpy of the ideal gas at $T_r = 0.8$ determined
    by the equations [4.77] and [4.79]   [kJ/kg]

H_l = enthalpy of the liquid at $T_r < 0.8$ determined
    by the equation [4.49]   [kJ/kg]

H'_l = enthalpy of the liquid at $T_r = 0.8$ determined
    by the equation [4.49]   [kJ/kg]

dH_{lgp} = enthalpy correction between the liquid and
    ideal gas calculated at $T_r < 0.8$ and P_{r_s} [kJ/kg]

dH'_{lgp} = enthalpy correction between the liquid and
    ideal gas calculated at $T_r = 0.8$ and P_{r_s} [kJ/kg]

The enthalpy correction is expressed as:

$$dH_{lgp} = \frac{R\,T_{c_f}}{M_f}\,dH_r \qquad\qquad [4.84]$$

$$dH_r = dH_{r_1} + 2.5138\,\omega_f\left(dH_{r_2} - dH_{r_1}\right)$$

where

M_f = molecular weight of the petroleum fraction f [kg/kmol]

R = ideal gas constant [8.314 kJ/(kmol·K)]

T_{c_f} = pseudocritical temperature of the fraction f [K]

ω_f = acentric factor of the fraction f

dH_{r_i} = reduced correction calculated by the equations of the Lee and Kesler model, given in. 4.3.1.3.b ($i = 1$ for methane and $i = 2$ for n-octane)

For mixtures, the enthalpy of an ideal gas is expressed as:

$$H_{gp_m} = \sum \left(x_{w_i} H_{gp_i}\right) \qquad\qquad [4.85]$$

where

H_{gp_m} = enthalpie of the mixture in the ideal gas state [kJ/kg]

H_{gp_i} = enthalpie of the component i in the ideal gas state [kJ/kg]

x_{w_i} = weight fraction of the component i

b. Enthalpy of a Real Gas

The enthalpy of a gas mixture is obtained by equation [4.82]. The correction dH_r is calculated by the Lee and Kesler model described in 4.3.1.3.b.

The terms dH_{r_1} and dH_{r_2} are determined after having calculated the reduced volumes V_{r_1} and V_{r_2} and compressibilities z_1 and z_2 of gases 1 and 2 at T_r and P_r.

This method has an average error of 5 kJ/kg except in the critical region where the deviations can be up to 30 kJ/kg.

The relation applies to mixtures of non-polar components such as hydrocarbons in this range:

$$0.3 < T_r < 4 \quad \text{and} \quad P_r < 10.$$

4.4.2 Thermophysical Properties of Gases

Calculation of thermophysical properties of gases relies on the principle of corresponding states. Viscosity and conductivity are expressed as the sum of the ideal gas property and a function of the reduced density:

$$\mu_g = \mu_{gp} + \mu_{sc}\,f\left(\rho_r\right) \qquad\qquad [4.86]$$

$$\lambda_g = \lambda_{gp} + \lambda_{sc}\,g\left(\rho_r\right) \qquad\qquad [4.87]$$

where

μ_g	= viscosity of the gas at T and P	[mPa·s]
μ_{gp}	= viscosity of the ideal gas at T	[mPa·s]
μ_{sc}	= scaling coefficient for viscosity	[mPa·s]
$f(\rho_r)$	= function of the reduced density	
λ_g	= conductivity of the gas	[W/(m·K)]
λ_{gp}	= conductivity of the ideal gas	[W/(m·K)]
λ_{sc}	= scaling coefficient for conductivity	[W/(m·K)]
$g(\rho_r)$	= function of the reduced density	

It is necessary to determine first the properties of each component in the ideal gas state, next to weight these values in order to obtain the property of the mixture in the ideal gas state.

4.4.2.1 **Viscosity of Gases**

a. *Viscosity of a Pure Hydrocarbon in the Ideal Gas State*

This viscosity can be calculated by the equation proposed by Yoon and Thodos (1970):

$$\mu_{gp} = \mu^0 f(T_r) \qquad [4.88]$$

$$\mu^0 = M^{1/2} P_c^{2/3} T_c^{-1/6}$$

$$f(T_r) = 10^{-5} \left[1 + 46.1 T_r^{0.618} - 20.4 \exp\left(-0.449 T_r\right) + 19.4 \exp\left(-4.058 T_r\right)\right]$$

where

T_r	= reduced temperature	
T_c	= critical temperature	[K]
P_c	= critical pressure	[bar]
M	= molecular weight	[kg/kmol]
μ_{gp}	= viscosity of the ideal gas	[mPa·s]
μ^0	= viscosity coefficient for the ideal gas	[mPa·s]

The viscosity coefficient μ^0 can also be derived from experimental data.

This method does not apply for hydrogen. For hydrocarbons, the average error is about 5%.

b. *Viscosity of an Ideal Gas Mixture*

The viscosity is obtained by weighting the viscosities of each component. The recommended method is that of Bromley and Wilke (1951):

$$\mu_{gp} = \sum_{i=1}^{n} \left[\frac{\mu_{gp_i} \, x_i}{\sum\limits_{j=1}^{n} \left(\phi_{ij} \, x_j\right)}\right] \qquad [4.89]$$

$$\phi_{ij} = \frac{\left[1 + \left(\dfrac{\mu_{gp_i}}{\mu_{gp_j}} \right)^{1/2} \left(\dfrac{M_i}{M_j} \right)^{1/2} \right]^2}{\sqrt{8} \left(1 + \dfrac{M_i}{M_j} \right)^{1/2}}$$

where

μ_{gp_m} = viscosity of the mixture in the ideal gas state   [mPa·s]

μ_{gp_i} = viscosity of the component i in the ideal gas state   [mPa·s]

x_i = mole fraction of component i

M_i = molecular weight of the component i [kg/kmol]

n = number of components

This method is applicable for all gas mixtures. The average error being about 3%.

c. Viscosity of a Gas Mixture at High Pressure

The Dean and Stiel (1965) method is based on the principle of corresponding states:

$$\mu_{g_m} = \mu_{gp_m} + \mu_{sc_m} f\left(\rho_r \right) \qquad [4.90]$$

$$\mu_{sc_m} = \mu_{a_m} M_m^{1/2} P_{c_m}^{2/3} T_{c_m}^{-1/6}$$

$$\mu_{a_m} = \Sigma \left(\mu_{a_i} x_i \right)$$

where

μ_{g_m} = viscosity of the gas at T and P [mPa·s]

μ_{gp_m} = viscosity of the ideal gas at T [mPa·s]

μ_{a_i} = viscosity coefficient of the component i, adjusted from experimental values, in general equal to 1

μ_{a_m} = viscosity coefficient of the mixture

T_{c_m} = pseudocritical temperature of the mixture   [K]

P_{c_m} = pseudocritical pressure of the mixture   [bar]

M_m = molecular weight of the mixture   [kg/kmol]

ρ_r = reduced density

x_i = mole fraction of component i

$$f(\rho_r) = 1.07 \cdot 10^{-4} \left(\exp \left(1.439 \, \rho_r \right) - \exp \left(- \, 1.11 \, \rho_r^{1.858} \right) \right)$$

This method applies only to mixtures of hydrocarbons. The average error is about 5%.

4.4.2.2 **Thermal Conductivity of Gases**

a. *Conductivity of a Pure Hydrocarbon in the Ideal Gas State*

The conductivity of a pure hydrocarbon in the ideal gas state is expressed as a function of reduced temperature T_r according to the equation of Misic and Thodos (1961):

$$\lambda_{gp} = C_{pgp} \, \lambda_{sc} \, f(T_r) \tag{4.91}$$

$$\lambda_{sc} = P_c^{2/3} \, M^{-1/2} \, T_c^{-1/6}$$

For methane and cyclic compounds: $f(T_r) = 4.56 \cdot 10^{-4} \, T_r$

For other compounds: $f(T_r) = (14.52 \, T_r - 5.14)^{2/3} \cdot 10^{-4}$

where

λ_{gp}	= conductivity of the ideal gas	[W/(m·K)]
C_{pgp}	= specific heat of the ideal gas	[kJ/(kg·K)]
P_c	= critical pressure	[bar]
M	= molecular weight	[kg/kmol]
T_c	= critical temperature	[K]

b. *Conductivity of a Mixture in the Ideal Gas State*

The conductivity of a gas mixture in the ideal gas state can be calculated by the Lindsay and Bromley method (1950):

$$\lambda_{gp_m} = \sum_{i=1}^{n} \left[\frac{\lambda_{gp_i} \, y_i}{\sum_{j=1}^{n} \left(A_{ij} \, y_j \right)} \right] \tag{4.92}$$

$$A_{ij} = \frac{1}{4} \frac{C_{ij}}{C_i} \left(1 + \sqrt{B_{ij}} \right)^2$$

$$C_{ij} = 1 + \frac{S_{ij}}{T} \qquad C_i = 1 + \frac{S_i}{T}$$

$$S_{ij} = \sqrt{S_i S_j} \qquad S_i = 1.5 \, T_{b_i}$$

$$B_{ij} = \left(\frac{\mu_{gp_i}}{\mu_{gp_j}} \right) \left(\frac{M_j}{M_i} \right)^{3/4} \left(\frac{C_i}{C_j} \right)$$

where

λ_{gp_m}	= conductivity of the mixture in the ideal gas state at T	[W/(m·K)]
λ_{gp_i}	= conductivity of the component i in the ideal gas state at T	[W/(m·K)]
y_i	= mole fraction of component i	

T_{b_i} = normal boiling point of the component i [K]

μ_{gp_i} = viscosity of the component i in the ideal gas
state at T [mPa·s]

M_i = molecular weight of the component i [kg/kmol]

The average error is about 5%.

c. Conductivity of Real Gases

The conductivity of a real gas can be calculated by the Stiel and Thodos method, already used for liquids and given in article 4.3.2.2.a:

$$\lambda_{g_m} = \lambda_{gp_m} + \lambda_{sc_m} f(\rho_r) \tag{4.66}$$

where

λ_{g_m} = conductivity of the gas mixture at T and P [W/(m·K)]

The average error is about 5%.

4.4.2.3 Diffusivity of Gases

a. Diffusivity of Gases at Low and Medium Pressure

For a binary mixture of two components A and B in the gas phase, the mutual diffusion coefficient D_{AB} such as defined in 4.3.2.3, does not depend on composition. It can be calculated by the Fuller (1966) method:

$$D_{AB} = \frac{1.43 \cdot 10^{-9} \, T^{1.75}}{P(M_{AB})^{1/2} \left[\left(\sum dV \right)_A^{1/3} + \left(\sum dV \right)_B^{1/3} \right]^2} \tag{4.93}$$

$$M_{AB} = \frac{2}{\dfrac{1}{M_A} + \dfrac{1}{M_B}}$$

where

T = temperature   [K]

P = pressure   [bar]

M_A = molecular weight of the component A   [kg/kmol]

$\left(\sum dV \right)_A$ = sum of the atomic volumes of diffusion
for component A $[m^3/kmol]$

D_{AB} = diffusivity $[m^2/s]$

The dV are as follows:

- carbon atom   0.0159 $[m^3/kmol]$
- hydrogen atom   0.00231 $[m^3/kmol]$
- nitrogen atom   0.00454 $[m^3/kmol]$
- sulfur atom   0.0229 $[m^3/kmol]$

For cyclic compounds, it is necessary to reduce the volume obtained by 0.0183 m^3/kmol. For example, benzene has a volume of diffusion of:

$$6 \, (0.0159 + 0.00231) - 0.0183 = 0.09096 \qquad m^3/kmol$$

Simple molecules have the following volumes of diffusion:

- hydrogen   0.00612 $[m^3/kmol]$
- nitrogen   0.0185 $[m^3/kmol]$
- carbon monoxide   0.0180 $[m^3/kmol]$
- CO_2 0.0269 $[m^3/kmol]$
- H_2O 0.0131 $[m^3/kmol]$

The error of this method is about 10% at atmospheric pressure. The accuracy becomes lower as the pressure increases.

b. Diffusivity of Gases at High Pressure

Fuller's equation, applied for the estimation of the coefficient of diffusion of a binary gas mixture, at a pressure greater than 10 bar, predicts values that are too high. As a first approximation, the value of the coefficient of diffusion can be corrected by multiplying it by the compressibility of the gas:

$$D_{AB} = D'_{AB} \, z$$

where

D_{AB} = mutual coefficient of diffusion $\qquad\qquad$ $[m^2/s]$

D'_{AB} = mutual coefficient of diffusion calculated
$\qquad$ by Fuller's method $\qquad\qquad\qquad\qquad\qquad$ $[m^2/s]$

z $\quad$ = compressibility factor of the gas

4.5 Estimation of Properties Related to Phase Changes

In the petroleum refining and natural gas treatment industries, mixtures of hydrocarbons are more often separated into their components or into narrower mixtures by chemical engineering operations that make use of phase equilibria between liquid and gas phases such as those mentioned below:

- distillation
- absorption
- condensation or vaporization.

When the fluids being treated contain water, the equilibria most often involve three phases (liquid-liquid-vapor).

Liquid-liquid and liquid-solid equilibria also find industrial applications in liquid-liquid extraction and fractional crystallization operations.

Finally, gas-solid equilibria should be studied to avoid plugging problems due in particular to hydrate formation.

4.5.1 **Phase Equilibria for a Pure Hydrocarbon**

At a given temperature and pressure, a pure compound can exist in one, two or three states. The compound exists at three different states at the triple point and at two different states along the curves of vaporization, freezing and sublimation. Refer to Figure 4.6.

The relations which permit us to express equilibria utilize the Gibbs free energy, to which we will give the symbol G and which will be called simply free energy for the rest of this chapter. This thermodynamic quantity is expressed as a function of enthalpy and entropy. This is not to be confused with the Helmholtz free energy which we will note as F:

$$G = H - TS$$
$$F = U - TS$$

where

G	= Gibbs molar free energy	[kJ/kmol]
S	= molar entropy	[kJ/(kmol·K)]
F	= Helmholtz free molar energy	[kJ/kmol]
H	= molar enthalpy	[kJ/kmol]
U	= molar internal energy	[kJ/kmol]

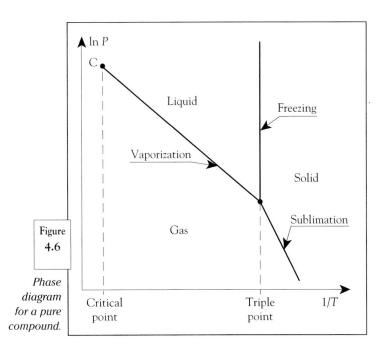

Figure 4.6

Phase diagram for a pure compound.

At the triple point, the free energies of each phase are equal:

$$G_g = G_l = G_s$$

The liquid -vapor equilibria, along the vaporization curve is expressed as:

$$G_g = G_l < G_s$$

Likewise, the liquid-solid equilibria along the fusion curve is:

$$G_l = G_s < G_g$$

And the gas-solid equilibria has this formula:

$$G_g = G_s < G_l$$

where

G_l	= molar free energy of the liquid	[kJ/kmol]
G_g	= molar free energy of the gas	[kJ/kmol]
G_s	= molar free energy of the solid	[kJ/kmol]

The determination of equilibria is done theoretically via the calculation of free energies. In practice, the concept of fugacity is used for which the unit of measurement is the bar. The equation linking the fugacity to the free energy is written as follows:

$$RT \ln \frac{f}{P^0} = G - G^0$$

$$f = P^0 \exp\left(\frac{G - G^0}{RT}\right)$$

where

R	= ideal gas constant	[8.314 kJ/(kmol·K)]
T	= temperature	[K]
f	= fugacity	[bar]
G	= molar free energy at T and P	[kJ/kmol]
G^0	= molar free energy of the ideal gas at T and P^0	[kJ/kmol]
P^0	= reference pressure	[bar]

The equilibrium relations are written as below:

- at the triple point: $\quad f_l = f_g = f_s$
- in vaporization: $\quad f_l = f_g < f_s$
- in freezing: $\quad f_l = f_s < f_g$
- in sublimation: $\quad f_g = f_s < f_l$

where

f_l	= fugacity of the liquid	[bar]
f_g	= fugacity of the gas	[bar]
f_s	= fugacity of the solid	[bar]

4.5.2 Phase Equilibria for Mixtures

Figure 4.7 gives liquid and vapor phase envelopes for a hydrocarbon mixture.

The region of coexistence between liquid and gas phases is delimited by three curves:

- the bubble point curve which ends at the critical point and which marks the transition between the liquid and the two-phase liquid-vapor regions
- the lower dew point curve which ends at the cricondentherm and delimits the gas and the two-phase liquid-vapor regions
- the higher dew point curve which joins cricondentherm to the critical point in passing by the cricondenbar and which separates the two-phase region from the supercritical gas.

The conditions of liquid-vapor equilibrium for a component i are expressed below:

$$G_{l_i} = \left(\frac{\partial G_l}{\partial n_i}\right)_{T,P,n_j} = \left(\frac{\partial G_g}{\partial n_i}\right)_{T,P,n_j} = G_{g_i}$$

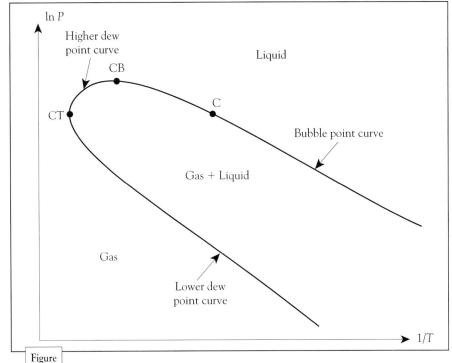

Figure 4.7   *Liquid and vapor phase envelope for a hydrocarbon mixture.*
C: critical point   CB: Cricondenbar   CT: Cricondentherm

where

G_{g_i} = partial free energy of component i in the gas
phase at temperature T and pressure P [kJ/kmol]

G_l = free energy of the liquid at T and P [kJ/kmol]

G_{l_i} = partial free energy of component i in the liquid
phase at T and P [kJ/kmol]

G_g = free energy of the gas phase at T and P [kJ/kmol]

T = temperature   [K]

P = pression   [bar]

n_j = quantity of species j different from i [kmol]

n_i = quantity of species i [kmol]

Utilizing the concept of partial fugacity which is defined by the relation:

$$RT \ln \left(\frac{f_i}{P^0 y_i} \right) = \left[\frac{\partial \left(G - G^0 \right)}{\partial n_i} \right]_{T, P, n_j}$$

the conditions for liquid-vapor equilibrium are written as:

$$f_{l_i} = f_{g_i}$$

for liquid-solid and gas-solid equilibria, one has similarly:

$$f_{s_i} = f_{l_i} \quad \text{and} \quad f_{s_i} = f_{g_i}$$

where

R = ideal gas constant   [8.314 kJ/(kmol·K)]

T = temperature   [K]

f_i = partial fugacity of component i in the phase
under consideration   [bar]

G = free energy of the phase considered at the
temperature T and pressure P [kJ/kmol]

G^0 = free energy of the mixture in the ideal gas
state at T and P^0 [kJ/kmol]

P^0 = reference pressure   [bar]

y_i = mole fraction of component i

f_{l_i} = partial fugacity of component i in the liquid phase   [bar]

f_{g_i} = partial fugacity of component i in the gas phase   [bar]

f_{s_i} = partial fugacity of component i in the solid phase   [bar]

The ratio f_i/Py_i is written as ϕ_i and called the partial fugacity coefficient; in the liquid phase, this coefficient is written as ϕ_{l_i} and in the gas phase ϕ_{g_i}.

4.5.3 **Vapor-Liquid Equilibria**

4.5.3.1 **Calculation of Vapor-Liquid Equilibria for a Mixture**

The general problem is to determine at given conditions of temperature and pressure, the quantities and compositions of the two phases in equilibrium starting from an initial quantity of material of known composition and to resolve the system of the following equations:

$$Az_i = Lx_i + Vy_i \qquad i \in \{1, n\}$$
$$f_{l_i} = f_{g_i} \qquad i \in \{1, n\}$$

where

A	= initial molar quantity	[kmol]
L	= molar quantity of the liquid phase	[kmol]
V	= molar quantity of the vapor phase	[kmol]
z_i	= mole fraction of component i in the feed	
x_i	= mole fraction of component i in the liquid phase	
y_i	= mole fraction of component i in the vapor phase	
f_{l_i}	= partial fugacity of component i in the liquid phase	[bar]
f_{g_i}	= partial fugacity of component i in the vapor phase	[bar]

The calculation of partial fugacities requires knowing the derivatives of thermodynamic quantities with respect to the compositions and to arrive at a mathematical model reflecting physical reality.

Utilization of equations of state derived from the Van der Waals model has led to spectacular progress in the accuracy of calculations at medium and high pressure.

In the calculation of vapor phase partial fugacities the use of an equation of state is always justified. In regard to the liquid phase fugacities, there is a choice between two paths:

- utilize an equation of state
- calculate the partial fugacity starting from the total fugacity of the component in the pure state.

a. *Calculation of the Partial Fugacity of a Component i in the Liquid Phase, starting from the Total Fugacity of Component i in the Pure Component State*

The partial fugacity of component i in the liquid phase is expressed as a function of the total fugacity of this same component in the pure liquid state, according to the following relation:

$$f_{l_i} = f_{l_i}^0 x_i \gamma_i \qquad [4.94]$$

where

f_{l_i} = partial fugacity of component i in the mixture
in the liquid state at T and P [bar]

$f^0_{l_i}$ = total fugacity of the pure component i as liquid
at T and P [bar]

x_i = mole fraction of component i in the liquid phase

γ_i = activity coefficient of the component i in the liquid phase

• Calculation of the total fugacity of the pure component in the liquid phase.

The total fugacity, if the liquid is considered to be incompressible, is calculated as a function of the vapor pressure by the expression:

$$f^0_{l_i} = \pi_i\,\phi_{ls_i}\exp\frac{V_{l_i}\left(P - \pi_i\right)}{RT}$$ [4.95]

where

π_i = vapor pressure of component i at T [bar]

ϕ_{ls_i} = fugacity coefficient of component i as saturated liquid at temperature T and pressure π_i

V_{l_i} = molar volume of component i as liquid at $T[\text{m}^3/\text{kmol}]$

P = pressure exerted on the liquid [bar]

T = temperature [K]

R = ideal gas constant $\left[0.08314\ \text{bar·m}^3/(\text{K·kmol})\right]$

The vapor pressure will be calculated according to one of the methods described later.

The molar volume of the liquid is easily estimated by one of the equations given in article 4.3.1.1.

The fugacity coefficient of component i at saturation is obtained after the calculation of the vapor fugacity at saturation, by the relation:

$$\phi_{ls_i} = \frac{f_{gs_i}}{\pi_i}$$ [4.96]

f_{gs_i} = fugacity of component i, in the gas phase at temperature T and pressure π_i [bar]

• Estimation of the activity coefficient in the liquid phase γ_i.

Hydrocarbon mixtures can be assumed to be regular solutions; it is thus possible to estimate the activity coefficient using a relation published by Hildebrandt (1950):

$$\ln \gamma_i = \frac{V^0_{l_i}\left(\delta_i - \delta_m\right)^2}{RT}$$ [4.97]

$$\delta_m = \frac{\displaystyle\sum_{j=1}^{n}\left(x_j\,V^0_{l_j}\,\delta_j\right)}{\displaystyle\sum_{j=1}^{n}\left(x_j\,V^0_{l_j}\right)}$$

where

δ_i = solubility parameter for component i at 25°C $\qquad$ $[(\text{kJ/m}^3)^{1/2}]$

γ_i = activity coefficient of the component

$V_{l_i}^0$ = molar volume of component i in the liquid state
at 25°C $\qquad$ $[\text{m}^3/\text{kmol}]$

R = ideal gas constant $\qquad$ $[8.314 \text{ kJ}/(\text{kmol·K})]$

T = temperature $\qquad$ $[\text{K}]$

The solubility parameters calculated at 25°C have been tabulated for numerous hydrocarbons in the API technical data book. They can also by calculated by the relationship:

$$\delta_i^2 = \frac{U_{gp_i} - U_{l_i}}{V_{l_i}^0} \qquad [4.98]$$

where

U_{gp_i} = internal molar energy of component i at 25°C
and in the ideal gas state $\qquad$ $[\text{kJ/kmol}]$

U_{l_i} = internal molar energy of component i at 25°C
and in the liquid state $\qquad$ $[\text{kJ/kmol}]$

$V_{l_i}^0$ = molar volume of the component i as liquid at 25°C $\quad$ $[\text{m}^3/\text{kmol}]$

δ_i = solubility parameter at 25°C $\qquad$ $[(\text{kJ/m}^3)^{1/2}]$

The values of common hydrocarbon solubility parameters vary between 300 and 600 $(\text{kJ/m}^3)^{1/2}$. Several tables are available where the solubility parameters are shown as $(\text{cal/cm}^3)^{1/2}$. To convert these values, it is necessary to multiply by 64.69. Thus a solubility parameter value of 10 $(\text{cal/cm}^3)^{1/2}$ is equal to 646.9 $(\text{kJ/m}^3)^{1/2}$.

This method applies only to liquids whose constituents have reduced temperatures less than 1. The average error is about 10%; the most important differences are observed in mixtures of components belonging to different chemical families.

b. Calculation of Partial Fugacities Using an Equation of State

In 1972, Soave published a method of calculating fugacities based on a modification of the Redlich and Kwong equation of state which completely changed the customary habits and became the industry standard. In spite of numerous attempts to improve it, the original method is the most widespread. For hydrocarbon mixtures, its accuracy is remarkable. For a mixture, the equation of state is:

$$P = \frac{RT}{V - b_m} - \frac{a_m}{V(V + b_m)} = \frac{z\,RT}{V} \qquad [4.99]$$

a_m and b_m are obtained by conventional equations as follows:

$$a_m = \sum_{j=1}^{n} \left[\sum_{i=1}^{n} \left(x_i\, x_j \sqrt{a_i\, a_j} \left(1 - k_{ij}\right) \right) \right]$$

$$b_m = \sum_{i=1}^{n} b_i\, x_i$$

The coefficients of each component are calculated by these expressions:

$$a_i = \frac{0.42747 R^2\, T_{c_i}^2\, f\!\left(T_{r_i}\right)}{P_{c_i}} \qquad b_i = 0.08664 \frac{R T_{c_i}}{P_{c_i}}$$

$$f\!\left(T_{r_i}\right) = \left(1 + m_i\left(1 - \sqrt{T_{r_i}}\right)\right)^2$$

$$T_{r_i} = \frac{T}{T_{c_i}}$$

where

z = compressibility factor

R = ideal gas constant $\qquad\qquad$ [0.08314 m³·bar/(K·kmol)]

T = temperature $\qquad\qquad$ [K]

P = pressure $\qquad\qquad$ [bar]

V = molar volume of the mixture $\qquad$ [m³/kmol]

T_{c_i} = critical temperature of the component i $\quad$ [K]

P_{c_i} = critical pressure of the component i $\qquad$ [bar]

m_i = constant ajustable for each component i

k_{ij} = constant ajustable for each binary $\{i, j\}$

The Soave equation of state, which is a third order equation for V, and which can be put in the form: $V^3 + \alpha V^2 + \beta V + \gamma = 0$, has, according to the value of its discriminant, one or three solutions.

When there are three solutions, the smallest in value corresponds to the liquid phase and the highest value corresponds to the vapor phase.

Fugacity is expressed as a function of the molar volume, the temperature, the parameters for pure substances a_i and b_i, and the binary interaction coefficients k_{ij}:

$$RT \ln\left[\frac{f_i}{P x_i}\right] = -RT \ln\left[\frac{P\left(V - b_m\right)}{RT}\right] + \frac{b_i\, RT}{\left(V - b_m\right)} - \xi_i \ln\left[\frac{V + b_m}{V}\right] - \frac{a_m\, b_i}{b_m\left(V + b_m\right)}$$

$$\xi_i = \frac{-a_m\, b_i}{b_m^2} + \frac{2}{b_m} \sum_{j=1}^{n} \left[x_j \sqrt{a_i} \sqrt{a_j}\left(1 - k_{ij}\right)\right]$$

where

f_i = fugacity of component i of the phase in question   [bar]

V = molar volume of the phase being considered $[m^3/kmol]$

Soave also proposed a formula to calculate the constant, m_i, as a function of the acentric factor:

$$m_i = 0.48508 + 1.55171 \, \omega_i - 0.15613 \, \omega_i^2 \qquad [4.100]$$

ω_i = acentric factor of the component i

In practice, however, it is recommended to adjust the coefficient m_i in order to obtain either the experimental vapor pressure curve or the normal boiling point. The function $f(T_r)$ proposed by Soave can be improved if accurate experimental values for vapor pressure are available or if it is desired that the Soave equation produce values estimated by another correlation.

The constants k_{ij} enable the improved representation of binary equilibria and should be carefully determined starting from experimental results. The API Technical Data Book has published the values of constants k_{ij} for a number of binary systems. The use of these binary interaction coefficients is necessary for obtaining accurate calculation results for mixtures containing light components such as:

- methane
- hydrogen
- H_2S
- CO_2.

The use of k_{ij} is equally necessary for binary systems where the relative volatility is needed with an error of better than 10%.

4.5.3.2 **Estimation of Hydrocarbon Vapor Pressures**

The true vapor pressure of hydrocarbons is expressed in bar and represents the vapor pressure directly above a saturated liquid in equilibrium with the vapor that rises over it.

For pure hydrocarbons and petroleum fractions the vapor pressure can be calculated by three methods which are:

- the Soave equation
- the Lee and Kesler method
- the Maxwell and Bonnel method.

For mixtures, the calculation is more complex because it is necessary to determine the bubble point pressure by calculating the partial fugacities of the components in the two phases at equilibrium.

The "Reid" vapor pressure characterizes the light petroleum products; it is measured by a standard test (refer to Chapter 7) which can be easily simulated.

a. Determination of Vapor Pressure Using the Soave Equation

The calculation of vapor pressure of a pure substance consists of finding the pressure for which the fugacities of the liquid and vapor are equal.

The objective of Soave being to represent equilibria at high pressure, the form of the function $f(T_r)$ was chosen in order to calculate accurately vapor pressures at pressures greater than 1 bar.

At low temperatures, using the original function $f(T_r)$ could lead to greater error. In Tables 4.11 and 4.12, the results obtained by the Soave method are compared with fitted curves published by the DIPPR for hexane and hexadecane. Note that the differences are less than 5% between the normal boiling point and the critical point but that they are greater at low temperature. The original form of the Soave equation should be used with caution when the vapor pressure of the components is less than 0.1 bar. In these conditions, it leads to underestimating the values for equilibrium coefficients for these components.

Temperature, °C	Vapor pressure DIPPR, bar abs	Vapor pressure Soave, bar abs	Difference SRK – DIPPR, %
5	0.079	0.077	−2.36
15	0.130	0.127	−2.28
25	0.203	0.199	−1.98
35	0.308	0.303	−1.52
45	0.452	0.447	−0.95
55	0.644	0.642	−0.32
65	0.895	0.899	0.35
68.75	1.007	1.013	0.57
75	1.217	1.230	1.01
85	1.623	1.650	1.66
95	2.125	2.173	2.26
105	2.738	2.815	2.81
115	3.480	3.593	3.27
125	4.366	4.525	3.65
135	5.415	5.628	3.92
145	6.649	6.920	4.08
155	8.089	8.423	4.12
165	9.761	10.154	4.03
175	11.691	12.135	3.80
185	13.909	14.387	3.43
195	16.450	16.930	2.92
205	19.351	19.788	2.26
215	22.655	22.982	1.45
225	26.407	26.536	0.49

Table
4.11   *Comparison of calculated vapor pressures for n-hexane.*

Temperature, °C	Vapor pressure DIPPR, bar abs	Vapor pressure Soave, bar abs	Difference SRK – DIPPR, %
100	0.0010	0.0009	−9.23
110	0.0018	0.0017	−8.96
120	0.0032	0.0029	−8.55
130	0.0054	0.0050	−8.03
140	0.0089	0.0082	−7.44
150	0.0140	0.0131	−6.78
160	0.0217	0.0203	−6.12
170	0.0326	0.0308	−5.43
180	0.0479	0.0456	−4.74
190	0.0689	0.0661	−4.07
200	0.0971	0.0938	−3.44
210	0.1344	0.1306	−2.83
220	0.1828	0.1787	−2.25
230	0.2449	0.2407	−1.72
240	0.3234	0.3194	−1.22
250	0.4212	0.4179	−0.80
260	0.5419	0.5398	−0.39
270	0.6892	0.6888	−0.05
280	0.8671	0.8692	0.24
286.85	1.0090	1.0133	0.42
290	1.0802	1.0855	0.49
300	1.3332	1.3424	0.68
310	1.6314	1.6450	0.83
320	1.9803	1.9989	0.94
330	2.3860	2.4098	0.99
340	2.8549	2.8836	1.01
350	3.3939	3.4269	0.97
360	4.0104	4.0461	0.89
370	4.7122	4.7484	0.77
380	5.5079	5.5409	0.60
390	6.4066	6.4312	0.38
400	7.4180	7.4272	0.12
410	8.5525	8.5372	−0.18
420	9.8215	9.7697	−0.53
430	11.2370	11.1337	−0.92
440	12.8121	12.6380	−1.36

Table 4.12   *Comparison of calculated vapor pressures for n-hexadecane.*

b. Determination of the Vapor Pressure by the Lee and Kesler Method

When the critical constants for a pure substance or the pseudocritical constants for a petroleum fraction are known, the vapor pressure for hydrocarbons and petroleum fractions can be calculated using the Lee and Kesler equations:

$$\pi = P_c \exp\left(f\left(T_r, \omega\right)\right) \qquad [4.33]$$

$$f\left(T_r, \omega\right) = \ln P_{r_0} + \omega \ln P_{r_1}$$

$$\ln P_{r_0} = 5.92714 - \frac{6.09648}{T_r} - 1.28862\ln T_r + 0.169347\,T_r^{\,6}$$

$$\ln P_{r_1} = 15.2518 - \frac{15.6875}{T_r} - 13.4721\ln T_r + 0.43577\,T_r^{\,6}$$

where

T_r = reduced temperature

ω = acentric factor

P_r = reduced pressure

P_c = critical pressure   [bar]

π = vapor pressure   [bar]

T_c = critical temperature   [K]

T = temperature   [K]

This method is based on the expression proposed by Lee and Kesler in 1975. It applies mainly to light hydrocarbons. The average error is around 2% when the calculated vapor pressure is greater than 0.1 bar.

For heavy hydrocarbons, it is preferable to use the Maxwell and Bonnel method described below.

c. Determination of Vapor Pressure by the Method of Maxwell and Bonnel

Maxwell and Bonnel (1955) proposed a method to calculate the vapor pressure of pure hydrocarbons or petroleum fractions whose normal boiling point and specific gravity are known. It is iterative if the boiling point is greater than 366.5 K:

- the factor f is determined starting with the boiling point temperature T_b
- T_b' is determined using the Watson characterization factor K_W and an estimation of the vapor pressure π_e
- the vapor pressure π is then determined after having calculated the parameter, X
- it is then possible to re-estimate T_b' by substituting π in π_e.

$$f = 0 \quad \text{if} \quad T_b < 366.5\,\text{K} \qquad\qquad [4.101] \text{ approaches } [4.24]$$
$$f = 1 \quad \text{if} \quad T_b > 477.6\,\text{K}$$

$$f = \frac{T_b - 366.5}{111.1} \quad \text{if} \quad 366.5\,\text{K} < T_b < 477.6\,\text{K}$$

$$T_b' = T_b - 1.39 f \left(K_W - 12\right) \log \frac{\pi_e}{1.013}$$

$$X = \frac{\dfrac{T_b'}{T} - 5.16 \cdot 10^{-4}\, T_b'}{748.1 - 0.386\, T_b'}$$

$$\log \pi = \frac{a X - b}{c X - d} - 2.8750$$

The values coefficients a, b, c and d depend on X:

$X < 0.0013$	$a = 2770.085$	$b = 6.412631$	$c = 36$	$d = 0.989679$
$0.0013 < X < 0.0022$	$a = 2663.129$	$b = 5.994296$	$c = 95.76$	$d = 0.972546$
$X > 0.0022$	$a = 3000.538$	$b = 6.761560$	$c = 43$	$d = 0.987672$

where

K_W = Watson characterization factor

T_b = normal boiling point   [K]

π = vapor pressure   [bar]

This method is applicable to heavy fractions whose boiling point is greater than 200°C. The average error is around 10% for pressures between 0.001 and 10 bar.

4.5.3.3 Estimation of the Reid Vapor Pressure

The ASTM D 323 standard describes a method for determining the vapor pressure employing two chambers, A and B; the volume of chamber A is four times that of chamber B.

Chamber B is filled with partially degassed sample material at 0°C. Chamber A is filled with air at 37.8°C and at atmospheric pressure.

The two chambers are opened to each other and the apparatus is brought to 37.8°C. The relative pressure resulting is called the "Reid vapor pressure".

This procedure can be easily simulated by a program if the composition of the petroleum material and the pure hydrocarbons or petroleum fractions composing the sample are known. Air can be considered as nitrogen without changing the results.

The "Reid" vapor pressure is generally barely different from the true vapor pressure at 37.8°C if the light gas content — methane, ethane, propane, and butane— of the sample is small, which is usually the case with petroleum products. The differences are greater for those products containing large quantities of dissolved gases such as the crude oils shown in Table 4.13.

"Reid" vapor pressure, bar	Vapor pressure crude oil, bar	Vapor pressure gasoline, bar	R crude oil	R stabilized gasoline
0.15	0.161	0.153	1.07	1.02
0.20	0.218	0.206	1.09	1.03
0.28	0.311	0.291	1.11	1.04
0.35	0.399	0.367	1.14	1.05
0.42	0.491	0.443	1.17	1.055
0.48	0.581	0.509	1.21	1.06
0.55	0.699	0.583	1.27	1.06
0.62	0.818	0.657	1.32	1.06
0.69	0.938	0.731	1.36	1.06
0.76	1.064	0.806	1.40	1.06
0.83	1.195	0.880	1.44	1.06
1.00	1.550	1.060	1.55	1.06

Table 4.13	*Relation between the true vapor pressure at 37.8°C and the "Reid" vapor pressure. R = True vapor pressure / "Reid" vapor pressure.*

4.5.3.4 **Estimation of the Flash Point**

The flash point measures the tendency of a petroleum material to form a flammable mixture with air. It is one of the properties to be considered when evaluating the flammability of a petroleum cut.

The flash point is measured in the laboratory following procedures that depend on the sample being tested.

The flammability limits of a hydrocarbon depend on its chemical nature and its molecular weight. Table 4.14 gives values for some common hydrocarbons.

Generally speaking, paraffins, naphthenes, and aromatics have a potential for exploding under the following conditions:

$$85 < P_{HC} M_v < 700$$

where

P_{HC} = partial pressure of hydrocarbons in the vapor phase   [bar]

M_v = average molecular weight of the hydrocarbons in the vapor phase   [kg/kmol]

For olefins, the limits are greater by about 30%. At ambient temperatures, heavy materials have a vapor pressure too low to cause an explosive mixture with air.

Calculating the flash point starting from the mixture's composition is not very accurate; however an estimation can be obtained if T_f is determined as the temperature for which the following relation holds true:

$$P_b M_v = 100$$

Component	Lower limit, mole %	Upper limit, mole %
Methane	5.0	15.0
Ethane	2.9	13.0
Propane	2.0	9.5
i-Butane	1.5	9.0
n-Butane	1.8	8.5
i-Pentane	1.4	8.3
n-Pentane	1.3	8.0
n-Hexane	1.1	7.7
n-Heptane	1.0	7.0
Cyclopentane	1.4	9.4
Cyclohexane	1.3	8.0
Benzene	1.4	7.1
Toluene	1.2	7.1
Ethylene	2.7	36.0
Propylene	2.0	11.0
1-Butene	1.6	9.3
Acetylene	2.5	80.0
Propadiene	2.1	13.0

Table 4.14   *Explosive limits of light hydrocarbons.*

where

P_b = bubble pressure of the mixture at T_f [bar]
M_v = molecular weight of the vapor at the bubble point   [kg/kmol]
T_f = flash point temperature   [K]

The relation developed by the API can also be used:

$$T_f = \frac{1}{-0.02421 + \dfrac{2.84947}{T_{10}} + 0.0034254 \ln T_{10}}$$ [4.102]

where

T_{10} = temperature at the 10% volume distilled point from ASTM D 86   [K]

The average error for these two methods is about 5°C.

The flash points obtained experimentally according to the different procedures differ slightly. The present estimation refers to the flash point called the "closed cup" method.

4.5.3.5 Estimation of a Petroleum Cut Flash Curve

The flash curve of a petroleum cut is defined as the curve that represents the temperature as a function of the volume fraction of vaporised liquid, the residual liquid being in equilibrium with the total vapor, at constant pressure.

The calculation of the flash curve is achieved using the models given earlier.

The flash curve at atmospheric pressure can be estimated using the results of the ASTM D 86 distillation by a correlation proposed by the API. For the same volume fraction distilled one has the following relation:

$$T' = a \, T^b \, S^c \qquad [4.103]$$

where

T' = temperature of the flash curve [K]

T = temperature of the ASTM D 86 test [K]

S = standard specific gravity of the petroleum cut

a, b, c = coefficients given in Table 4.15, which also supplies an example

% volume distilled or vaporized	Coefficient a	Coefficient b	Coefficient c	ASTM D 86, °C	Specific gravity	Flash curve, °C
0	2.9748	0.8466	0.4208	43	0.746	71
10	1.4459	0.9511	0.1287	66	0.746	82
30	0.8506	1.0315	0.0817	93	0.746	93
50	3.2680	0.8274	0.6214	113	0.746	103
70	8.2873	0.6871	0.9340	131	0.746	116
90	10.6266	0.6529	1.1025	163	0.746	134
100	7.9950	0.6949	1.0737	182	0.746	138
					example	

Table 4.15 *Coefficients for converting an ASTM D 86 curve to an atmospheric flash curve and an application for a petroleum cut whose standard specific gravity is 0.746.*

The average accuracy of this method is mediocre since the deviations can be greater than 10°C.

The curve obtained can be transformed into a curve at a different pressure by the equations of Maxwell and Bonnel (see article 4.5.3.2.c).

4.5.3.6 **Estimation of the TBP, ASTM D 86, ASTM D 1160 Distillation Curves**

The normalized distillation curves, TBP, ASTM D 86, and ASTM D 1160, provide a way to judge the quality of a fractionation performed on petroleum cuts.

Moreover, certain commercial products should meet specifications including those for fractions distilled at certain temperatures.

In the distillation units for producing petroleum cuts, the curves are determined in the laboratory from samples taken at regular intervals.

For the units for which a calculation is made to simulate operations, or for which a calculation is made for sizing purposes, the compositions are known; it is necessary then to calculate the distillation curves starting from the characteristics of the components.

a. Construction of the TBP Distillation Curve

Calculation of the atmospheric TBP is rapid if it can be assumed that this distillation is ideal (which is not always the case in reality). It is only necessary to arrange the components in order of increasing boiling points and to accumulate the volumes determined by using the standard specific gravity.

The atmospheric TBP can be transformed into a TBP at 10 mm Hg by the Maxwell and Bonnel relations (refer to article. 4.5.3.2.c).

The results of the calculation are close to those of the laboratory for medium and heavy cuts. They are somewhat different for light cuts, notably those containing dissolved gases.

Daubert has recently published a new method that correlates D 86 and TBP results. Refer to 4.1.3.1.

Table 4.16 shows the results of this new method for an example. They differ significantly from those obtained with the method proposed by Riazi for the initial and 10% volume distilled points. The accuracy is improved and the deviations observed at small distilled fractions are corrected.

b. Construction of an ASTM D 86 Distillation Curve

This calculation starts from the TBP at atmospheric pressure. The API recommends a relation established by Riazi (1982):

$$T' = a\,T^b \qquad \text{[4.104] similar to [4.19]}$$

where

T' = temperature of the ASTM D 86 test [K]

T = temperature of the TBP test [K]

a, b = coefficients that depend on the fraction distilled. See Table 4.16.

The accuracy depends on the fraction distilled; it deviates particularly when determining the initial and final boiling points; the average error can exceed 10°C. When calculating the ASTM D 86 curve for gasoline, it is better to use the Edmister (1948) relations. The Riazi and Edmister methods lead to very close results when they are applied to ASTM D 86 calculations for products such as gas oils and kerosene.

Generally speaking, these correlations for refined products lead to too-low results for the points at small distilled fractions and to too-high results for those of large distilled fractions.

% volume distilled	Coefficient *a*	Coefficient *b*	TBP, °C	ASTM D 86, °C
0	1.08947	0.99810	35	59
10	1.71243	0.91743	55	75
30	1.29838	0.95923	95	103
50	1.10755	0.98270	135	134
70	1.13047	0.97790	175	170
90	1.04643	0.98912	215	204
95	1.21455	0.96572	225	216
				example

Table 4.16a *Coefficients for converting a TBP curve into an ASTM D 86 curve and an example applied to a petroleum cut (Riazi's method).*

% volume distilled	Coefficient *A*	Coefficient *B*	Temperature TBP, °C	Temperature D 86, °C	Index *i*
0	7.40120	0.60244	35	74	1
10	4.90040	0.71644	55	82	2
30	3.03050	0.80076	95	105	3
50	0.87180	1.02580	135	134	4
70	2.52820	0.82002	175	167	5
90	3.04190	0.75497	215	204	6
95	0.11798	1.66060	225	216	7

Table 4.16b *Coefficients for converting a TBP curve into an ASTM D 86 curve and an example applied to a petroleum cut (Daubert's method).*

c. Construction of the ASTM D 1160 Distillation Curve

The TBP at atmospheric pressure is transformed to a TBP at 0.0133 bar (10 mmHg) by the Maxwell and Bonnel relations and the latter is converted to a D 1160 at 0.0133 bar.

The API recommends using the Edmister and Okamoto method (1959).

For distilled fractions greater than 50% in volume, the two curves are considered identical. For distilled fractions at 30, 10 and 0 volume %, the temperatures are obtained using the following relations:

$$T_{50} = T'_{50} \qquad \text{[4.105] similar to [4.22]}$$
$$T_{30} = T_{50} - g_1\left(T'_{50} - T'_{30}\right)$$
$$T_{10} = T_{30} - g_1\left(T'_{30} - T'_{10}\right)$$
$$T_i = T_{10} - g_2\left(T'_{10} - T'_i\right)$$

where

T_{50} = temperature at the ASTM D 1160 50% distilled point   [°C]

T_{30} = temperature at the ASTM D 1160 30% distilled point   [°C]

T_{10} = temperature at the ASTM D 1160 10% distilled point   [°C]

T_i = temperature at the ASTM D 1160 initial point, 0% distilled   [°C]

T'_{50} = temperature at the TBP 50% distilled point   [°C]

T'_{30} = temperature at the TBP 30% distilled point   [°C]

T'_{10} = temperature at the TBP 10% distilled point   [°C]

T'_i = temperature at the TBP initial point, 0% distilled   [°C]

The functions $g_1(dT)$ and $g_2(dT)$ are calculated by interpolating between the values shown in Table 4.17.

The average error of this method is about 5°C. It is about 7°C for the 10% distilled points and around 10°C for the initial point.

dT, °C	$g_1(dT)$, °C	$g_2(dT)$, °C
0	0.0	0.0
10	7.5	5.0
20	16.5	10.0
30	25.5	16.0
40	35.5	23.5
50	46.5	32.5
60	57.0	44.0
70	68.0	60.0
80	78.5	76.0
90	89.0	88.0
100	100.0	100.0

Table 4.17

Values of the functions g(dT) for converting a TBP into an ASTM D 1160 at 10 mmHg

dT   = *TBP temperature interval*
$g_1(dT)$ = *ASTM D 1160 temperature interval*
$g_2(dT)$ = *ASTM D 1160 temperature interval*

4.5.3.7 **Estimation of the Interfacial Tension**

The interfacial tension is usually expressed in mN/m or dynes/cm. It measures the tendency for a liquid to form an interface having the least area. The interfacial tension decreases as the temperature increases.

a. *Estimation of the Interfacial Tension of a Pure Hydrocarbon*

The surface tension is calculated starting from the parachor and the densities of the phases in equilibrium by the Sugden method (1924):

$$\sigma = \left[\frac{Pa\left(\rho_l - \rho_g\right)}{1000M}\right]^4 \tag{4.106}$$

where

Pa = parachor $\qquad\qquad$ $[10^3 \, (\text{mN/m})^{1/4} \, (\text{m}^3/\text{kmol})]$

M = molecular weight $\qquad$ [kg/kmol]

ρ_l = density of the liquid $\qquad$ $[\text{kg/m}^3]$

ρ_g = density of the gas $\qquad$ $[\text{kg/m}^3]$

σ = interfacial tension $\qquad$ [mN/m]

This method has an average accuracy of about 5% and applies only to reduced temperatures less than 0.9. The parachor can be estimated by the method described in 4.1.1.2.c.

b. Estimation of the Interfacial Tension of Petroleum Fractions

The interfacial tension of a petroleum fraction at 20°C can be estimated by the following formula, cited by the API:

$$\sigma_f = \frac{673.7\left(1 - \dfrac{293.15}{T_{c_f}}\right)^{1.232}}{K_W} \qquad\qquad [4.107]$$

where

σ_f = the interfacial tension of a petroleum fraction f at 20°C [mN/m]

K_W = Watson characterization factor

T_{c_f} = pseudocritical temperature of the fraction f $\qquad$ [K]

The error of this method is about 10%. The interfacial tension obtained is used in calculating the parachor by equation [4.106].

c. Estimation of the Interfacial Tension of Mixtures

For mixtures, the concept of weighting is employed:

$$\sigma_m = \sum_{i=1}^{n}\left[\frac{Pa_i\left(\dfrac{\rho_l x_i}{M_l} - \dfrac{\rho_g y_i}{M_g}\right)}{1000}\right]^4 \qquad\qquad [4.108]$$

where

σ_m = interfacial tension of the mixture $\qquad$ [mN/m]

Pa_i = parachor of component i

x_i = mole fraction of component i in the liquid phase

y_i = mole fraction of component i in the gas phase

ρ_l = density of the liquid $\qquad\qquad$ $[\text{kg/m}^3]$

ρ_g = density of the gas $\qquad\qquad$ $[\text{kg/m}^3]$

M_l = molecular weight of the liquid $\qquad$ [kg/kmol]

M_g = molecular weight of the gas $\qquad$ [kg/kmol]

4.5.4 **Estimation of the Properties Used in Determining the Liquid-Liquid Equilibria**

4.5.4.1 **Solubility of Water in Hydrocarbons**

The solubility of water in hydrocarbons increases strongly with temperature.

For temperatures less than 200°C, it can be estimated by the Hibbard and Schalla (1952) formula:

$$\log x = \left(0.0016 - \frac{1}{T}\right)(4200y + 1050) \qquad [4.109]$$

where

   y = weight ratio of hydrogen to carbon

   T = temperature   [K]

   x = solubility of water as mole fraction

   $\log$ = common logarithm (base 10)

The average error is around 30%. This formula applies to pure substances and mixtures. For pure hydrocarbons, it is preferable to refer to solubility charts published by the API if good accuracy is required.

4.5.4.2 **Solubility of Hydrocarbons in Water**

The hydrocarbons are in general only slightly soluble in water. The solubility depends especially on the chemical nature of the hydrocarbons.

For similar molecular weights, the following are ranked by order of decreasing solubility:

- polynuclear aromatics
- aromatics
- acetylenics
- diolefins
- olefins
- naphthenes
- paraffins.

The solubility of hydrocarbon liquids from the same chemical family diminishes as the molecular weight increases. This effect is particularly sensitive; thus in the paraffin series, the solubility expressed in mole fraction is divided by a factor of about five when the number of carbon atoms is increased by one. The result is that heavy paraffin solubilities are extremely small. The polynuclear aromatics have high solubilities in water which makes it difficult to eliminate them by steam stripping.

Solubility increases with temperature. This increase is, however, smaller than the increase in hydrocarbon vapor pressure.

Table 4.18 shows the solubility trend for some common hydrocarbons.

Component	Solubility at 25°C ppm mol	Solubility at 75°C ppm mol	Solubility at 125°C ppm mol
Propane	200		
n-Butane	60	95	180
i-Butane	50	70	140
n-Pentane	10	13	30
n-Hexane	2	35	95
n-Heptane	0.4	0.7	3
Cyclopentane	40	55	120
Cyclohexane	6	8	25
Methylcyclohexane	3	5	15
Ethylcyclohexane	1	2	7
Propene	1000	1000	
1-Butene	200	300	500
1,3-Butadiene	600	800	
Propyne	7000		
Benzene	400	700	1400
Toluene	130	200	450
Ethylbenzene	55	110	350
Xylene	38	60	150
n-Propylbenzene	18	20	33
Naphthalene	50	400	

Table 4.18	*Solubilities of liquid hydrocarbons in water.*

If it is necessary to evaluate liquid-liquid equilibria where the hydrocarbon phase is a mixture, the following relation can be used:

$$x_i = x_i^0 \frac{f_{l_i}}{f_{ls_i}} \qquad [4.110]$$

where

x_i = molar solubility of component i in the aqueous phase   [ppm]

x_i^0 = molar solubility of pure component i in the
aqueous phase   [ppm]

f_{l_i} = partial fugacity of component i in the mixture of
hydrocarbons in the liquid phase at T and P [bar]

f_{ls_i} = fugacity of pure component i, at T and at saturation
pressure P_s [bar]

The x_i^0 can be interpolated using Table 4.18 and the fugacities calculated by the Soave model.

4.5.4.3   Three-Phase Equilibria for Liquid Water, Hydrocarbon Liquid, and Hydrocarbon Gas

In the case of three-phase equilibria, it is also necessary to account for the solubility of hydrocarbon gases in water. This solubility is proportional to the partial pressure of the hydrocarbon or, more precisely, to its partial fugacity in the vapor phase. The relation which ties the solubility expressed in mole fraction to the fugacity is the following:

$$f_{g_i} = H_i \, x_i \qquad\qquad [4.111]$$

where

f_{g_i} = partial fugacity of component i in the vapor phase   [bar]

H_i = Henry's constant for component i [bar]

x_i = mole fraction of component i in the liquid phase

Table 4.19 gives the Henry constants for a few common gaseous components. The chemical nature is also a dominant factor. The effect of temperature is moderate; note that the solubility passes through a minimum that depends on the hydrocarbon in question and that it is around 100°C.

A general model has been developed and published by Kabadi and Danner (1985) to calculate the equilibria in the presence of water. It is related to an adaptation of the Soave model where water is treated in a particular fashion.

The model is predictive and uses a method of contributing groups to determine the parameters of interaction with water. It is generally used by simulation programs such as HYSIM or PRO2. Nevertheless the accuracy of the model is limited and the average error is about 40%. Use the results with caution.

Component	*H* at 25°C, bar	*H* at 50°C, bar	*H* at 75°C, bar	*H* at 100°C, bar	*H* at 125°C, bar	*H* at 150°C, bar
Methane	58,000	77,000	91,000	91,000	77,000	67,000
Ethane	67,000	100,000	110,000	110,000	95,000	77,000
Propane	38,000	77,000	105,000	133,000	125,000	105,000
n-Butane	46,000	69,000	100,000	125,000	133,000	133,000
Ethylene	15,000	20,000	28,000	24,000	23,000	
Propylene	11,000	22,000	36,000	36,000		
1-Butene	18,000	24,000	30,000	45,000	56,000	
Acetylene	1400					
Propyne	770	1400	2100	3500		
Nitrogen	62,000	71,000	74,000	74,000	69,000	61,000
CO_2	1800	3400	4600	6600	9,500	9,500
H_2S	590	1020	1380	1720	1970	2300
Hydrogen	71,000	77,000	77,000	71,000		

Table **4.19**   *Solubility of hydrocarbon gases in water expressed as Henry's constants.*

4.5.4.4 **Equilibrium between a Hydrocarbon Liquid and a Partially Miscible Liquid**

There are of course liquid-liquid equilibria between hydrocarbons and substances other than water. In practice these equilibria are used in solvent extraction processes. The solvents most commonly used are listed as follows:

- furfuraldehyde (furfural)
- N-methyl pyrrolidone (NMP)
- N-dimethyl formamide (DMF)
- sulfolane
- dimethyl sulfoxyde (DMSO)
- methanol.

The extraction processes are based on the property of these solvents to dissolve preferentially hydrocarbons of a certain chemical nature.

Propane deasphalting uses propane as an anti-solvent for asphaltenes.

Other kinds of liquid-liquid equilibria are encountered in processes such as alkylation, where anhydrous hydrofluoric acid (HF) is partially soluble in hydrocarbons.

Sufficiently accurate thermodynamic models used for calculating these equilibria are not available in simulation programs. It is generally not recommended to use the models proposed. Only a specific study based on accurate experimental results and using a model adapted to the case will succeed.

4.5.5 **Estimation of Properties needed to Determine Liquid-Solid Equilibria**

4.5.5.1 **Estimation of the Freezing Point**

The solubility of a solid in the liquid phase of a mixture depends on the properties of the two phases; for the components that crystallize, the equilibrium is governed by the following equation:

$$f_{l_i} = f_{s_i} \quad \text{for others} \quad f_{l_j} < f_{s_j}$$

where

f_{l_i} = partial fugacity of component i in the liquid phase   [bar]

f_{s_i} = partial fugacity of component i in the solid phase   [bar]

i = component participating in the crystallization equilibrium

j = component not participating in the crystallization equilibrium

The fugacity in the liquid phase is determined by methods we have seen previously.

The solid phase can be considered as a pure substance or a solid solution.

When pure component i constitutes the solid phase, the liquid-solid equilibrium obeys the following equation:

$$\frac{f_{l_i}}{f^0_{l_i}} = -dH_{f_i}\left(\frac{1-\dfrac{T}{T_T}}{RT}\right) + dC_{p_i}\left(\frac{T_T-T}{RT}\right) - \frac{dC_{p_i}}{R}\frac{T_T}{T}\ln\frac{T_T}{T}$$ [4.112]

$$dH_{f_i} = H^0_{l_i} - H^0_{s_i}$$
$$dC_{p_i} = C_{pl_i} - C_{ps_i}$$

where

R	= ideal gas constant	[8,314 kJ /(kmol·K)]
T	= temperature	[K]
T_T	= temperature of the triple point	[K]
f_{l_i}	= partial fugacity of component i in the liquid phase at T et P	[bar]
$f^0_{l_i}$	= total fugacity of pure component i liquid at T and P	[bar]
dH_{f_i}	= enthalpy of fusion at the triple point of component i	[kJ/kmol]
dC_{p_i}	= the difference of specific heats for component i between the liquid and the solid	[kJ/(kmol·K)]
$H^0_{l_i}$	= enthalpy of pure liquid component i at the triple point	[kJ/kmol]
$H^0_{s_i}$	= enthalpy of pure solid component i at the triple point	[kJ/kmol]
C_{pl_i}	= average specific heat of the pure liquid between the triple point and temperature in question, T	[kJ/(kmol·K)]
C_{ps_i}	= average specific heat of the pure solid component i between the triple point and temperature in question, T	[kJ/(kmol·K)]

The properties of the solids most commonly encountered are tabulated. An important problem arises for petroleum fractions because data for the freezing point and enthalpy of fusion are very scarce. The MEK (methyl ethyl ketone) process utilizes the solvent's property that increases the partial fugacity of the paraffins in the liquid phase and thus favors their crystallization. The calculations for crystallization are sensitive and it is usually necessary to revert to experimental measurement.

4.5.5.2 **Estimation of the Pour Point**

The pour point of a cut or petroleum fraction can be derived from the characteristic properties by an expression published by the API:

$$T_{EC} = 130.47\, S^{2.971}\, M^{(0.612-0.474\,S)}\, \nu_{100}^{(0.31-0.333\,S)}$$ [4.113]

where

S = standard specific gravity

M = molecular weight   [kg/kmol]

ν_{100} = kinematic viscosity at 100°F $[\text{mm}^2/\text{s}]$

T_{EC} = pour point temperature   [K]

Average error is about 5°C. The method should not be used for pour points less than 60°C.

4.5.5.3 **Estimation of the Hydrate Formation Temperature**

Calculating the hydrate formation temperature is essential when one needs to guard against equipment and line plugging that can result when wet gas is cooled, intentionally or not, below 30°C.

Hydrates are solid structures composed of water molecules joined as crystals that have a system of cavities. The structure is stable only if at least one part of the cavities contains molecules of small molecular size. These molecules interact weakly with water molecules. Hydrates are not chemical compounds; rather, they are "clathrates".

There are two hydrate structures, types I et II, each composed of two cavity sizes. Only the lightest of the hydrocarbons can form hydrates. Table 4.20 gives the hydrates formed by the most common compounds.

Hydrate formation is possible only at temperatures less than 35°C when the pressure is less than 100 bar. Hydrates are a nuisance; they are capable of plugging (partially or totally) equipment in transport systems such as pipelines, filters, and valves; they can accumulate in heat exchangers and reduce heat transfer as well as increase pressure drop. Finally, if deposited in rotating machinery, they can lead to rotor imbalance generating vibration and causing failure of the machine.

Type	I		II	
Cavity size	**7.95 Å**	**8.60 Å**	**7.82 Å**	**9.46 Å**
Nitrogen	**	*	*	*
Methane	**	*	*	*
Ethane		**		*
Propane				**
i-Butane				**
Ethylene	**	*	*	*
Propylene				**
CO_2	**	*	*	*
H_2S	**	*	*	*

| Table 4.20 | *Common compounds that form type I and II hydrates.*
** Cavities occupied in first priority   * Cavities occupied in second priority |

To avoid hydrate formation, it is necessary either to dry the stream, or to inject a substance that, dissolving the water, lowers its partial fugacity and, consequently, the temperature of hydrate formation.

Calculation of the conditions of hydrate formation is generally accomplished by software employing the Parrish and Prausnitz (1972) model. It is difficult to predict the conditions by a simple method.

Nevertheless, for pressures less than 100 bar, temperatures less than 30°C, and if the gas is saturated with water, an estimation can be made using the following formula:

$$\sum_{i=1}^{n}\left(\frac{y_i}{K_i(T_h)}\right) = 1 \qquad [4.114]$$

$$K_i(T) = \frac{K_i^0}{P}\exp\left[\alpha_i(T-T^0)\right]\exp\left[\sum_{j=1}^{n}(\beta_{ij}\,y_j)\right]$$

where

y_i = mole fraction of component i in the gas phase

K_i^0, α_i = parameters of the simplified model (see Table 4.21)

P = pressure   [bar]

T_h = temperature of hydrate formation   [K]

β_{ij} = binary interaction parameter of the simplified model

T^0 = reference temperature = 273.15   [K]

n = number of components capable of forming hydrates present in the mixture

i, j = components in the mixture capable of forming hydrates

Component	K_i^0	α_i	β propane	β i-butane	β propylene
Methane	24.55	0.110	−15.5	−18.7	−10.2
Ethane	4.71	0.127	−4.0	−1	0
Propane	1000.00	0.000	0.0	0	0
i-Butane	1000.00	0.000	0.0	0	0
Ethylene	5.37	0.120	0.0	0	0
Propylene	1000.00	0.000	0.0	0	0
Nitrogen	157.00	0.106	−24.5	−26.8	−17.2
CO$_2$	11.17	0.156	−7.3	−9.4	0
H$_2$S	1.05	0.096	−5.0	−6.2	−2.4

Table 4.21   *Parameters of the simplified model for calculating hydrate formation temperatures.*

The average error of this simplified method is about 3°C and can reach 5°C. Table 4.22 shows an application of this method calculating the temperature of hydrate formation of a refinery gas at 14 bar. Table 4.23 gives an example applied to natural gas at 80 bar. Note that the presence of H_2S increases the hydrate formation temperature.

Component	Y_i	K_i	Y_i / K_i
Methane	0.175	0.395	0.4428
Ethane	0.071	2.131	0.0333
Propane	0.044	71.429	0.0006
i-Butane	0.060	71.429	0.0008
Ethylene	0.078	2.743	0.0284
Propylene	0.146	71.429	0.0020
Nitrogen	0.023	0.353	0.0652
CO_2	0.005	4.247	0.0012
H_2S	0.060	0.141	0.4255
Other	0.338		0.0000
Total	1.000		1.0000

Table 4.22 | *Example calculation of the hydrate formation temperature for an FCC gas at 14 bar abs. Result:* $T_h = 16.4°C$.

Component	Y_i	K_i	Y_i / K_i
Methane	0.680	4.570	0.1488
Ethane	0.030	2.159	0.0139
Propane	0.020	12.500	0.0016
i-Butane	0.010	12.500	0.0008
Ethylene	0.000	2.198	0.0000
Propylene	0.000	12.500	0.0000
Nitrogen	0.010	20.042	0.0005
CO_2	0.100	10.241	0.0098
H_2S	0.150	0.182	0.8247
Other	0.000		0.0000
Total	1.000		1.0000

Table 4.23 | *Example calculation of the hydrate formation temperature for a natural gas at 80 bar abs. Result:* $T_h = 29.1°C$.

5

Characteristics of Petroleum Products for Energy Use (Motor fuels – Heating fuels)

Jean-Claude Guibet

In this chapter, we will discuss petroleum products used for energy purposes, that is, motor fuels and heating fuels. Chapter 6 will be devoted to other products such as special gasolines, lubricants, petrochemical bases, and asphalts.

First of all, a technical clarification is necessary; in the wider sense, motor fuels are chemical compounds, liquid or gas, which are burned in the presence of air to enable thermal engines to run: gasoline, diesel fuel, jet fuels. The term heating fuel is reserved for the production of heat energy in boilers, furnaces, power plants, etc.

Thus, according to the definitions, diesel fuel (or gas oil) is not a heating fuel but a motor fuel. Incidentally, heavy fuel can be considered a heating fuel or a motor fuel depending on its application: in a burner or in a marine diesel engine.

It is useful to specify at the start the principal quality criteria for each type of product (motor fuels, heating fuels), imposed by the requirements for the different kinds of energy converters: motors, turbines, burners.

The regular and premium motor fuels used in spark ignition motors must have physical properties such as density, distillation curve, and vapor pressure that allow their atomization in air by carburetion or injection before entering the cylinder. Furthermore, the need to assure satisfactory operation of vehicles in extremely variable atmospheric conditions (in France, for example, from $-20°C$ to $+40°C$) implies distinct volatility specifications for gasoline according to the season. Finally, the gasoline engine requires that its fuel offer strong resistance to auto-ignition which is expressed as the octane number. This characteristic determines the compression ratio of the motor which itself affects the performance of the automobile (consumption of fuel, power).

However, in practice the octane number has a ceiling imposed by refining industry constraints such as composition, lead reduction or elimination, cost, and demand volume and distribution.

Diesel fuel should have, in contrast to gasoline, a strong tendency to auto-ignite since the principle of operation of diesel engines is based on the ignition of the fuel injected at high pressure in previously compressed air. This quality of diesel is expressed by the cetane number which, as opposed to the octane number, does not directly affect the motor performance but rather acts on driving comfort factors such as cold starting, noise, and exhaust emissions.

Another essential diesel characteristic is its behavior in cold climates; paraffin crystal formation in the fuel at winter temperatures (0 to $-20°C$ in France, for example) can plug the fuel filter. The refiner adjusts the cetane number and the cold characteristics of the diesel fuel by modifying the nature and composition of the basic fuel and adding certain additives. In this area too, optimization is difficult for many reasons: the constantly increasing demand for diesel motor fuel, the increasing share of the diesel engine in the passenger car fleet and the development of heavy duty vehicles, and the new refining structure that is supplying middle distillates of low quality from catalytic cracking and other conversion processes.

The properties required by jet engines are linked to the combustion process particular to aviation engines. They must have an excellent cold behavior down to $-50°C$, a chemical composition which results in a low radiation flame that avoids carbon deposition on the walls, a low level of contaminants such as sediment, water and gums, in order to avoid problems during the airport storage and handling phase.

With respect to fuels utilized as heating fuels for industrial furnaces, or as motor fuels for large diesel engines such as those in ships or power generation sets, the characteristics of primary importance are viscosity, sulfur content and the content of extremely heavy materials (asphaltenes) whose combustion can cause high emissions of particulates which are incompatible with antipollution legislation.

This chapter comprises three parts wherein we will examine the specifications of motor and heating fuels imposed by combustion, storage and distribution, and protection of the environment.

5.1 **Properties Related to the Optimization of Combustion**

All petroleum energy products, as distinct and dissimilar as they can be, are subjected to the process of flame combustion. It is helpful at this point to bring to mind some definitions and general laws of thermochemistry.

5.1.1 **Fundamentals of Thermochemistry**

5.1.1.1 **Stoichiometric Equation for Combustion Stoichiometric Ratio**

With regards to the overall balance of combustion, the chemical structure of the motor or heating fuel, e.g., the number of carbon atoms in the chain and the nature of the bonding, does not play a direct role; the only important item is the overall composition, that is, the contents of carbon, hydrogen, and — eventually — oxygen in the case of alcohols or ethers added to the fuel.

Thus the quantitative elemental analysis of the fuel establishes an overall formula, $(CH_yO_z)_x$, where the coefficient x, related to the average molecular weight, has no effect on the fuel-air ratio.

The chemical equation for combustion is then written as:

$$CH_yO_z + \left(1 + \frac{y}{4} - \frac{z}{2}\right)(O_2 + 3.78\,N_2) \longrightarrow CO_2 + \frac{y}{2}H_2O + 3.78\left(1 + \frac{y}{4} - \frac{z}{2}\right)N_2$$

It is known that air contains 20.9% O_2 and 79.1% N_2 by volume; argon whose content is 0.93% and the other rare gases present in trace quantities are included with nitrogen.

Stoichiometry is the composition of the air-fuel mixture required to obtain complete combustion. The stoichiometric ratio, r, is the quotient of the respective masses, m_a and m_c, of air and fuel arranged in the stoichiometric conditions:

$$\left(r = m_a / m_c\right)$$

r is generally between 13 and 15 for hydrocarbons, more exactly 14 and 14.5 for conventional liquid fuels such as gasolines, diesel, and jet fuel; r increases with the H/C ratio, moving from 11.49 to 34.46 from carbon to pure hydrogen.

5.1.1.2 **The Expression, Calculation and Importance of the Equivalence Ratio in Different Combustion Systems**

In practice, for motors, turbines or furnaces, the conditions of combustion are frequently far from those corresponding to stoichiometry and are characterized either by an excess or by an insufficiency of fuel with respect to oxygen. The composition of the fuel-air mixture is expressed by the equivalence ratio, φ, defined by the relation:

$$\varphi = \frac{m_{ce}/m_{ae}}{m_c/m_a}$$

where m_{ce} and m_{ae} are, respectively, the amounts of fuel and air actually utilized. The equivalence ratio is then written as a function of the stoichiometric ratio:

$$\varphi = \frac{m_{ce}}{m_{ae}}\,r$$

The equivalence ratio refers to the more "noble" reactant, that is, the fuel, and the mixture is rich or lean according to whether the fuel is in excess or deficient with respect to the stoichiometry.

In a general manner, diesel engines, jet engines, and domestic or industrial burners operate with lean mixtures and their performance is relatively insensitive to the equivalence ratio. On the other hand, gasoline engines require a fuel-air ratio close to the stoichiometric. Indeed, a too-rich mixture leads to an excessive exhaust pollution from CO emissions and unburned hydrocarbons whereas a too-lean mixture produces unstable combustion (reduced driveability and misfiring).

The relation defining the equivalence ratio shows that for a given mechanical setting of the motor (m_{ce} and m_{ae} fixed), the equivalence ratio varies with the stoichiometric ratio. It is recommended to have a relatively narrow margin of variation for the composition of gasolines in order not to miss the optimum adjustment zone. The forgoing considerations also explain that fuels too rich in oxygen are not interchangeable with conventional fuels for a given automotive fleet. This is the reason for the European Directive of 5 December 1985 establishing the maximum oxygen content of 2.5% for gasoline.

5.1.1.3 **Heating Value or Calorific Value**

a. *Definition of Heating Value*

The mass or volume heating value represents the quantity of energy released by a unit mass or volume of fuel during the chemical reaction for complete combustion producing CO_2 and H_2O. The fuel is taken to be, unless mentioned otherwise, at the liquid state and at a reference temperature, generally 25°C. The air and the combustion products are considered to be at this same temperature.

A distinction is made between the greater (or gross) heating value (GHV) and the net heating value (NHV), according to whether water obtained from combustion is assumed to be in the liquid or vapor state. The only really useful value in practice is the net heating value because the combustion products rejected by engines or burners contain water in the vapor state.

b. *Methods for Determining the Heating Value*

Measuring the gross heating value (mass) is done in the laboratory using the ASTM D 240 procedure by combustion of the fuel sample under an oxygen atmosphere, in a bomb calorimeter surrounded by water. The thermal effects are calculated from the rise in temperature of the surrounding medium and the thermal characteristics of the apparatus.

The calorimetric techniques give the weight gross heating value (GHV_w); to determine the net heating value (NHV_w), the weight hydrogen content W_H of the fuel must be known. If the latter is expressed as a percentage, the mass of water M produced during combustion of one kilogram of fuel is written as:

$$M = \frac{W_H}{100} \cdot \frac{18}{2}$$

Taking into account the mass enthalpy of vaporization at 25°C, ($\Delta H = 2443$ kJ/kg),

$$NHV_w = GHV_w - 220 \, W_H$$

where NHV_w and GHV_w are expressed in kJ/kg.

The error attributed to the determination of the NHV_w (calorimetric measurement, determination of hydrogen content, and final calculation) is satisfactory with a repeatability of 0.3% and a reproducibility of 0.4%.

For pure organic materials, it is also possible to calculate the heating value starting from the heats of formation found in tables of thermodynamic data. The NHV is obtained using the general relation of thermochemistry applicable to "standard" conditions of pressure and temperature (1 bar and 25°C).

$$\Delta H° = \underset{\text{combustion}}{\sum} \left(\Delta H°_{f, T} \right) - \underset{\text{reactants}}{\sum} \left(\Delta H°_{f, T} \right)$$
$$\text{products}$$

where

$\Delta H°$ combustion = change in mass enthalpy or thermal effect of the combustion reaction, equal in absolute value but of opposite sign as the NHV_w

$\sum \left(\Delta H°_{f, T} \right)$ products = sum of the mass enthalpy changes of formation for the products

$\sum \left(\Delta H°_{f, T} \right)$ reactants = sum of the mass enthalpy changes of formation for the reactants

Table 5.1 gives a sample calculation of the NHV_w for toluene, starting from the molar enthalpies of formation of the reactants and products and the enthalpies of changes in state as the case requires.

This calculational method is nevertheless only of limited interest and concerns only certain pure organic products for which direct measurement of the NHV_w is difficult owing to a variety of reasons such as cost and availability.

There also exist relatively simple correlations between the heating value of motor fuels and certain characteristics such as density and composition by chemical family supplied by FIA analysis. Refer to Chapter 3.

As an example, we cite two formulas proposed by Sirtori et al. (1974):

$$NHV_w = 4.18 \, (106.38 \, PAR + 105.76 \, OL + 95.55 \, ARO)$$

and

$$NHV_v = 4.18 \, (10,672.05 \, \rho - 8.003 \, ARO)$$

where NHV_w and NHV_v are the mass and volume net heating values in kJ/kg and kJ/l, PAR, OL and ARO are the paraffin, olefin and aromatic contents resulting from the FIA analysis and ρ is the density (kg/l) at 15°C.

$$\Delta H^\circ_{v\ 298}\left(C_6H_5\ CH_3\right) = \quad 38,003 \text{ J/mol}$$

$$\Delta H^\circ_{f\ 298}\left(C_6H_5\ CH_3\right) = \quad 49,999 \text{ J/mol}$$

$$\Delta H^\circ_{f\ 298}\left(H_2O\right) \quad = -241,818 \text{ J/mol}$$

$$\Delta H^\circ_{f\ 298}\left(CO_2\right) \quad = -393,510 \text{ J/mol}$$

$$C_6H_5\ CH_{3l} + 9\ O_{2g} + n\ N_{2g} \longrightarrow 7\ CO_{2g} + 4\ H_2O_g$$

$$\Delta H = 7\left(-393,510\right) + 4\left(-241,818\right) - 49,999 + 38,003 = -3,733,838 \text{ J}$$

$$M\left(C_6H_5\ CH_3\right) = 0{,}09213 \text{ kg/mol}$$

$$NHV = \frac{3,733,838}{0.09213} = 40,528 \text{ kJ/kg}$$

Table
5.1
Example of NHV calculation for toluene based on thermodynamic data from "Thermodynamic Tables – Hydrocarbons" edited by TRC (Thermodynamic Research Center, The Texas A&M University System College Station, Texas, USA).
Nota: in thermochemistry 1 cal = 4.184 J.

c. Numerical Values for Heating Values

Table 5.2 lists NHV_w values for a certain number of pure organic compounds while Table 5.3 gives values of NHV_w and NHV_v for conventional motor and heating fuels.

For a large number of petroleum fractions, the NHV_w is between 40,000 and 45,000 kJ/kg: generally, it increases with the H/C ratio, in conjunction with the fact that hydrogen and carbon heating values are, respectively, 120,000 and 32,760 kJ/kg. Moreover, the presence of oxygen in the molecule tends to reduce the NHV. Thus, the NHV_w of methane is 50,030 kJ/kg, that of the methanol 19,910 kJ/kg. Finally, the NHV_w does not only depend on the weighted composition; it varies according to the energy of formation of the motor fuel in question. Note, for example, that cycloparaffins have lower NHV_w values than olefins having the same number of carbon atoms and overall composition (H/C = 2).

The volume heating value (NHV_v) is obtained from the density ρ at 25°C and the NHV_w by the relationship:

$$NHV_v = \rho\, NHV_w$$

For hydrocarbons, the NHV_v is generally between 27,000 and 36,000 kJ/l: It is much lower for alcohols; methanol, for example, is 15,870 kJ/l.

The magnitude of the NHV_v has economic importance because the consumption and cost of motor fuels are frequently expressed in liters/100 km and in Francs/liter in France. From the technical viewpoint, the NHV_v establishes the maximum range for a transport system with a given load. This is a decisive criterion for applications like aviation.

Compound		NHV$_w$ kJ/kg	Compound		NHV$_w$ kJ/kg
n-Paraffins			**Acetylenics**		
Methane	(g)	50,030	Acetylene	(g)	48,277
Ethane	(g)	47,511	Methylacetylene	(g)	46,152
Propane	(g)	46,333	1-Butyne	(g)	45,565
Butane	(g)	45,719	1-Pentyne	(l)	44,752
Pentane	(l)	44,975	**Aromatics**		
Hexane	(l)	44,735	Benzene	(l)	40,142
Heptane	(l)	44,556	Toluene	(l)	40,525
Octane	(l)	44,420	o-Xylene	(l)	40,811
Nonane	(l)	44,321	m-Xylene	(l)	40,802
Decane	(l)	44,237	p-Xylene	(l)	40,811
Undecane	(l)	44,166	Ethylbenzene	(l)	40,924
Dodecane	(l)	44,110	1,2,4-Trimethylbenzene	(l)	41,023
Isoparaffins			n-Propylbenzene	(l)	41,218
Isobutane	(g)	45,560	Cumene	(l)	41,195
Isopentane	(l)	44,900	**Alcohols**		
2-Methylpentane	(l)	44,665	Methanol	(l)	19,910
2,3-Dimethylbutane	(l)	44,648	Ethanol	(l)	26,820
2,3-Dimethylpentane	(l)	44,517	n-Propanol	(l)	30,680
2,2,4-Trimethylpentane	(l)	44,343	Isopropanol	(l)	30,450
Naphthenes			n-Butanol	(l)	33,130
Cyclopentane	(l)	43,785	Isobutanol	(l)	33,040
Methylcyclopentane	(l)	43,656	Tertiarybutanol	(l)	32,700
Cyclohexane	(l)	43,438	n-Pentanol	(l)	34,720
Methylcyclohexane	(l)	43,358	**Ethers**		
Olefins			Dimethyloxide	(l)	28,840
Ethylene	(g)	47,165	Diethyloxide	(l)	33,780
Propylene	(g)	45,769	Di n-propyloxide	(l)	36,460
1-Butene	(g)	45,284	Di n-butyloxide	(l)	37,990
cis 2-Butene	(g)	45,161	**Aldehydes and ketones**		
trans 2-Butene	(g)	45,097	Formaldehyde	(g)	17,300
Isobutene	(g)	44,989	Acetaldehyde	(g)	25,070
1-Pentene	(l)	44,624	Propionaldehyde	(l)	29,020
2-Methyl 1-pentene	(l)	44,224	n-Butyraldehyde	(l)	31,950
1-Hexene	(l)	44,431	Acetone	(l)	28,570
Diolefins			**Other compounds**		
1-3 Butadiene	(g)	44,154	Carbon (graphite)	(s)	32,760
Isoprene	(l)	43,811	Hydrogen	(g)	120,000
Nitrogen compounds			Carbon monoxide	(g)	10,100
Nitromethane	(l)	10,540	Ammonia	(g)	18,600
1-Nitropropane	(l)	20,860	Sulfur	(s)	9257

Table 5.2	*Weight net heating values NHV$_w$ of pure organic compounds at 25°C and 1 bar abs. Sources: for hydrocarbons: TRC, for others: DIPPR (see chapter 4).*

Product	NHV$_w$ kJ/kg at 25°C	Density kg/l at 25°C	Volume NHV$_v$ kJ/l at 25°C
LPG-motor fuel	46,000	0.550	25,300
Regular gasoline	43,325	0.735	31,845
Premium gasoline	42,900	0.755	32,390
Jet fuel	42,850	0.795	34,065
Diesel fuel	42,600	0.840	35,785

Table 5.3   *Weight and volume net heating values of commercial motor fuels (average values).*

It is interesting to compare the respective variations of net heating values (mass and volume) of hydrocarbons as a function of their chemical structure. Figure 5.1 shows very distinct differences between aliphatic components such as paraffins and olefins on one hand, and aromatic components on the other. The former have a higher NHV$_w$ but a lower NHV$_v$. Generally, the NHV$_v$ increases with density, that is, for aliphatic hydrocarbons, with the number of carbon atoms in the molecule. For aromatics it is the simplest compound, benzene, that is the most dense. However, all members of this family have a high density, greater than 0.850 kg/l which goes with a high NHV$_v$.

Table 5.3 shows that for gasolines, the NHV$_v$ correlates closely with the density which itself depends on the aromatic content. The NHV$_v$ increases from regular gasoline to conventional leaded premium gasoline and from the latter to unleaded premium gasoline: in fact, in the majority of production schemes, obtaining high octane numbers means using large quantities of aromatic-rich reformate produced at high severity.

Table 5.3 also shows that diesel fuel has a net heating value slightly lower than gasoline: but, taking into account its much higher density, on the order of 0.840 kg/l instead of an average of 0.750 kg/l, the NHV$_v$ of the diesel is in all cases appreciably higher than for gasoline: the difference, close to 10%, should always be considered when the fuel consumptions in l/100 km are compared for gasoline and diesel vehicles.

In the expression for heating value, it is useful to define the physical state of the motor fuel: for conventional motor fuels such as gasoline, diesel fuel, and jet fuels, the liquid state is chosen most often as the reference. Nevertheless, if the material is already in its vapor state before entering the combustion system because of mechanical action like atomization or thermal effects such as preheating by exhaust gases, an increase of useful energy results that is not previously taken into consideration.

For hydrocarbons, this gain remains moderate since their mass enthalpy of vaporization, between 300 and 500 kJ/kg, usually represents only 0.8% to 1% of the NHV. It is thus not extremely useful to know in this case whether the motor fuel is introduced as a vapor or a liquid. The same is not true for

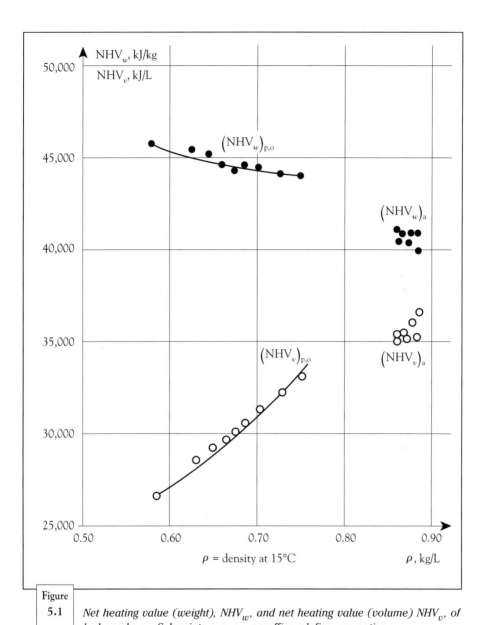

Figure
5.1

Figure
5.1   *Net heating value (weight), NHV$_w$, and net heating value (volume) NHV$_v$, of hydrocarbons. Subscripts p, o, a = paraffins, olefins, aromatics.*

certain special motor fuels such as the alcohols. The mass enthalpy of vaporization of methanol at 25°C (1100 kJ/kg) reaches 5.5% of its NHV in the liquid state. A previously vaporized methanol in the inlet of the motor, for example, due to an exchange of sensible heat with the exhaust, constitutes an elegant and efficient method of increasing the performance of this type of motor fuel.

d. Specific Energy

In the applications where the compactness of the energy conversion system is the determining factor as in the case of engines, it is important to know the quantity of energy contained in a given volume of the fuel-air mixture to be burned. This information is used to establish the ultimate relations between the nature of the motor fuel and the power developed by the motor; it is of prime consideration in the development of fuels for racing cars.

If one imagines that the fuel is used in the liquid state in the form of droplets — as in the case of fuel injection — the specific energy of the motor fuel (SE) is expressed in kilojoules per kilogram of air utilized, under predetermined conditions of equivalence ratio (stoichiometry for example). The SE is none other than the NHV_w/r quotient where r represents the previously defined stoichiometric ratio.

Table 5.4 gives the specific energies of selected organic liquid compounds. Compared with the isooctane chosen as the base reference, the variations from one compound to another are relatively small, on the order of 1 to 5%, with the exception of some particular chemical structures such as those of the short chain nitroparaffins (nitromethane, nitroethane, nitropropane) that are found to be "energetic". That is why nitromethane, for example, is recommended for very small motors such as model airplanes: it was also used in the past for competitive auto racing, for example in the Formula 1 at Le Mans before being forbidden for safety reasons.

Compound	Specific energy	
	kJ/kg of air	**Relative value**
Isooctane	2932	1.000
Decane	2940	1.003
Hexane	2938	1.002
Butane	2961	1.010
Cyclohexane	2942	1.003
1-Hexene	3008	1.026
Benzene	3032	1.034
Toluene	3011	1.027
o and m-Xylene	3000	1.023
Methanol	3086	1.054
Ethanol	2982	1.019
Isopropanol	2945	1.004
Tertiarybutanol	2915	0.994
Nitromethane	6221	2.122
Nitropropane	5010	1.709

Table 5.4 *Specific energy for selected organic compounds.*

For fuels occurring naturally in the gaseous state such as methane and LPG, or used in this form as in the case of total vaporization ahead of the engine's combustion chambers, the available energy per unit volume of the gaseous fuel mixture no longer depends only on the NHV/r ratio but also on the molecular weight of the motor fuel. In this case one observes a significant loss, compared with isooctane, that is a function of the material's volatility (9.4% with methane, 3.8% with ethane, 2% with butane). The alcohols likewise show a small loss (1.2% for methanol), while the nitroparaffins bring, once again in this scenario, a considerable gain.

5.1.2 Gasoline Combustion and Corresponding Quality Criteria

We distinguish on one hand the physical properties that establish the correct fuel feed for a vehicle (starting up, warm up, cold and hot operation) and on the other hand the chemical properties essentially related to the octane numbers, acting to achieve optimum performance for the motor without risking abnormal combustion. The first items are largely the result of each country's climatic conditions and can not therefore be made uniform over a wide geographical area: on the other hand, they are identical for each type of gasoline (regular, premium, with or without lead). The second group of properties establish the classification of gasolines by types such as regular or premium; they depend mostly on a compromise taking into account refining capacities, the options taken by the automobile manufacturers, and traffic conditions, etc. We observe that in this case, slightly different situations in the United States, Europe, and Japan for example.

5.1.2.1 Physical Properties of Gasolines

The density and the volatility, expressed by the distillation curve and the vapor pressure, constitute the most important physical characteristics of motor fuels for obtaining satisfactory operation of a vehicle in all circumstances.

a. Density of Gasolines

Density is generally measured at 15°C using a hydrometer in accordance with the NF T 60-101 method: it is expressed in kg/l with an error of 0.0002 to 0.0005 according to which category of hydrometer is utilized. However, in practice only three decimal places are usually retained.

The density varies with the temperature according to the relationship:

$$\rho_T = \rho_{15} - k\,(T - 15)$$

where T represents the temperature in °C, ρ_T and ρ_{15} the density at T°C and 15°C respectively; k is a numerical coefficient that, for gasolines, is close to 0.00085. Thus when the temperature increases from 15°C to 25°C, for example, ρ decreases by 0.008, about 1%.

Although they are small in absolute value, these fluctuations must be taken into account in the various commercial transactions related to storage and distribution of gasoline.

The French specifications set for each category of motor fuel (regular gasoline, conventional premium gasoline with or without lead) the lower and upper limits to follow for the density at 15°C (see Table 5.5); the acceptable margin for variation is about 0.030 kg/l with the possibility for overlapping zones from one product to another, theoretically allowed but seldom occurring. Regular gasolines are almost always lighter than conventional premium gasolines, which are themselves less dense than unleaded premium gasoline. For overall calculations, the average values of 0.720, 0.750 and 0.760 kg/l, respectively, can be used for each type of gasoline.

Maintaining a certain density range within the same category of motor fuel is required for satisfactory operation of the vehicle. Indeed the automobile manufacturers take this into account in the development of motor fuel feed systems and consequently select the flows for different mechanical parts. Afterwards, in actual operation, a too-high change in density between different motor fuels could disrupt the settings by changing the fuel equivalence ratio.

Several parameters come into the relation between density and equivalence ratio. Generally, the variations act in the following sense: a too-dense motor fuel results in too lean a mixture causing a potential unstable operation; a motor fuel that is too light causes a rich mixture that generates greater pollution from unburned material. These problems are usually minimized by the widespread use of closed loop fuel-air ratio control systems installed on new vehicles with catalytic converters.

Product	Density at 15°C (kg/l)	
	Minimum	Maximum
Regular gasoline	0,700	0,750
Premium leaded gasoline	0,720	0,770
Premium unleaded gasoline	0,725	0,780

Table
5.5   *Density specifications for gasolines in France.*

In practice, the user prefers the densest possible motor fuel that is compatible with the specifications, for it gives him the best volume NHV and highest fuel economy. It is estimated that an increase in density of 4 to 5% brings a reduction in consumption of 3 to 5%. Finally for the refiner, a margin of 50 thousandths accorded for the density of each type of gasoline, makes an acceptable compromise, while a tightening of the specification would be too constraining.

b. Vapor Pressure of Gasolines

The criterion retained up to now in the specifications is not the true vapor pressure, but an associated value called the Reid vapor pressure, RVP. The procedure is to measure the relative pressure developed by the vapors from a sample of motor fuel put in a metallic cylinder at a temperature of 37.8°C. The variations characteristic of the standard method are around 15 millibar in repeatability and 25 millibar in reproducibility.

In the standard method, the metal enclosure (called the air chamber) used to hold the hydrocarbon vapors is immersed in water before the test, then drained but not dried. This mode of operation, often designated as the "wet bomb" is stipulated for all materials that are exclusively petroleum. But if the fuels contain alcohols or other organic products soluble in water, the apparatus must be dried in order that the vapors are not absorbed by the water on the walls. This technique is called the "dry bomb"; it results in RVP values higher by about 100 mbar for some oxygenated motor fuels. When examining the numerical results, it is thus important to know the technique employed. In any case, the dry bomb method is preferred.

Note that the RVP is a relative pressure: that is a difference compared to the atmospheric pressure. The RVPs for gasolines are generally between 350 and 1000 millibar. The level corresponding to European specifications are shown in Table 5.6; the fuel must be simultaneously within minimum and maximum limits, identical for each type of fuel, gasoline and premium, but

Class number	Vapor pressure (mbar)		FVI**
	Minimum	Maximum	Maximum
1	350	700	900
2	350	700	950
3	450	800	1000
4	450	800	1050
5	550	900	1100
6	550	900	1150
7	600	950	1200
8	650	1000	1250

Table 5.6   *Vapor pressure specifications for gasolines in Europe* (Distribution by class).*

---

* As of 1993, France has chosen classes 1, 3 and 6, according to the following periods of the year:
- from 20 June to 9 September: class 1
- from 10 April to 19 June: class 3
- from 10 September to 31 October: class 3
- from 1 November to 9 April: class 6.

** Fuel volatility index FVI = RVP + 7 E70.

different according to the season. Indeed the volatility of the motor fuel should be sufficient in cold weather to ensure the quick starting and satisfactory warm-up of the vehicle. Conversely, during warm operation, it is best to limit the volatility to avoid certain incidents such as a reduced driveability, stalling due to vapor locking in the fuel system, difficulty or failure to start after stopping in summer while parking or at expressway toll booths.

Current requirements for vehicles are more pronounced for warm conditions than for cold for many reasons e.g., improved aerodynamics, transversal placement of the motor, generally higher temperatures under the hood, such that the automobile manufacturers prefer a reduction, rather than increase in RVP.

To these technological constraints are added an increasing preoccupation with limiting the evaporation losses that are an important source of atmospheric pollution (Mc Arragher et al., 1990).

Finally, the present (and increasing) trend is to a reduction of gasoline vapor pressures. This does not please the refiners who see themselves as having to limit the addition of light fractions to the gasoline pool. Note that an addition of 1% by weight of a C_4 cut (butanes, butenes) brings about an average increase in RVP of 50 mbar. Incorporating alcohols such as methanol, and ethanol also causes a noticeable increase in RVP (about 150 mbar for additions of methanol in excess of 1%, 50 mbar for the same quantity of ethanol). In order to meet volatility specifications, it follows that fuels containing alcohol must have limits on their light hydrocarbon contents.

We believe to have shown here that the RVP of gasoline is a primary characteristic for quality resulting from a delicate compromise between the demands for vehicle performance, optimization of refinery operations and environmental protection.

c. Distillation Curve for Gasoline

The distillation curve shows how the volume fraction distilled at atmospheric pressure changes as a function of temperature in an appropriate apparatus (NF M 07-002). Most often, several points along the curve are defined: the initial point IP, the end point EP, distilled fractions as a volume per cent at 70, 100, 180 and 210°C designated respectively as E70, E100, E180, and E210. The specifications relative to European unleaded gasoline are shown in Table 5.7. In practice the E70 and E100 values are the subject of particular concern. They should be within a precise range, that is, between 15% and 45% for E70 and between 40% and 65% for E100 in order to allow both satisfactory cold start-ups and acceptable warm operation. The French automobile manufacturers, who are concerned mainly with warm operation, require in their specifications that the E70 always remain under 40% during summer.

The gasoline end point should not exceed a given value, currently established for Europe at 215°C. In fact the presence of too-heavy fractions leads to incomplete combustion and to a number of accompanying problems:

Class	E70		E100		E180	EP (°C)
number	Min.	Max.	Max.	Min.	Min.	Max.
1, 2, 3, 4	15	45	40	65	85	215
5, 6, 7	15	47	43	70	85	215
8	20	50	43	70	85	215

Table 5.7   *Distillation specifications* for unleaded gasoline in Europe**.*

high fuel consumption, fouling of the combustion chamber and increased octane requirement, lubricant dilution, and premature motor wear. In practice, the gasoline distillation end point falls between 170 and 200°C which is sufficiently far from the maximum limit. However, this could be a problem of considerable importance in the case of the addition of certain stocks to the gasoline pool, such as light olefin dimers and trimers, that contain traces of heavy components.

d. Fuel Volatility Index for Gasoline

This term, often called the Fuel Volatility Index (FVI), is expressed by the relation:

$$FVI = RVP + 7 E70$$

in which the Reid Vapor Pressure, RVP, is in millibar and E70 in volume per cent distilled at 70°C.

Tests on vehicles have shown that the volatility index as defined expresses satisfactorily the fuel contribution during hot operation of the engine (Le Breton, 1984). In France, specifications stipulate that its value be limited to 900, 1000 and 1150, respectively, according to the season (summer, spring/fall, winter). The automobile manufacturers, being even more demanding, require in their own specifications that the FVI not be exceeded by 850 in summer.

e. Other Characteristics Related to Volatility

Among several other gasoline volatility criteria, we will cite the *V/L* ratio and the relations existing between the various ways of expressing the vapor pressure.

The *V/L* ratio is a volatility criterion seldom used in France but is used in Japan and in the United States where it has been standardized as ASTM D 2533. At a given temperature and pressure, the *V/L* ratio represents the volume of vapor formed per unit volume of liquid taken initially at 0°C.

---

* Characteristics coupled with those of vapor pressure (see Table 5.6).
** As of 1993, France has selected classes 1, 3 and 6, corresponding to the season: summer, fall/spring, winter.

The volatility of the fuel is expressed then by the temperature levels for which the V/L ratio is equal to certain particular values: for example $V/L = 12$, $V/L = 20$, $V/L = 36$. There are correlations between the temperatures corresponding to these vaporization ratios and the conventional volatility parameters such as the RVP and the distillation curve. Consider, for example, the following relations:

$$T_{(V/L)12} = 88.5 - 0.19\,\text{E70} - 42.5\,\text{RVP}$$
$$T_{(V/L)20} = 90.6 - 0.25\,\text{E70} - 39.2\,\text{RVP}$$
$$T_{(V/L)36} = 94.7 - 0.36\,\text{E70} - 32.3\,\text{RVP}$$

where $T_{(V/L)x}$ designate the temperature, in °C at which $V/L = x$.

E70 and RVP are expressed as per cent distilled and bar, respectively.

Concerning the vapor pressure more specifically, the current tendency is to substitute gradually more modern and meaningful techniques for the RVP. In fact, the RVP is not the true vapor pressure of the motor fuel, in the sense that it covers the contribution of gases dissolved in the fuel and desorbed during the increase in temperature from 0°C to 37.8°C. The more sophisticated instruments, described by Grabner's method and ASTM D 4953, avoid this unwanted phenomenon and give the true vapor pressure (TVP) at various temperatures. The graph in Figure 5.2 gives correlations for TVP and RVP.

5.1.2.2 **Chemical Properties of Gasolines – Octane Numbers**

For many years the development of refining processes and the formulation of gasolines has centered around the octane number. It is therefore appropriate to explain briefly what is the current situation and what are the prospects in this area.

a. *Knocking Phenomenon*

In induced ignition engines, still called "explosion" or "spark", several types of combustion are possible.

The normal process is a rapid-but-smooth combustion of the fuel-air mixture in the engine due to the propagation of a flame front emanating from the spark created between the electrodes of the spark plug.

One undesirable phenomenon is knocking. It is caused by a sudden and massive auto-ignition of a part of the fuel not yet burned and brought to the elevated temperature and pressure by movement of the piston and by the release of energy coming from the flame front. The result is a local pressure increase followed by vibrations in the gas phase which are attenuated once the pressure is equalized in all parts of the combustion chamber. This is accompanied by a characteristic noise similar to a series of metallic clicking or knocking sounds which explains the origin of the term. The basic corresponding frequency is about 5000 to 8000 Hz.

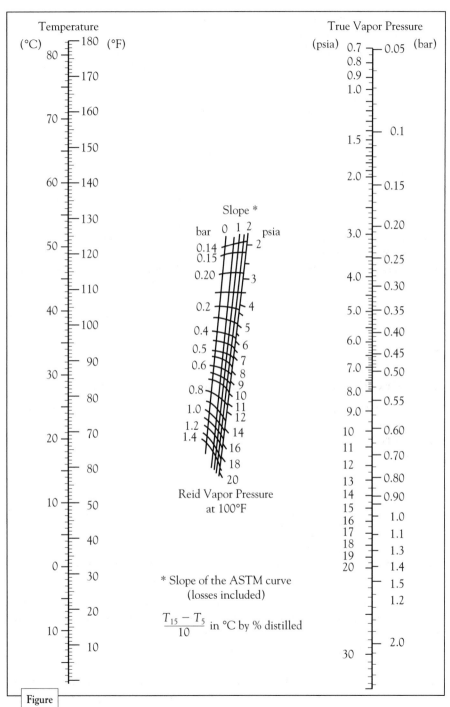

Figure 5.2 *Correlation between Reid Vapor Pressure (RVP) and True Vapor Pressure (TVP).*

Figure 5.3 gives an example of a combustion diagram recorded during knocking conditions. This is manifested by intense pressure oscillations which continue during a part of the expansion phase.

The term detonation often employed to describe knocking is incorrect because the phenomenon can not be attributed to the propagation of a flame in the supersonic region, accompanied by a shock wave..

Knocking should be absolutely avoided because if allowed to continue it will cause severe mechanical and thermal problems (Eyzat et al., 1982), which can

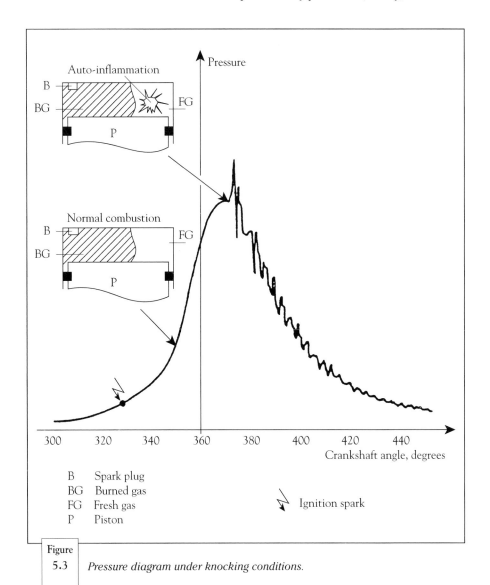

| Figure 5.3 | *Pressure diagram under knocking conditions.* |

cause destructive incidents such as rupture of the cylinder head gasket, failure or partial melting of pistons, or deterioration of the cylinder head and valves.

b. Definition of Octane Numbers – Standard Procedures

The preceding considerations show that the formulation or motor fuels should be directed towards the selection of products offering a very good resistance to auto-ignition. Qualitatively, one knows for this matter what the most preferred hydrocarbon structures are: highly branched paraffins and olefins and aromatics: benzene, toluene, xylenes. Inversely, the long, straight-chain paraffins and olefins having more than 4 carbon atoms, are shown to be the most susceptible to knocking.

In order to characterize the behavior of motor fuels or their components with regard to knocking resistance but without involving chemical composition criteria which are complex and not easy to quantify, the traditional method that has been universally employed for more than 50 years consists of introducing the concept of octane number.

The motor fuel under test is compared to two pure hydrocarbons chosen as references. The first is 2, 2, 4-trimethylpentane or isooctane which is very resistant to auto-ignition and to which is arbitrarily given the number 100:

$$CH_3 - \overset{\overset{\displaystyle CH_3}{|}}{\underset{\underset{\displaystyle CH_3}{|}}{C}} - CH_2 - \overset{\overset{\displaystyle CH_3}{|}}{CH} - CH_3$$

The other compound is n-heptane which has little resistance to knocking and is given the number, 0:

$$CH_3 - \left(CH_2\right)_5 - CH_3$$

A motor fuel has an octane number X if it behaves under tightly defined experimental conditions the same as a mixture of X volume % of isooctane and $(100 - X)\%$ of n-heptane. The isooctane-heptane binary mixtures are called primary reference fuels. Octane numbers higher than 100 can also be defined: the reference material is isooctane with small quantities of tetraethyl lead added: the way in which this additive acts will be discussed later.

Measurement of octane numbers is carried out using a reference motor called CFR (Cooperative Fuel Research), referring to a series of studies conducted in 1928 in the United States in order to standardize the methods for characterizing motor fuels.

The CFR engine has a single cylinder designed to withstand continued knocking without damage. It operates at full throttle and at low rotating speed (600 or 900 rpm, according to one of two standard procedures described below). The variable compression ratio can be adjusted during operation by moving the cylinder vertically by means of a rack-and-pinion crank. There is also a mechanism to adjust the fuel-air ratio, consisting of varying the fuel level in the carburetor tank.

The basic procedure is to increase the compression ratio of the CFR engine just to the point of obtaining a "standard" knocking intensity indicated by a pressure detector in the combustion chamber. The critical compression ratio thus recorded is bracketed by two values taken along with two binary heptane-isooctane of neighboring compositions. For each operation, the fuel-air ratio is taken to be that corresponding to the strongest tendency to knock. The octane number is calculated by linear interpolation determining the primary reference mixture having similar behavior as the fuel being tested.

There are two standard procedures for determining the octane numbers: "Research" or F1 and the "Motor" or F2 methods. The corresponding numbers are designated as RON (Research Octane Number) and MON (Motor Octane Number) which have become the international standard.

The distinctions between the two procedures RON and MON concern essentially the engine speed, temperature of admission and spark advance. See Table 5.8.

Operating parameters	RON	MON
Engine speed, rpm	600	900
Ignition advance, degrees before Top Dead Center	13	14 to 26*
Inlet air temperature, °C	48	–
Fuel mixture temperature, °C	–	149
Fuel-air ratio	**	**

Table 5.8 *Test conditions for the determination of the RON and MON in the CFR engine.*

During the determination of the RON, the CFR engine operates at 600 rpm with a timing advance set at 13° TDC and with no fuel mixture preheating. The MON by contrast operates at 900 rpm, with an advance from 14 to 26° depending on compression ratio and a fuel mixture temperature of 149°C.

The measurement error for conventional motor fuels is around 0.3 points and 0.7 points for the RON and the MON respectively. The RON is the characteristic more often used and more widespread than the MON; moreover, when the octane number is used without reference either procedure, it is taken to be the RON.

A gasoline's MON is always lower than its RON: their difference, an average of 10 to 12 points, is called the sensitivity, S. This is an indication of a fuel's "sensitivity" to a modification of the experimental conditions and more

---

* Variable with the compression ratio.
** Adapted in each case to obtain maximum knocking intensity, it is usually between 1.05 and 1.10.

particularly to an increase in temperature such as happens during the MON procedure.

Most conventional regular and premium fuels have an RON between 90 and 100, while their MON is between 80 and 90.

c. Required Levels – Specifications

Table 5.9 shows the octane specifications for the different types of fuels distributed in France.

Type of product	Required values		
	RON		MON
	Min.	Max.	Min.
Regular gasoline	89	92	–
Premium gasoline with lead	97	–	86
Eurosuper	95		85
Superplus*	98		88

Table
5.9 *Specifications for octane numbers of motor fuels in France.*

For regular gasoline, the specification concerns only the RON which should be between 89 and 92. This is a product on the way to extinction in France, for in 1993, it had only about 0.3% of the total gasoline market. This is a recent trend; in 1980, for example, the share of regular gasoline was 17% of the total French market. The only western European country where regular gasoline is still used to a large extent is Germany with a 30% share of the total market in 1993.

The conventional leaded premium gasoline is characterized by minimum RON and MON values of 97 and 86, respectively. In 1993, it was the principal product in the gasoline pool in France having 60% of the sales, but its share is diminishing and, in 2000, it will play mostly a minor role.

The unleaded premium fuels appeared in Europe and France in 1988. In 1993, they represented 40% of the French and 47% of the European Community markets. They are divided into three types.

• The Eurosuper as defined by the European Directive of 16 December 1985, offering a minimum RON of 95 and a minimum MON of 85. Found throughout the European countries, it still is fairly scarce in France with 10% of the total demand for unleaded fuels.

---

* These are not official specifications, but rather "Quality" levels specified by French automobile manufacturers.

- The Superplus with its RON and MON greater than or equal to 98 and 88 respectively. These two values do not correspond to official specifications but to "quality" specification sheets drawn up by French auto manufacturers. Strictly according to the rules, Superplus is only a variety of a "top quality" Eurosuper.

In every part of the world, the same type of classification as above is found for fuels: premium or regular, with or without lead. The octane numbers can be different from one country to another depending on the extent of development of their car populations and the capabilities of their local refining industries. The elimination of lead is becoming the rule wherever there are large automobile populations and severe anti-pollution requirements. Thus the United States, Japan and Canada no longer distribute leaded fuels.

In the United States, the different categories of motor fuels are not defined as they are in Europe by minimum RON and MON values, but by a combination of the two numbers: more exactly by the relation $\left(\dfrac{RON + MON}{2}\right)$. There are thus three kinds of motor fuels: regular, intermediate, premium with average values of 87, 90 and 93 for $\dfrac{RON + MON}{2}$, respectively.

The fact that the sensitivity is often close to 10, makes it rather easy to calculate the individual values of RON and MON.

d. Gasoline Octane Numbers and Vehicle Performance

In order to adapt an engine to a given fuel of a given octane number, the automobile manufacturer must consider the design and control parameters in order to prevent knocking in all possible operating conditions; the variables at hand are essentially the compression ratio and ignition advance which in turn determine the motor performance (thermal efficiency and specific horsepower). Horsepower can always be maintained by technological devices such as cylinder displacement and transmission ratios but the thermal efficiency always remains closely tied to the octane number. This is illustrated by the following example: a 6-point increase in octane number (RON or MON) — corresponding to an average difference between a premium gasoline and a regular gasoline— enables a one-point gain in compression ratio (from 9 to 10, for example), which results in an efficiency improvement of 6%. An average 1% efficiency gain per point of octane number increase is thereby obtained. This approach has led to the concept of Car Efficiency Parameter (CEP). For an engine with a compression ratio exactly adapted to the fuel used, the CEP represents the weight per cent change in consumption resulting from a one-point change in octane number. In the preceding example, the CEP equals 1. That is the value most often used in economic evaluation of the technology. Now if the manufacturer changes the system acting on not the compression ratio but the ignition advance, the preceding tendency still applies but with a lower CEP, between 0.5 and 1. As a conclusion, we have seen that an increase in octane number will always be an important factor for the thermal efficiency, thereby reducing the fuel consumption of motor vehicles. This fact alone justifies all the processes employed by the refiner to produce high octane gasolines.

e. Importance and Significance of RON and MON

Knowledge of RON and MON, or their combination, is not enough to predict the real behavior of a motor fuel in a mass-production engine. In fact for this case, the change in pressure and temperature as a function of time in gases under knocking conditions is usually very different from that observed in the CFR engine. Complementary experiments on the vehicle are necessary to find the correlations between the characteristics of the motor fuel (RON, MON, composition by chemical family) and its real behavior. This approach has caused the concept of "Road" octane number — a discussion of which would be too lengthy to describe here. However we will give the general trends which can be brought to light.

For vehicles, special attention is most often focused on the knocking potential encountered at high motor speeds in excess of 4000 rpm for which the consequences from the mechanical point of view are considerable and lead very often to mechanical failure such as broken valves or pistons, and rupture of the cylinder head gasket. Between RON and MON, it is the latter which better reflects the tendency to knock at high speeds. Conversely, RON gives the best prediction of the tendency to knock at low engine speeds of 1500 to 2500 rpm.

The two numbers each have their own utility which explains why both are taken into account when setting specifications. Nevertheless, in the context of refining today, it is the minimum MON that is the most difficult constraint for the refiner. For example, to obtain an 85 MON for Eurosuper it often happens that the RON is greater than 95: in actual practice it is 96 or even 97. Likewise, Superplus can represent, for an 88 MON, an RON of 99 or 100.

f. Other Methods for Characterizing Knocking Resistance

The observation of possible knocking in certain vehicles undergoing acceleration brings to light phenomena totally ignored by the standards covering the CFR engine. During the acceleration phase of a carbureted motor vehicle, the inertia of the gasoline droplets compared to air can cause a segregation of the mixture in time and space. The engine intakes the more volatile fractions of the fuel preferentially. These too should offer good resistance to knocking.

To evaluate the real behavior of fuels in relation to the segregation effect, the octane numbers of the fuel components can be determined as a function of their distillation intervals In this manner, new characteristics have been defined, the most well-known being the "delta R 100" ($\Delta R100$) and the "Distribution Octane Number" (DON). Either term is sometimes called the "Front-End Octane Number".

The $\Delta R100$ corresponds to the difference between the RON of the motor fuel and that of the fraction distilled at 100°C. The determination is carried out in three stages:

- Conventional determination of the motor fuel RON
- ASTM Distillation up to 100°C and recovery of the distillate
- Measurement of RON of this volatile fraction.

The final error of the $\Delta R100$ is on the order of one point in repeatability.

The $\Delta R100$ for commercial motor fuels is most often between 5 and 15. A value less than 10 is considered satisfactory. There is no official specification in France; on the other hand Germany requires a maximum $\Delta R100$ of 8. For this characteristic, trying to meet the low values has been, until recently, a severe constraint concerning the formulation of motor fuels. Developments in refining flow schemes, most notably in the increasing production of gasolines from cracking processes, offer lower, thus more satisfactory, $\Delta R100$s and have solved the problem almost completely. The residual problem is that faced by simple refineries that have no cracking units and whose major gasoline component is reformate having higher $\Delta R100$s. Furthermore, widespread use of fuel injection engines will eliminate the possibility of segregation such as described.

Another characteristic similar to $\Delta R100$ is the Distribution Octane Number (DON) proposed by Mobil Corporation and described in ASTM 2886. The idea is to measure the heaviest fractions of the fuel at the inlet manifold to the CFR engine. For this method the CFR has a cooled separation chamber placed between the carburetor and the inlet manifold. Some of the less volatile components are separated and collected in the chamber. This procedure is probably the most realistic but less discriminating than that of the $\Delta R100$; likewise, it is now only of historical interest.

g. Desirable Gasoline Compositions – Characteristics Available at the Refinery

The RON and MON of hydrocarbons depend closely on their chemical structure. Figure 5.4 shows how the RON varies with the boiling point of each family of hydrocarbons.

For the *n*-paraffins, the RON is very high for the lighter components but decreases as the chain length increases reaching zero by definition for *n*-heptane. The RON always increases with branching (ramification), i.e., with the number and complexity of the lateral branches. The MON for paraffins is generally lower — by 2 to 3 points— than the RON. There are, however, important exceptions: for example, for ethyl-3 hexane, the MON of 52.4 is much higher than its RON of 33.5. As in the case of RON, branching increases the MON.

The octane numbers for olefins, as for paraffins, depends on the length and ramification of their chains. Olefin RONs are generally higher than those for paraffins having the same carbon skeleton. The displacement of the

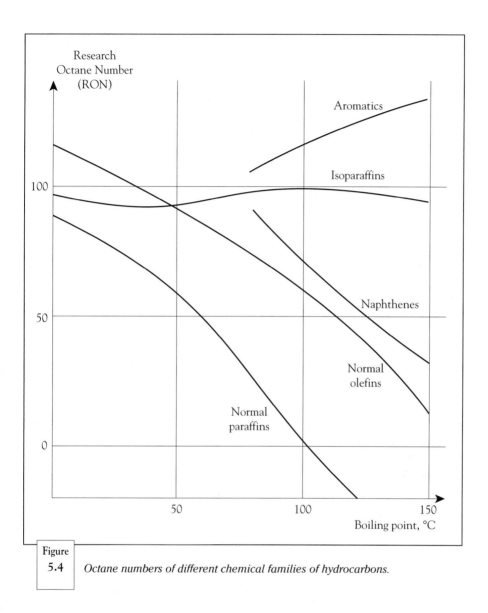

Figure 5.4 *Octane numbers of different chemical families of hydrocarbons.*

double bond towards the center of the molecule tends to improve the RON, at least for the first members of the series. Olefin MONs are always lower than their RON by an average of 10 to 15 points, explained by a high activation energy in the kinetic process of auto-ignition. In practice, the commercial motor fuel MONs always correlates strongly with the olefin content.

Naphthenes always have higher octane numbers — RON and MON — than their non-cyclic homologs; for example, the RON of n-hexane is 24.8 while cyclohexane attains 83.

The aromatic RONs are always greater than 100 (up to 115-120); these values are only orders of magnitude considering the poor accuracy in measuring octane number in this range. Aromatic MONs are usually high, greater than 100, but lower than their RONs by around 10 points. It is almost impossible to discern the major octane number differences among the components found in gasoline: benzene, toluene, xylenes, ethylbenzene, cumene. As a matter of fact, all have excellent behavior in this area.

Outside of hydrocarbons, certain organic oxygenated compounds such as the alcohols and ethers are henceforth utilized in the formulation of gasolines. These are mostly methanol, ethanol, propanols and butanols, as well as methyl and ethyl ethers obtained from C_4 and C_5 olefins: methyltertiarybutylether (MTBE) ethyltertiarybutylether (ETBE), tertiaryamylmethylether (TAME). All these compounds — alcohols and ethers — are characterized by very high RONs, up to 120-130 for methanol and ethanol. The MONs are also much higher than 100, but the sensitivity (RON-MON) is high, on the order of 15 to 20 points. These are generalized estimates; in reality the CFR engine procedures are not adapted at characterizing such substances that are quite different from conventional hydrocarbons with regards to certain properties such as heat of vaporization and heating value. Usually, it is more important to know the behavior of these components in a mixture than in their pure state. Incorporating them into gasoline always results in substantial gains in octane number (Unzelman, 1989).

The preceding information indicates the paths to follow in order to obtain stocks of high octane number by refining. The orientation must be towards streams rich in aromatics (reformate) and in isoparaffins (isomerization, alkylation). The olefins present essentially in cracked gasolines can be used only with moderation, considering their low MONs, even if their RONs are attractive.

Table 5.10 gives octane number examples for some conventional refinery stocks. These are given as orders of magnitude because the properties can vary according to process severity and the specified distillation range.

Figure 5.5 can be used to place the different product streams with respect to the objectives required for commercial octane numbers for Eurosuper and Superplus. It is clearly evident that the preparation of Superplus (RON 98, MON 88) will require careful screening of its components.

h. Formulation of Motor Fuels – Blending Index

Formulation consists of mixing the effluent streams coming from the different refining units in order to obtain products conforming to the specifications. It is also at this point that additives are added, the reasons for which and whose action will be described later. One can easily see that as far as octane numbers are concerned, or for that matter any other parameter, the characteristics of a mixture are not always identical to those predicted by a linear addition. To take into account the deviation from ideality, the concept of blending index M, as defined below, is introduced.

Stream type	RON	MON
Butane	95	92
Isopentane	92	89
Light straight run gasoline	68	67
Medium pressure reformate	94	85
Low pressure reformate	99	88
Heavy reformate*	113	102
Total FCC gasoline	91	80
Light FCC gasoline	93	82
Heavy FCC gasoline*	95	85
Alkylate	95	92
Isomerisate	85	82
Dimersol (oligomerization of light olefins)	97	82
MTBE	115	99
ETBE	114	98

Table 5.10 | *Octane numbers (RON and MON) of some conventional refinery streams (orders of magnitude).*

In a system consisting of two stocks, A and B, the blending index M_A of one of the constituents, A for example, is calculated using the relation:

$$RON_{AB} = xM_A + (1 - x)\, RON_B$$

where RON_B and RON_{AB} represent respectively the RON of constituent B and of the final mixture, AB; x is the volume fraction of component A in the mixture. Generally, the blending index refers to the minor component.

The preceding definition is applicable to other characteristics such as MON, vapor pressure and volatility characteristics such as E70 and E100.

In reality, the blending index of a compound varies according to its concentration and the nature of the product receiving it; it is not, therefore, an intrinsic characteristic. In spite of this problem, refiners have long used the concept of blending index to predict and establish their refining flow sheets based on data drawn from their own experience. This approach is disappearing except in certain cases, for example, concerning the addition of oxygenates. In this manner, Table 5.11 gives estimated blending values for different alcohols and ethers when they are added in small quantities to an unleaded fuel close to the specifications for Eurosuper (RON 95, MON 85). Taking into account the diversity of situations encountered in regards to the composition of the receiving product stream, one does not retain a unique value for the blending value, but on the contrary, a margin for possible variation.

---

* Initial boiling point about 110°C.

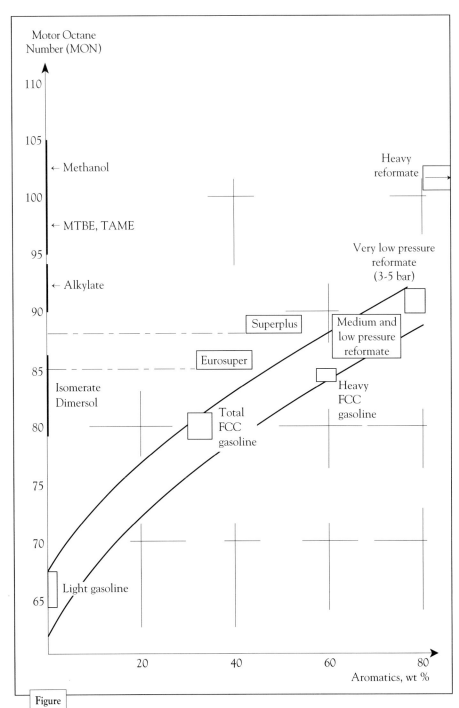

Figure 5.5 *Motor Octane Number (MON) as a function of aromatics content for stocks from different refining processes.*

Compound	Blending index, M	
	RON	**MON**
Methanol	125-135	100-105
MTBE	113-117	95-101
Ethanol	120-130	98-103
ETBE	118-122	100-102
Tertiarybutyl alcohol (TBA)	105-110	95-100
TAME	110-114	96-100

Table
5.11 *Blending values of some alcohols and ethers.*

Modern methods for formulating motor fuels rely on chemical analysis and composition-octane number correlations. A technique proposed by BP Corporation (Descales et al., 1989) consists of predicting the octane numbers of gasolines based on their infrared spectra. The method has the advantage of being very fast, taking about one minute. It has been shown to be very useful for adjusting a formulation by slight changes in a composition-type and is applied as a first priority in real refinery situations at a given site.

More general techniques covering a wider range employ gas chromatography (Durand et al., 1987). This enables identification and analysis of the nearly 200 gasoline components whose octane numbers are known.

From these data, a first approach is to develop linear models using a relation of the following type:

$$ON = \Sigma \left(ON_{ppi} + K_i\right) C_i$$

in which ON is the octane number (RON or MON) of the gasoline, ON_{ppi} the octane number of the pure component i, C_i its weight per cent concentration, and K_i a coefficient representing a deviation from ideality — plus or minus — taking into account its behavior in the mixture.

In these calculations, the terms K_i are identical for the components belonging to the same hydrocarbon family with the same degree of branching and the same number of carbon atoms. Chromatographic analyzers associated with a program for determining the octane number are commercialized. A good example is Chromoctane®, built under license by IFP and distributed by the company, Vinci-Technologie. This on-line instrument determines the octane numbers of reformates automatically.

To predict the octane numbers of more complex mixtures, non-linear models are necessary; the behavior of a component i in these mixtures depends on its hydrocarbon environment.

The octane number is expressed then by the equation:

$$ON = \sum_{1}^{n} \left(ON_{ppi} + K_i^*\right) C_i$$

with the same preceding notations, but where K_i^* is a function of the concentrations of different hydrocarbon families:

$$K_i^* = f\left(C_{f1}, C_{f2}, C_{f1} \cdot C_{f2} \ldots\right)$$

The more-or-less pronounced interactions between the added component and the receiving system are identified, for example, as follows:

- in an aromatic-olefinic environment: the addition of normal paraffins and isoparaffins with a methyl group will be generally favorable
- in an aromatic environment: the addition of normal olefins and iso-olefins with a methyl group will also be generally favorable
- in an olefinic environment: the addition of light aromatics (C_6 to C_8) creates a negative effect and adding heavy aromatics (C_{9+}), a positive effect
- in a paraffinic environment: all aromatics have a negative effect.

From such models the prediction of octane numbers can be considerably refined. Figure 5.6 also shows for one hundred gasoline samples having a large range of composition the differences between the measured and calculated values for the RON and MON. For 70% of the sample population, these deviations are dispersed within a margin of $\pm$ 0.3 point. The preceding information is of course only an example, revealing however the refining needs towards the year 2000, where the constitution of a more and more complex motor fuel pool will require very precise modelling tools.

i. Action of Lead Alkyls

These compounds have been incorporated in the gasolines of the entire world since 1922. Their use will disappear with the general acceptance of catalytic converters with which they are totally incompatible.

It is useful, nevertheless, to bring to mind their composition and their means of action (Goodacre, 1958). Several components of the same family can in reality be utilized: tetraethyl lead, $Pb\left(C_2H_5\right)_4$ or TEL, tetramethyl lead, $Pb\left(CH_3\right)_4$ or TML, mixtures of these products or yet mixed chemical components including various combinations of the groups C_2H_5 and CH_3: $Pb\left(C_2H_5\right)_2\left(CH_3\right)_2$, $Pb\left(C_2H_5\right)_3 CH_3$, $Pb\left(C_2H_5\right)\left(CH_3\right)_3$.

The base products, TEL and TML, are liquids having boiling points of 205°C and 110°C respectively. The contents of additives used are usually expressed in grams of lead per liter of fuel; in the past they have reached 0.85 g Pb/l. These concentrations are still found in some of the countries of Africa. Elsewhere, when part or all of the motor fuel pool contains lead, the concentrations are much smaller. Thus in Western Europe they no longer exceed 0.15 g Pb/l.

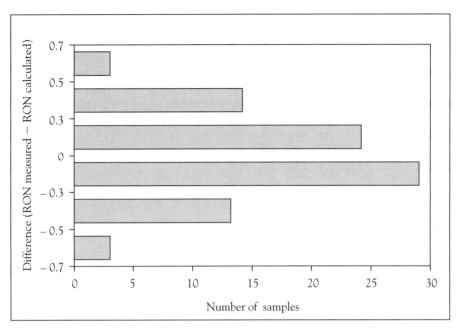

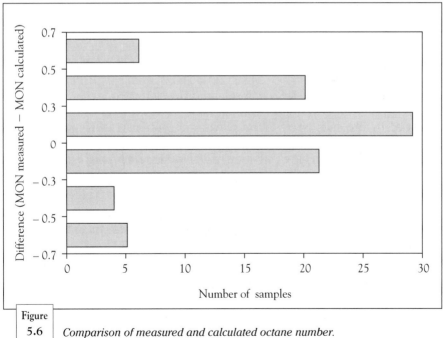

Figure
5.6

Comparison of measured and calculated octane number.

Type of Measurement	RON	MON
Standard deviation for all samples	*0.34*	*0.42*
Standard deviation for samples whose RON > 95	*0.30*	*0.40*

The lead alkyls inhibit auto-ignition which, by intermediary of lead oxide, PbO, increases the resistance to knocking.

Adding lead to a fuel increases octane numbers by several points. From an RON of around 92, the increase is on the order of 2 to 3 points for 0.15 g Pb/l and of 5 to 6 points for 0.4 g Pb/l. For higher concentrations the effect of saturation appears and additional improvement in the octane number becomes more modest. The preceding values concern the RON as well as the MON. Nevertheless, one more often observes slightly larger increases for the RON. In other words, lead addition tends to increase the sensitivity slightly (on an order of one point for 0.4 g Pb/l).

Finally, one should note that the increases in octane numbers due to lead alkyls are higher when the octane levels in the receiving stream are low. Frequently the term "susceptibility" is used to express the efficiency of the lead alkyls. A stock is deemed more "susceptible" when it shows higher gains in octane number for a given lead content. Some examples of susceptibility of different refinery stocks are given in Figure 5.7. One sees notably that in a stock whose RON approaches 70, therefore essentially paraffinic, the incorporation of lead up to 0.6 g/l gives a gain of more than 10 points. It is understandable that for a country not having a very elaborate refining industry, the addition of lead remains a very effective means to obtain relatively satisfactory octane levels.

Introducing lead alkyls in the fuel can cause heavy deposition of solids in the combustion chamber and exhaust systems. To avoid this problem, additives called "scavengers" are added to keep these substances from appearing. Dibromoethane, $C_2H_4Br_2$, and dichloroethane, $C_2H_4Cl_2$, transform the lead derivatives into lead chloride and bromide. These compounds are gases at the temperatures prevailing inside the cylinder and are largely removed by the exhaust gas. The scavenger content is expressed often in terms of "theories". A theory represents the quantity of additive strictly necessary to convert all the lead into lead chloride and bromide in the motor fuel. In practice, 1.5 theories of scavenger are added (a theory of dichloroethane and 0.5 theory of dibromoethane).

However, in some countries such as Germany there is considerable reservation to adding scavengers because of their possible contribution to dioxin emissions. Furthermore, for lead contents of 0.15 g/l, the need for scavengers is questionable. It is possible that the leaded fuels sold in the coming years will contain neither chlorine nor bromine.

j. Unleaded Fuels

The lead alkyls and scavengers contained in fuels cause rapid poisoning of exhaust gas catalytic converters. They are tolerated only in trace quantities in fuels for vehicles having that equipment. The officially allowed content is 0.013 g Pb/l, but the contents observed in actual practice are less than 0.005 g Pb/l.

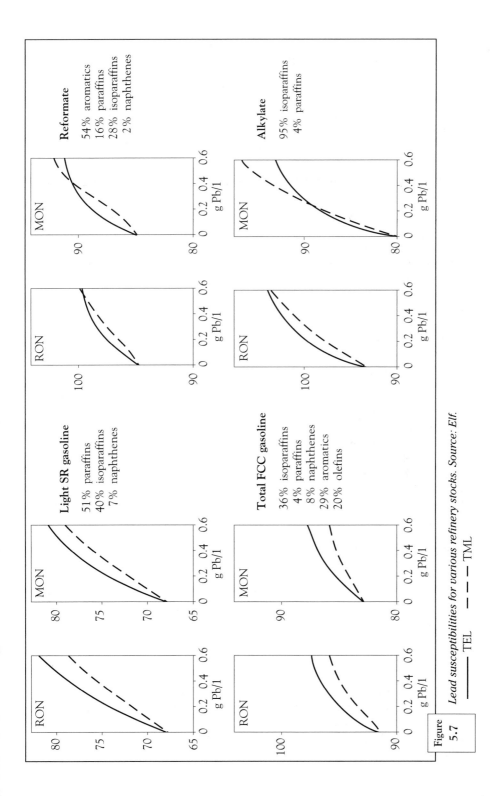

Figure 5.7 *Lead susceptibilities for various refinery stocks. Source: Elf.*

— TEL   – – – TML

The gradual reduction and ultimate elimination of lead has seen considerable effort by the refiner to maintain the octane numbers at satisfactory levels. In Europe, the conventional unleaded motor fuel, Eurosuper, should have a minimum RON of 95 and a minimum MON of 85. These values were set in 1983 as the result of a technical-economic study called RUFIT (Rational Utilization of Fuels in Private Transport). A compromise was then possible between refining energy expenses and vehicle fuel consumption (Anon., 1983).

As of 1 July 1991 all new vehicles registered in Europe should have been adapted for using Eurosuper. However, another quality of unleaded fuel had already appeared in 1988 and has since been well established in some countries, notably France. It is called Superplus, the highest octane level, RON 98, MON 88. Superplus is suitable for vehicles having catalytic converters as well as for a large part of the former automotive fleet requiring an RON of 97 and higher.

Note that for certain older engines, small quantities lead deposits from combustion could have a positive effect as a solid lubricant and prevent exhaust valve recession. For these motors which still represented in 1993 from 20 to 30% of the French automotive fleet, the use of unleaded fuel is not possible.

In France, Superplus represented in 1993 about 90% of the demand for unleaded gasoline which itself has attained 40% of the total sales. Remember that this is an unusual situation for Europe because in most other countries Eurosuper has the major share.

The penetration of unleaded fuels will continue rapidly in Europe in the coming years. Figure 5.8 portrays a scenario predicting the complete disappearance of leaded motor fuels by 2000-2005. Regular gasolines will also be eliminated very soon. The remaining uncertainty for unleaded fuels is the distribution between Eurosuper and Superplus. From a strictly regulatory point of view, the predictions lean towards a predominance of Eurosuper, defined by the European Directive of 20 March 1985, still in force. Nevertheless, the automobile manufacturers desire to have available, abundantly if possible, a motor fuel of high quality so that they can more easily reach their objective of reducing fuel consumption in the early years following 2000.

To the refiner, the question of octane numbers in future gasolines is of primary importance because it determines the course of operations, the development or on the contrary the stagnation of such and such a process. Table 5.12 thus gives an example of the typical composition by origin and concentration of different base constituents of three grades of the most common motor fuels distributed today in Europe: conventional premium gasoline at 0.15 g Pb/l, Eurosuper and Superplus.

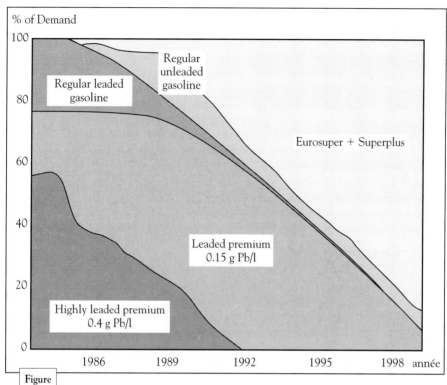

Figure 5.8 *Trends in consumption for various gasolines in Europe.*

Stock	Composition range, vol. %		
	Super with 0.15 g Pb/l	**Eurosuper**	**Superplus**
C_4 cut	2-4	2-4	2-4
Straight run gasoline	5-10	0-8	0-5
FCC gasoline	20-40	15-35	10-25
Reformate	30-60	35-60	45-80
Isomerate	0-5	0-5	0-8
Alkylate	0-10	0-15	0-20
MTBE	0-3	0-5	0-9
RON clear	95*	96.5	99.5
MON clear	84*	85.5	88

Table 5.12 *Typical compositions of gasolines according to blending stock origin.*
* Before lead addition.

5.1.3 **Diesel Fuel Characteristics Imposed by its Combustion Behavior**

This fuel is used for the diesel engines in passenger cars and in utility vehicles ranging from light delivery vans to heavy trucks carrying 38 tons. Such diversity of application can complicate the search for quality criteria, but in each situation, it will be the most demanding type of application, very often as it would be, the private automobile that determines the fuel specifications. Note that other types of non-highway diesel engines —fishing boats, construction equipment, farm tractors, and large ships— use motor fuels distinctly different from diesel fuel such as marine diesel, home-heating oil, and heavy fuel oil. Certain characteristics specified or recommended for these products take into account that they are ultimately going to be used for a diesel-type of combustion.

5.1.3.1 **Particularities of Diesel Engines**

All properties required by diesel fuel are justified by the characteristics of the diesel engine cycle, in particular the following:

- the means of fuel introduction
- the process of ignition and maintaining the combustion
- the means of regulating the horsepower output.

The diesel engine takes in and compresses the air. The fuel is injected into the cylinder in atomized form at the end of the compression stroke and is vaporized in the air. Ignition begins by auto-ignition in one or several zones in the combustion chamber where the conditions of temperature, pressure and concentration combine to enable combustion to start.

To promote auto-ignition, especially under marginal operating conditions —cold starting, for example— a high compression ratio is necessary, generally between 15 and 22 according to the type of technology, e.g., direct or prechamber. This distinction along with other purely thermodynamic considerations such as average specific heat of the gases present in the cylinder, explain the generally high efficiency of the diesel engine.

Power output is controlled, not by adjusting the quantity of fuel/air mixture as in the case of induced spark ignition engines, but in changing the flow of diesel fuel introduced in a fixed volume of air. The work required to aspirate the air is therefore considerably reduced which contributes still more to improve the efficiency at low loads.

The diesel engine operates, inherently by its concept, at variable fuel-air ratio. One easily sees that it is not possible to attain the stoichiometric ratio because the fuel never diffuses in an ideal manner into the air; for an average equivalence ratio of 1.00, the combustion chamber will contain zones that are too rich leading to incomplete combustion accompanied by smoke and soot formation. Finally, at full load, the overall equivalence ratio in a diesel engine is somewhere between 0.70 and 0.85 according to the type of technology. This situation explains why diesel engines have a low specific horsepower that can be surmounted, nevertheless, by supercharging (Guibet et al., 1987).

It can seem surprising at first glance that auto-ignition, so undesirable in gasoline engines, should be so highly desirable and preferred in the diesel cycle. This is because auto-ignition occurs here in very localized zones and that it concerns only a small fraction of the total energy introduced. There are thus one or more ignition centers instead of the auto-ignition of one massive homogeneous system which characterizes knocking. Furthermore, the sturdy structure of the diesel engine provides good resistance to a very rapid release of energy at the moment of auto-ignition. The principal exterior manifestation of this phenomenon is the very characteristic noise associated with the diesel cycle.

5.1.3.2 **Physical Properties of Diesel Fuel**

The density, distillation curve, viscosity, and behavior at low temperature make up the essential characteristics of diesel fuel necessary for satisfactory operation of the engine.

a. Density*

The density should be between 0.820 and 0.860 kg/l according to the European specifications (EN 590).

Imposing a minimum density is justified by the need to obtain sufficient maximum power for the engine, by means of an injection pump whose flow is controlled by regulating the volume.

Moreover, a limit to maximum density is set in order to avoid smoke formation at full load, due to an increase in average equivalence ratio in the combustion chamber.

b. Distillation Curve*

The necessity of carrying out injection at high pressure and the atomization into fine droplets using an injector imposes very precise volatility characteristics for the diesel fuel. French and European specifications have established two criteria for minimum and maximum volatility; therefore, the distilled fraction in volume % should be:

- less than 65% for a temperature of 250°C
- greater than 85% for a temperature of 350°C
- greater than 95% for a temperature of 370°C.

---

* See note on the bottom of page 214.

The distillation initial and end points are not specified because their determination is not very accurate; the values obtained for commercial products are found to be between 160 and 180°C for the initial point and between 350 and 385°C for the end point.

c. Viscosity*

This property should also be within precise limits. In fact, a too-viscous fuel increases pressure drop in the pump and injectors which then tends to diminish the injection pressure and the degree of atomization as well as affecting the process of combustion. Inversely, insufficient viscosity can cause seizing of the injection pump.

For a long time the official specifications for diesel fuel set only a maximum viscosity of 9.5 mm^2/s at 20°C. Henceforth, a range of 2.5 mm^2/s minimum to 4.5 mm^2/s maximum has been set no longer for 20°C but at 40°C which seems to be more representative of injection pump operation. Except for special cases such as "very low temperature" very fluid diesel fuel and very heavy products, meeting the viscosity standards is not a major problem in refining.

d. Low Temperature Characteristics

Low temperature characteristics of a diesel fuel affect more its fuel feed system than its behavior when burning. However, we will examine them here because of their strong impact on refinery flow schemes.

The diesel fuel must pass through a very fine mesh filter (a few μm) before entering the injection pump, a precision mechanical device whose operation might be jeopardized by contaminants and suspended particles in the liquid. It happens that certain paraffinic hydrocarbons in diesel fuel can partially crystallize at low temperature, plug the fuel filter and immobilize the vehicle. These considerations justify adopting strict specifications for the cold behavior of diesel fuel even though technical devices such as fuel filter heaters installed in recent vehicles have contributed to reducing the risk of such incidents occurring.

The characteristics of diesel fuel taken into account in this area are the cloud point, the pour point, and the cold filter plugging point (CFPP).

The cloud point, usually between 0 and −10°C, is determined visually (as in NF T 07-105). It is equal to the temperature at which paraffin crystals normally dissolved in the solution of all other components, begin to separate and affect the product clarity. The cloud point can be determined more accurately by differential calorimetry since crystal formation is an exothermic phenomenon, but as of 1993 the methods had not been standardized.

---

* Specifications for density, distillation curve and viscosity shown above are for products distributed in "temperate" climates. Other limits are required for "arctic" regions, particularly the Scandinavian countries. See Tables 5.13 and 5.14.

At lower temperatures, the crystals increase in size, and form networks that trap the liquid and hinder its ability to flow. The pour point is attained which can, depending on the diesel fuel, vary between −15 and −30°C. This characteristic (NF T 60-105) is determined, like the cloud point, with a very rudimentary device (maintaining a test tube in the horizontal position without apparent movement of the diesel fuel inside).

The cold filter plugging point (CFPP) is the minimum temperature at which a given volume of diesel fuel passes through a well defined filter in a limited time interval (NF M 07-042 and EN 116 standards). For conventional diesel fuels in winter, the CFPP is usually between −15 and −25°C.

The experimental conditions used to determine the CFPP do not exactly reflect those observed in vehicles; the differences are due to the spaces in the filter mesh which are much larger in the laboratory filter, the back-pressure and the cooling rate. Also, research is continuing on procedures that are more representative of the actual behavior of diesel fuel in a vehicle and which correlate better with the temperature said to be "operability", the threshold value for the incident. In 1993, the CEN looked at two new methods, one called SFPP proposed by Exxon Chemicals (David et al., 1993), the other called AGELFI and recommended by Agip, Elf and Fina (Hamon et al., 1993).

In Europe, the classification of diesel fuels according to cold behavior is shown in Tables 5.13 and 5.14. The products are divided into ten classes, six for "temperate" climates, four for "arctic" zones.

Each country adopts such and such a class as a function of its climatic conditions. France has chosen classes B, E, and F, respectively for the summer, winter, and "cold wave" periods. The first is from 1 May to 31 October, the second is from 1 November to 30 April, while the third has

Characteristics		Units	Limiting values	
			Minimum	Maximum
CFPP	Classe A			+5°C
	Classe B			0°C
	Classe C			−5°C
	Classe D			−10°C
	Classe E			−15°C
	Classe F			−20°C
Density at 15°C		kg/m^3	820	860
Viscosity at 40°C		mm^2/s	2	4.5
Measured cetane number			49	
Calculated cetane number			46	
E250		% vol.		65
E350		% vol.	85	
E370		% vol.	95	

Table 5.13   *European diesel fuel specifications (EN 590 Standard). Requirements for "temperate" climatic zones.*

Characteristics	Units	Limiting values by class				
		0	**1**	**2**	**3**	**4**
CFPP	°C	−20	−26	−32	−38	−44
Cloud point	°C	−10	−16	−22	−28	−34
Density at 15°C	kg/m³ min.	800	800	800	800	800
	kg/m³ max.	845	845	845	840	840
Viscosity at 40°C	mm²/s min.	1.5	1.5	1.5	1.4	1.2
	mm²/s max.	4.0	4.0	4.0	4.0	4.0
Measured cetane number	min.	47	47	46	45	45
Calculated cetane number	min.	46	46	45	43	43
E180	% vol. max.	10	10	10	10	10
E340	% vol. min.	95	95	95	95	95

Table 5.14 *European diesel fuel specifications (EN 590 Standard). Requirements for "arctic" climatic zones.*

not been set officially. The petroleum companies thereby take advantage of the opportunity to promote their special brands.

The means available to the refiner to improve the cold characteristics are as follows:

- reduction of the distillation end point, taking into account the often present n-paraffins in the heaviest fractions of the diesel fuel
- reduction of the initial point, implying a more pronounced overlap with the kerosene cut
- selection of fractions being more naphthenic and aromatic than paraffinic: in this case, the crude oil origin exerts considerable influence.

One remaining possibility that is less costly from an energy point of view but needs to be carefully controlled is to incorporate additives called flow improvers. These materials favor the dispersion of the paraffin crystals and in doing so prevent them from forming the large networks which cause the filter plugging. The conventional flow improvers essentially change the CFPP and pour point, but not the cloud point. They are usually copolymers, produced, for example, from ethylene and vinyl acetate monomers:

$$[\cdots CH-CH_2-CH_2-CH_2-CH-CH_2-CH_2-CH_2-CH\cdots]_n$$

with pendant groups:

$$O \quad O \quad O$$
$$| \quad | \quad |$$
$$C=O \quad C=O \quad C=O$$
$$| \quad | \quad |$$
$$CH_3 \quad CH_3 \quad CH_3$$

Figure 5.9 shows an example of the efficiency of these products. The reductions of CFPP and pour point can easily attain 6 to 12°C for concentrations between 200 and 600 ppm by weight. The treatment cost is relatively low, on the order of a few hundredths of a Franc per liter of diesel fuel. In practice, a diesel fuel containing a flow improver is recognized by the large difference (more than 10°C) between the cloud point and the CFPP.

Finally, we can mention the existence of additives acting specifically on the cloud point. These are polymers containing chemical groups resembling paraffins in order to associate with the paraffins and solubilizing functions to keep the associations in solution. The gains are more modest than those described above being on the order of 2 to 4°C for concentrations between 250 and 1000 ppm. These are, however, appreciable effects for the refiner, considering the difficulty encountered in meeting the cloud point specification.

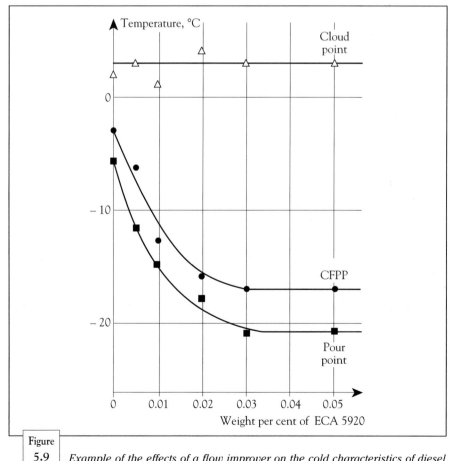

Figure
5.9
Example of the effects of a flow improver on the cold characteristics of diesel fuels.

5.1.3.3 **Diesel Cetane Number**

a. Definition and Measurement

For diesel engines, the fuel must have a chemical structure that favors auto-ignition. This quality is expressed by the cetane number.

The behavior of the diesel fuel is compared to that of two pure hydrocarbons selected as a reference:

- *n*-cetane or hexadecane $CH_3 - (CH_2)_{14} - CH_3$ which is given the number 100.
- α-methylnaphthalene which is given the number 0.

A diesel fuel has a cetane number X, if it behaves like a binary mixture of $X\%$ (by volume) *n*-cetane and of $(100 - X)\%$ α-methylnaphthalene.

In practice, the reference base is usually taken not as α-methylnaphthalene but as heptamethylnonane (HMN), a branched isomer of *n*-cetane. The HMN has a cetane number of 15. In a binary system containing $Y\%$ of *n*-cetane, the cetane number CN will be, by definition:

$$CN = Y + 0.15\,(100 - Y)$$

The standard measurement of cetane number is conducted on a CFR engine similar to that used for determining octane numbers, but obviously having a diesel-type of combustion chamber. For the diesel fuel being tested, the engine's compression ratio is varied so that auto-ignition, as shown on a diagram of pressure vs. crankshaft angle, °CA, is reproduced exactly at TDC (Top Dead Center) while the injection of fuel takes place at 13°CA before TDC. The measured compression ratio is in turn bracketed by two measured values at the same conditions of combustion, with primary reference mixtures whose cetane numbers differ by no more than 5 points. The cetane number of the diesel fuel is then calculated by linear interpolation. The procedure is not very precise with possible deviations for the same sample on the order of one point in repeatability and two points in reproducibility.

b. Cetane Number – Engine Requirements

The European specifications require a minimum cetane number of 49 for the temperate climatic zones and the French automotive manufacturers require at least 50 in their own specifications. The products distributed in France and Europe are usually in the 48-55 range. Nevertheless, in most Scandinavian countries, the cetane number is lower and can attain 45-46. This situation is taken into account in the specifications for the "arctic" zone (Table 5.14). In the United States and Canada, the cetane numbers for diesel fuels are most often less than 50.

The cetane number does not play the same essential role as does the octane number in the optimization of engines and motor fuels. In particular, it does not have a direct influence on the engine efficiency. However, a cetane number less than the required level could lead to operating problems: difficulties in starting, louder noise especially during idling while cold, and higher smoke emissions upon starting (refer to Figure 5.10). These tendencies, incidentally, are more pronounced in passenger cars than in heavy trucks. It is highly desirable to produce a diesel fuel with sufficiently high cetane number in order to maintain or improve the diesel engine's image for the customers. We will show eventually that obtaining a low level of pollution can not be done without an adequate diesel fuel cetane number.

c. Methods of Calculating the Cetane Number (Index)

The procedure for determining the cetane number in the CFR engine is not extremely widespread because of its complexity and the cost of carrying it out. There also exist several methods to estimate the cetane number of diesel fuels starting from their physical characteristics or their chemical structure.

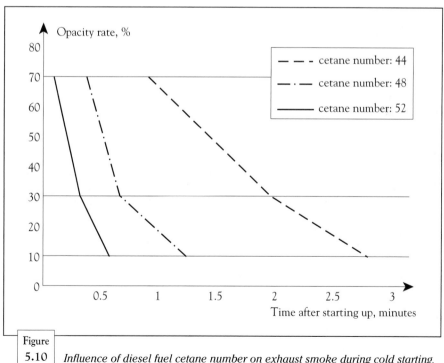

| Figure 5.10 | *Influence of diesel fuel cetane number on exhaust smoke during cold starting. Direct injection turbo-diesel engine (heavy duty vehicle).* *Start at −10°C* *Source: VROM Program – Ricardo* |

The most common formula was developed by the Ethyl Corporation (ASTM D 976) and is expressed by the relation:

$$CCI = 454.74 - 1641.416\, \rho + 774.74\, \rho^2 - 0.554\, (T_{50}) + 97.083\, (\log T_{50})^2$$

where

CCI = calculated cetane index

ρ = density at 15°C in kg/l

T_{50} = temperature, °C, corresponding to the ASTM D 86 50% distilled point.

The differences between the measured cetane numbers and the cetane indices calculated by the above formula are relatively small for cetane indices between 40 and 55. Nevertheless, for diesel fuels containing a pro-cetane additive, the CCI will be much less than the actual value (refer to paragraph **d**).

Another characteristic used for some time to measure the propensity of a diesel fuel for auto-ignition, is the Diesel Index (DI). This is defined by the relation:

$$DI = \frac{(PA) \cdot API}{100}$$

which is a function of API gravity and aniline point, AP. Remember that the aniline point is the temperature in °F at which equal volumes of aniline and diesel fuel are completely miscible.

Some empirical formulas have been devised to link the cetane index to the Diesel Index (D.I.) or even directly to the aniline point. We will cite two such formulas here in order to illustrate their comparative values:

- Cetane index = 0.72 D.I. + 10
- Cetane index = AP − 15.5

with AP in °C.

Other techniques for predicting the cetane number rely on chemical analysis (Glavinceski et al., 1984) (Pande et al., 1990). Gas phase chromatography can be used, as can NMR or even mass spectrometry (refer to 3.2.1.1.b and 3.2.2.2).

In gas chromatography, the columns employed do not give a complete separation of the diesel cut. There are a large number of unknown components starting from C_{12}-C_{13}. In spite of this problem, it is possible to extract enough information in order to estimate the final cetane number and the cetane profile along the distillation curve. Figure 5.11 gives an example for a straight-run diesel fuel.

The spectroscopic methods, NMR and mass spectrometry for predicting cetane numbers have been established from correlations of a large number of samples. The NMR of carbon 13 or proton (see Chapter 3) can be employed. In terms of ease of operation, analysis time (15 minutes), accuracy of prediction (1.4 points average deviation from the measured number), it is proton NMR which seems to be most advantageous. Figure 3.11 shows a very good correlation between the calculated and measured values for cetane numbers ranging between 20 and 60.

With regard to mass spectrometry, accuracy is not as high with an average error of 2.8 points, but on the other hand, the sample required is very small, being around 2 µl.

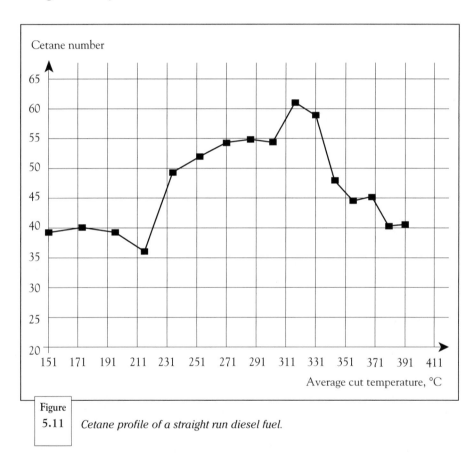

| Figure 5.11 | *Cetane profile of a straight run diesel fuel.* |

d. Improving the Cetane Number using Additives

The additives for improving the cetane number, called pro-cetane, are particularly unstable oxidants, the decomposition of which generates free radicals and favors auto-ignition. Two families of organic compounds have been tested: the peroxides and the nitrates. The latter are practically the only ones being used, because of a better compromise between cost-effectiveness and ease of utilization. The most common are the alkyl nitrates, more specifically the 2-ethyl-hexyl nitrate. Figure 5.12 gives an example of the

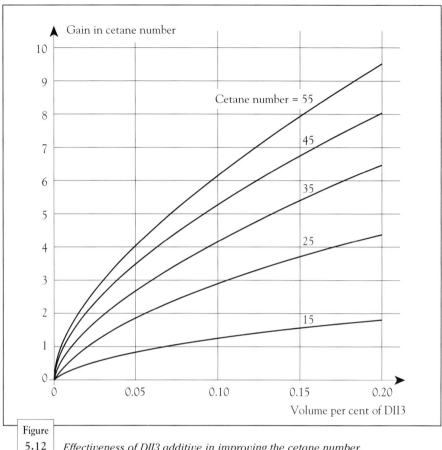

Figure
5.12 *Effectiveness of DII3 additive in improving the cetane number.*
Source: (Unzelman, 1984).

effectiveness of these products, sometimes designated as DII (Diesel Ignition Improvers). The gain in cetane number depends on the composition and characteristics of the recipient diesel fuel but they are just as small as the initial cetane number is low (Unzelman, 1984).

The desired improvements are usually on the order of 3 to 5 points, beginning with a base of 45-48. Under these conditions between 300 and 1000 ppm by weight of additive must be used. The treating costs are acceptable, being on the order of a centime (hundredth of a French Franc) per liter.

Improving the cetane number by additives results in better engine behavior, as would be predicted by the combustion mechanisms in the diesel engine (noise reduction, better operating characteristics, particularly when cold). Nevertheless, concerning certain items such as pollution emissions, it may be better to obtain a higher cetane number rather by modification of the chemical structure than by additives. Furthermore, a diesel fuel high in additives shows a large difference between the calculated cetane index and the measured cetane number, the first being much lower. This explains why the French automobile manufacturers have established minimum values for both the calculated cetane index (49) and the measured cetane number (50) for their quality specifications.

e. Characteristics of Different Refinery Stocks Used in the Formulation of Diesel Fuel

As we have shown previously, obtaining both good cold operation characteristics and sufficient cetane numbers constitutes the principal objective for the refiner in the formulation of diesel fuel. To this is added the need for deep desulfurization and, perhaps in the future, limitations placed on the chemical nature of the components themselves, e.g., aromatics content.

Table 5.15 gives some physical-chemical characteristics of selected main refinery streams capable of being added to the diesel fuel pool. Also shown is the weight per cent yield corresponding to each stock, that is, the quantity of product obtained from the feedstock.

The properties of straight run diesel fuels depend on both nature of the crude oil and selected distillation range. Thus the paraffinic crudes give cuts of satisfactory cetane number but poorer cold characteristics; the opposite will be observed with naphthenic or aromatic crudes. The increasing demand for diesel fuel could lead the refiner to increase the distillation end point, but that will result in a deterioration of the cloud point. It is generally accepted that a weight gain in yield of 0.5% could increase the cloud point by 1°C. The compromise between quantity and quality is particularly difficult to reconcile.

The gas oil cut from catalytic cracking called Light Cycle Oil (LCO), is characterized by a very low cetane number (about 20), high contents in aromatics, sulfur and nitrogen, all of which strongly limit its addition to the diesel fuel pool to a maximum of 5 to 10%.

Hydrotreating the LCO increases its cetane number to around 40 (Table 5.16), but this technique needs large amounts of hydrogen for rather mediocre results, the aromatics being converted into naphthenes which are still not easily auto-ignited. That is why LCO is sent to the domestic heating oil pool.

The gas oils from visbreaking and coking have better cetane numbers than LCO but they are unstable and need hydrotreatment before they can be used.

Hydrocracking makes very good quality diesel fuels concerning the cetane number, cold behavior, stability, and sulfur content. However this type of stock is only available in limited quantities since the process is still not widely used owing essentially to its high cost.

Feedstock	Paraffinic crude			Naphthenic crude		Vacuum distillate		Vacuum residue		Deasphalted atmospheric residue
Process	**Atmospheric distillation**			**Atmospheric distillation**		**FCC**	**Hydro-cracking**	**Vis-breaking**	**Coking**	**Hydrocracking**
Yield, weight %	30.3	32.8	36.7	29.2	47.2	10-15	30-40	5-15	35	20
Density at 15°C, kg/l	0.835	0.825	0.843	0.827	0.856	0.930	0.814	0.845	0.900	0.807
Distillation, °C IP EP	170 370	180 375	170 400	180 350	170 370	170 370	220 370	170 370	170 370	260 380
Cloud point,* °C	5	−2	+1	−10	−20	−5	−17	−4	−8	−13
Pour point, °C	−12	−9	−6	−18	−33	−14	−20	−18	−20	−18
Cetane number	50	51	54	54	43	24	64	40	28	70
Sulfur content, weight %	0.12	0.04	0.83	0.80	0.09	2.8	0.001	2.33	2.10	0.0005

Table 5.15   *Examples of stocks used in formulating diesel fuels.*

* *The results given here are on a laboratory basis. During formulation in the refinery, the cold characteristics are much less satisfactory with a penalty of around 6°C for cases where it is desired to keep the same yield from the crude oil.*
Source: Elf, Total.

Case	(1)	(2)	(3)	(4)	(5)
H2 pressure, bar	40	60	100	100	100
T, °C	360	360	360	380	380
LHSV, m^3/m^3h	2	2	2	2	0.5
Hydrogen consumption, m^3/m^3	180	230	270	290	410
weight %	1.8	2.3	2.7	2.9	4.1
HDS, %	93.0	96.4	98.3	99	99
HDN, %	75.0	98.3	99.5	99.5	99.5

Product 150°C$^+$	(1)	(2)	(3)	(4)	(5)
Specific gravity d_4^{15}	0.912	0.907	0.896	0.886	0.862
Viscosity at 20°C, mm^2/s	4.1	3.8	3.67	3.58	3.44
Sulfur, weight %	0.18	0.09	0.04	0.02	0.003
Total nitrogen, ppm	100	10	3	2	1
Cetane number	22	24	27.5	29	40

Composition by mass spectrometry	(1)	(2)	(3)	(4)	(5)
Paraffins, weight %	10.5	12.5	12.3	11.0	8.0
Naphthenes, weight %	9.5	12.5	22.3	35.0	66.0
Aromatics, weight %	80.0	75.0	65.4	54.0	26
of which: Monoaromatics	55	55	55.2	45.7	23
Polynuclear aromatics	12	6.0	4.6	4.9	1.7

Table
5.16
Gas oils from FCC (LCO).
Effect of operating conditions on hydrotreated gas oil characteristics.

Finally, other new processes can supply stocks for the diesel fuel pool. Oligomerization of light olefins followed by hydrogenation gives products having cetane numbers between 40 and 50, with neither sulfur nor aromatics. With regard to Fischer-Tropsch synthesis followed by hydroisomerization, totally paraffinic compounds result with high cetane numbers from 65 to 75, with good cold characteristics when the iso- to n-paraffin ratio is optimized.

5.1.4 **Combustion of Jet Fuels and Corresponding Quality Criteria**

As their name implies, these products are used essentially for jet aircraft; they can also be used in stationary turbines, which are themselves adaptable to a large variety of fuels such as natural gas, LPG, diesel fuel, and heating oil.

5.1.4.1 **Classification of Various Jet Fuels**

There are several types of jet fuels; according to their civilian or military application, their names can vary from one country to another.

The most widely used product is TRO (TR for turbo-reactor) or JP8 (JP for Jet Propulsion), still designated by the NATO symbols F34 and F35. In the United States, the corresponding fuel is called Jet A1. The military sometimes still uses a more volatile jet fuel called TR4, JP4, Jet B, F45 or F40. The preceding terms correspond to slight variations and it would be superfluous to describe them here.

There is finally another type of jet fuel somewhat heavier and less volatile than TRO, which allows safe storage on aircraft carriers. This is the TR5 or JP5. Among these products, TRO or Jet A1 have the most widespread acceptance because they are used for almost all the world's civil aviation fleet. The information that follows will concern essentially TRO, and very rarely TR4.

5.1.4.2 **Mode of Combustion for Jet Fuels**

Unlike piston engines, jet engines use a continuous gas stream, the motion of the aircraft being the result of a force associated with the kinetic energy of the exhaust gases. The chemical characteristics relating to auto-ignition do not have any particular involvement here. The important properties of the jet fuel are linked rather to preparation of the fuel-air mixture, flame radiation, and potential formation of carbon deposits (Odgers et al., 1986). Moreover, considering the high altitudes encountered, it is evident that the jet fuel must remain fluid at very low temperatures. Finally, under special conditions, particularly for supersonic aircraft, a high thermal stability is required.

5.1.4.3 **Required Characteristics for Jet Fuels**

a. *Physical Properties of Jet Fuels*

For optimum combustion, the fuel should vaporize rapidly and mix intimately with the air. Even though the design of the injection system and combustion chamber play a very important role, properties such as volatility, surface tension, and fuel viscosity also affect the quality of atomization and penetration of the fuel. These considerations justify setting specifications for the density (between 0.775 and 0.840 kg/l), the distillation curve (greater than 10% distilled at 204°C, end point less than 288°C) and the kinematic viscosity (less than 8 mm^2/s at $-20°C$).

b. *Chemical Properties of Jet Fuels*

In order to maintain high energy efficiency and ensure a long service life of the materials of construction in the combustion chamber, turbine and jet nozzle, a clean burning flame must be obtained that minimizes the heat exchange by radiation and limits the formation of carbon deposits. These qualities are determined by two procedures that determine respectively the smoke point and the luminometer index.

The smoke point corresponds to the maximum possible flame height (without smoke formation) from a standardized lamp (NF M 07-028). The values commonly obtained are between 10 and 40 mm and the specifications for TRO fix a minimum threshold of 25 mm. The smoke point is directly linked to the chemical structure of the fuel; it is high, therefore satisfactory, for the linear paraffins, lower for branched paraffins and much lower still for naphthenes and aromatics.

The luminometer index (ASTM D 1740) is a characteristic that is becoming less frequently used. It is determined using the standard lamp mentioned above, except that the lamp is equipped with thermocouples allowing measurement of temperatures corresponding to different flame heights, and a photo-electric cell to evaluate the luminosity. The jet fuel under test is compared to two pure hydrocarbons: tetraline and iso-octane to which are attributed the indices 0 and 100, respectively. The values often observed in commercial products usually vary between 40 and 70; the official specification is around 45 for TRO.

Figure 5.13 shows that the luminometer index depends directly on the mono-aromatic and di-aromatic contents. For this reason, the specifications

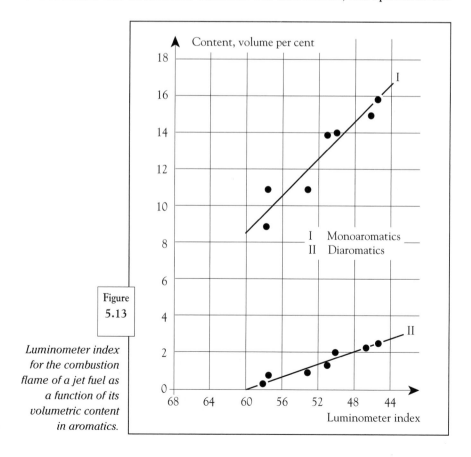

Figure 5.13

Luminometer index for the combustion flame of a jet fuel as a function of its volumetric content in aromatics.

have a maximum aromatics content of 20% in kerosene. Furthermore, certain American rules impose a maximum concentration of 3% for naphthalene compounds. In practice, the aromatics content is replacing the determination of the luminometer index which does not add really new information.

c. Cold Behavior of Jet Fuels

After a few hours of flight at high altitude, an aircraft's fuel tank will reach the same temperature as the outside air, that is, around −40 to −50°C. Under these conditions, it is important that the fuel remains sufficiently fluid to assure good flow to the jet engine. This property is expressed by the temperature at which crystals disappear or the Freezing Point (ASTM D 2386). It is the temperature at which the crystals formed during cooling disappear when the jet fuel is reheated. It should be around −50°C maximum for Jet A1, with a more and more frequent waiver to −47°C. Meeting such a cold service can be hindered by the presence of small quantities of water dissolved in the jet fuel.

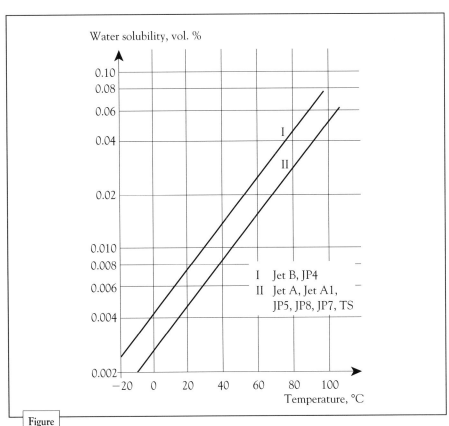

| Figure 5.14 | *Solubility of water in jet fuels as a function of temperature (Jet A is a variant of Jet A1, used in the USA for domestic flights. Jet A has a freezing point 3°C higher than that of Jet A1).* |

When the temperature decreases, water becomes less soluble (see Figure 5.15) and deposits as fine droplets that begin to freeze as the temperature reaches 0°C. To prevent this occurrence, it is possible to use anti-freeze additives that absorb the water and lower the freezing point. These products, used at maximum levels of 1500 ppm, are ethers-alcohols: for example, 2-methoxy ethanol:

$$CH_3 - O - CH_2 - CH_2 - OH$$

or β 2-methoxy ethoxy ethanol:

$$CH_3 - O - CH_2 - CH_2 - O - CH_2 - CH_2 - OH$$

d. Thermal Stability of Jet Fuels

Fuel passing through certain hot zones of an aircraft can attain high temperatures; moreover it is used to cool lubricants, hydraulic fluids, or air conditioning. It is therefore necessary to control the thermal stability of jet fuels, more particularly during supersonic flight where friction heat increases temperatures in the fuel tanks.

The most common technique for estimating thermal stability is called the Jet Fuel Thermal Oxidation Test (JFTOT). It shows the tendency of the fuel to form deposits on a metallic surface brought to high temperature. The sample passes under a pressure of 34.5 bar through a heated aluminum tube (260°C for Jet A1). After two and one-half hours, the pressure drop across a 17-micron filter placed at the outlet of the heater is measured (ASTM D 3241).

For Jet A1, the pressure drop should be less than 33 mbar, and the visual observation of the tube should correspond to a minimum of three on the scale of reference.

e. Formulation of Jet Fuels

The stocks used for jet fuel production come almost essentially from direct distillation of crude oil. They correspond to the fraction distilled between 145 and 240°C, more or less expanded or contracted according to the circumstances. The yield of such a cut depends largely on the nature of the crude but is always larger than the demand for jet fuel which reaches about 6% of the petroleum market in Europe. For the refiner, the tightest specifications are:

- the Freezing Point which is adjusted by changing the range of the cut
- the aromatics content which depends on the available selection of crudes
- other constraints such as sulfur content, acidity and anti-corrosion behavior which will be addressed later in conjunction with storage and distribution.

Finally, note that hydrocracking is ideal for obtaining middle distillate cuts that can be used in jet fuel formulation.

5.1.5 Characteristics of Special Motor Fuels

5.1.5.1 Liquefied Petroleum Gas (LPG) Used as Motor Fuel

Utilization of LPG as motor fuel represents only a tiny fraction of the market, on the order of 30,000 t/an in France out of a total 3,000,000 t; that is, about 1%. Globally, the fraction of LPG used in the transportation industry is on the order of 5%; the other applications are essentially petrochemicals (25%) and the heating fuel sectors (70%). However noticeable openings in the LPG-fuel markets in some countries such as Japan (1.8 Mt), the United States (1.3 Mt), the Netherlands (0.9 Mt), Italy (0.8 Mt) and in certain regions (Latin America, Africa, and the Far East) have been observed for specialized vehicles such as taxis.

LPG, stored as a liquid at its saturation pressure, is vaporized and introduced as vapor in conventional spark ignition motors. These motors are not modified with the exception of their feed system. Moreover, in the majority of cases, dual fuel capabilities have been adapted, that is, the vehicle can use either LPG or liquid fuel.

The potential advantages of LPG concern essentially the environmental aspects. LPG's are simple mixtures of 3- and 4-carbon-atom hydrocarbons with few contaminants (very low sulfur content). LPG's contain no noxious additives such as lead and their exhaust emissions have little or no toxicity because aromatics are absent. This type of fuel also benefits often enough from a lower taxation. In spite of that, the use of LPG motor fuel remains static in France, if not on a slightly downward trend. There are several reasons for this situation: little interest from automobile manufacturers, reluctance on the part of automobile customers, competition in the refining industry for other uses of C_3 and C_4 fractions, (alkylation, etherification, direct addition into the gasoline pool). However, in 1993 this subject seems to have received more interest (Hublin et al., 1993).

Regarding product characteristics, European specifications were established in 1992. They concern mainly the motor octane number (MON) that limits the olefin content and which should be higher than 89, and the vapor pressure, tied to the C_3/C_4 ratio which should be less than 1550 mbar at 40°C (ISO 4256). On the other hand, to ensure easy vehicle start-ups, a minimum vapor pressure for winter has been set which is different for each country and depends on climatic conditions. Four classes, A, B, C, and D, are thus defined in Europe with a minimum vapor pressure of 250 mbar, respectively, at $-10°C$ (A), $-5°C$ (B), $0°C$ (C) and $+10°C$ (D). France has chosen class A.

Finally, there are some limits regarding LPG fuels: butadiene content (0.5 wt. % maximum, ISO 7941), the absence of hydrogen sulfide (ISO 8819) and copper strip corrosion (class 1, ISO 6251) which are not usually problems for the refiner.

5.1.5.2 **Fuels for Two-Stroke Engines**

These products are used for motor scooters, outboard motors and other engines for domestic use such as power mowers and chain saws equipped with spark-ignition two-stroke engines. Their design will require either a conventional fuel, if the lubrication is separate, or mixtures of fuel and lubricating oil, with 2 to 6% oil depending on the manufacturer's specifications. In the latter case, up to now, the oil is mixed with regular gasoline containing lead. Yet, this will soon disappear. Its replacement by other products like conventional premium, Eurosuper, and Superplus will not present many particular problems concerning combustion; however, certain problems of engine failure due to insufficient lubrication have been brought to light. Currently the rules covering this area are not directed toward two-stroke engines, which would be unrealistic considering their limited market, but towards a more rigorous and better adapted lubricant specifications. Under these conditions, the two-stroke engine fuel that will be recommended worldwide will very likely be a conventional unleaded premium fuel like Eurosuper.

5.1.5.3 **The "Aviation" Gasolines**

The "aviation" gasolines are used for the small touring and pleasure aircraft still equipped with piston engines. The market is extremely small (32,000 t in 1992 in France) and the fuel is produced in only a few refineries. Table 5.17 shows the principal specifications for these products, distributed in three grades. Each grade is represented by two numbers: 80/87, 100/130 and 115/145 which are related to the octane numbers. The lowest values, 80, 100 and 115, are close to the MON (ASTM D 2700); the highest values are the performance numbers (ASTM D 909) measured on the CFR engine, according to the so-called "Supercharge" method or F4.

Characteristics		ASTM reference	Required value	
			minimum	maximum
Vapor pressure, mbar at 37.8°C		D 2551	385	490
Distillation end point, °C		D 86		170
Freezing point, °C		D 2386		−60
Sulfur content	(ppm weight)	D 1266		500

			Grade 80-87	Grade 100-130	Grade 115-145
Octane number or performance number (minimum)					
Lean mixture	(MON)	D 2700	80	100	115
Rich mixture	(supercharge)	D 909	87	130	145
Lead content	(g/l − maximum)	D 3341	0.14	0.56	1.28

Table 5.17 *Main specifications for "aviation" gasolines.*

"Civil" aviation gasoline corresponds to the grade, 100/130; it is still called 100 LL (LL signifies Low Lead) because of its low lead content of 0.56 g/l compared to military gasolines of the 115/145 type which can contain as much as 1.28 g/l.

Outside of their very high resistance to auto-ignition, the aviation gasolines are characterized by the following specifications: vapor pressure between 385 and 490 mbar at 37.8°C, a distillation range (end point less than 170°C), freezing point (−60°C) and sulfur content of less than 500 ppm.

Their production in a refinery begins with base stocks having narrow boiling ranges and high octane numbers; iso C_5 cuts (used in small concentrations because of their high volatility) or alkylates are sought for such formulations.

5.1.6 Home and Industrial Heating Fuels

This category comprises conventional LPG (commercial propane and butane), home-heating oil and heavy fuels. All these materials are used to produce thermal energy in equipment whose size varies widely from small heaters or gas stoves to refinery furnaces. Without describing the requirements in detail for each combustion system, we will give the main specifications for each of the different petroleum fuels.

5.1.6.1 Characteristics of LPG Used as Heating Fuel

LPG is divided into two types of products: commercial propane and commercial butane, each stored as liquid at ambient temperature and corresponding vapor pressure.

Their satisfactory combustion requires no particular characteristics and the specifications are solely concerned with safety considerations (vapor pressure) and the C_3 and C_4 hydrocarbon distribution.

Commercial propane is defined as a mixture containing about 90% C_3. Its density should be equal to or greater than 0.502 kg/l at 15°C (i.e., 0.443 kg/l at 50°C). The vapor pressure at 37.8°C is between 8.3 bar and 14.4 bar, which corresponds to a range of 11.5-19.3 bar at 50°C. Finally, sulfur is limited to 50 ppm by weight. The evaporation test, NF M 41-012, must result in an end point less than or equal to −15°C.

Commercial butane comprises mainly C_4 hydrocarbons, with propane and propylene content being less than 19 volume %. The density should be equal to or greater than 0.559 kg/l at 15°C (0.513 kg/l at 50°C). The maximum vapor pressure should be 6.9 bar at 50°C and the end point less than or equal to 1°C.

5.1.6.2 **Characteristics of Home-Heating Oil**

This product, given the abbreviation FOD (fuel-oil domestique) in France, still held a considerable market share there of 17 Mt in 1993. However, since 1973 when its consumption reached 37 Mt, FOD has seen its demand shrink gradually owing to development of nuclear energy and electric heating. FOD also faces strong competition with natural gas. Nevertheless, its presence in the French, European and worldwide petroleum balance will still be strong beyond the year 2000.

FOD is very similar to diesel fuel. Both are designated in economic studies by the generic term, "middle distillates".

We will give here just the main distinguishing characteristics of home heating oil with respect to diesel fuel.

a. Coloration and Tracing of Home-Heating Oil

Different taxation of diesel fuel and home-heating oil implies rigorous control in the end use of each product. That is the reason why home-heating oil contains a scarlet-red colorant (ortho-toluene, ortho-azotoluene, β-azonaphthol) in a concentration of 1 g/hl and two tracer compounds: diphenylamine and furfural with respective concentrations of 5 g/hl and 1 g/hl. The two latter components can be detected by relatively simple testing. In fact, aniline acetate gives a bright red color with furfural while chromium sulfide reagent leads to an intense blue color with diphenylamine. These control methods are very effective and reliable in detecting possible fraud.

b. Physical Properties of Home-Heating Oil

It is mainly in cold behavior that the specifications differ between home heating oil and diesel fuel. In winter diesel fuel must have cloud points of −5 to −8°C, CFPPs from −15 to −18°C and pour points from −18 to −21°C according to whether the type of product is conventional or for severe cold. For home-heating oil the specifications are the same for all seasons. The required values are +2°C, −4°C and −9°C, which do not present particular problems in refining.

For other physical properties, the specification differences between diesel fuel and home-heating oil are minimal. Note only that there is no minimum distillation end point for heating oil, undoubtedly because the problem of particulate emissions is much less critical in domestic burners than in an engine.

The winter period corresponds, of course, to the moment in the year where the diesel fuel and home-heating oil characteristics are noticeably different. Table 5.18 gives a typical example of the recorded differences; heating oil appears more dense and viscous than diesel fuel, while its initial and final boiling points are higher.

Characteristics			Product	
			Diesel fuel	Home-heating oil
Density		kg/l	0.824	0.860
Kinematic viscosity at 20°C		mm^2/s	3.47	6.20
Sulfur		weight %	0.206	0.300
Cloud point		°C	-8	$+3$
CFPP		°C	-23	-3
Pour point		°C	-27	-21
Distillation	IP	°C	167	185
	EP	°C	364	374
Cetane number			50.5	48.0
NHV$_w$		kJ/kg	42,825	42,220
NHV$_v$		kJ/l	35,285	36,310
Composition	Carbon	weight %	86.0	87.2
	Hydrogen	weight %	13.35	12.7

Table 5.18   *Characteristics of a diesel fuel and a home-heating oil (for severe winter conditions).*

c. Chemical Characteristics of Home-Heating Oil

To obtain satisfactory burner operation without forming deposits and with minimum pollutant emissions, heating oil must burn without producing residues. This property is expressed by a value called "Conradson Carbon" (NF T 60-116). It is obtained by recovering the heaviest distilled fraction of the sample (greater than 90 wt. % distilled, i.e., about 10 wt. % of the feed) and subjecting it to pyrolysis in a crucible at 550°C for thirty minutes. The residue is weighed and the Conradson Carbon is expressed as a weight per cent of the sample.

The specifications require a maximum Conradson Carbon of 0.35%. This limit is very easily met; in fact the values obtained on commercial products rarely exceed 0.1%. On the other hand, for heavy fuels, the Conradson Carbon can often reach 5 to 10%, as we will show later.

There exists, only in France, a specification requiring a minimum cetane number of 40 for home-heating oils. This rule is to assure satisfactory operation of diesel engines such as tractors, agricultural equipment, or civil construction equipment which use heating oil as fuel. In practice, the cetane numbers for this product are usually high enough, often greater than 45.

d. Sulfur Content in Home-Heating Oils

Until 1992, the maximum sulfur level, identical for diesel fuel and home-heating oil, was 0.3% for all of Europe. Nevertheless, certain countries had already established a limit of 0.2 weight %. A European directive of 23 March 1993 (93/12/EEC) set a maximum sulfur content of 0.2%, beginning 1 October 1994.

Following 1 October 1996, diesel fuel should be desulfurized to a level of 0.05% while the maximum sulfur content of home-heating oils will stay provisionally at 0.2%.

e. Home-Heating Oil Formulation

Formulation of home-heating oil is done without difficulty by the refiner who, selecting from available middle distillates, sends the streams whose characteristics are incompatible with the final specifications of diesel fuel to the home-heating oil pool. Examples of these characteristics are the cetane number and cold behavior. The segregation of diesel fuel and home-heating oil was not always done in the past but is now increasingly necessary, especially in winter for distribution of very fluid diesel fuels having cloud points less than $-5°C$. Furthermore, future specifications imposing different sulfur contents for each product will accentuate their differences.

5.1.7 Properties of Heavy Fuels

Heavy fuels are used for two kinds of applications: industrial combustion in power plants and furnaces, and fueling large ships having low-speed powerful diesel engines (Clark, 1988).

In 1993, French consumption of these products was around 6 Mt and 2.5 Mt respectively for use in burners and in diesel engines. The latter figure appears in the statistics under the heading, "marine bunker fuel". Its consumption been relatively stable for several years, whereas heavy industrial fuel use has diminished considerably owing to the development of nuclear energy. However, it seems that heavy fuel consumption has reached a bottom limit in areas where it is difficult to replace, e.g., cement plants.

5.1.7.1 Classification and Specifications of Heavy Fuels

In France there are four categories of heavy fuels whose specifications are given in Table 5.19; the different product qualities are distinguished essentially by the viscosity, equal to or less than 110 mm^2/s at 50°C for No. 1 fuel oil, equal to or greater than 110 mm^2/s for No. 2 fuel oil, and by the sulfur content varying from 4 wt. % (No. 2 fuel oil) to 1 wt. % (No. 2 TBTS – very low sulfur content fuel oil).

In the industrial combustion sector, for a total French market of 6 Mt in 1993, the distribution between the four product types was as follows:

- No. 1 fuel oil   0.2%
- Ordinary No. 2 fuel oil   73.9%

- No. 2 fuel oil BTS   (low sulfur content)   11.0%
- No. 2 fuel oil TBTS (very low sulfur content)   14.9%

The production of No. 1 fuel oil is thus quite marginal whereas the BTS and TBTS products will be undergoing important development in the coming years. In applications as diesel fuel, ordinary No. 2 fuel, and No. 2 BTS fuel are the most commonly used.

Characteristics	Fuel			
	No. 1	No. 2	No. 2 BTS	No. 2 TBTS
Kinematic viscosity, mm²/s				
50°C (min)	15	110	110	110
50°C (max)	110	–	–	–
100°C (max)	–	40	40	40
Flash point, °C, min	70	70	70	70
Distillation (% max)				
250°C	65	65	65	65
350°C	85	85	85	85
Maximum sulfur content, wt. %	2	4	2	1
Maximum water content, wt. %	0.75	1.5	1.5	1.5
Maximum insolubles, wt. %	0.25	0.25	0.25	0.25

Table
5.19   *Specifications for heavy fuels in France.*
BTS: Low sulfur content.   TBTS: Very low sulfur content.

5.1.7.2 Main Characteristics of Heavy Fuels

a. Density of Heavy Fuels

The density of heavy fuels is greater than 0.920 kg/l at 15°C. The marine diesel consumers focus close attention on the fuel density because of having to centrifuge water out of the fuel. Beyond 0.991 kg/l, the density difference between the two phases —aqueous and hydrocarbon— becomes too small for correct operation of conventional centrifuges; technical improvements are possible but costly. In extreme cases of fuels being too heavy, it is possible to rely on water-fuel emulsions, which can have some advantages of better atomization in the injection nozzle and a reduction of pollutant emissions such as smoke and nitrogen oxides.

b. Viscosity of Heavy Fuels

The heavy fuel should be heated systematically before use to improve its operation and atomization in the burner. The change in kinematic viscosity with temperature is indispensable information for calculating pressure drop and setting the preheating temperature. Table 5.20 gives examples of viscosity required for burners as a function of their technical design.

Atomization method	Required viscosity, mm^2/s
Mechanical Mechanical with air at low pressure	15 to 20
Mechanical with steam assistance	20 to 25
Rotating tip	60 to 70

Table 5.20	*Average kinematic viscosity required for the atomization of heavy fuels.*

Examination of the diagram in Figure 5.15 enables the temperature range to be found for various No. 2 fuel oils.

c. Heating Values of Heavy Fuels

The high C/H ratio for heavy fuels and their high levels of contaminants such as sulfur, water, and sediment, tend to reduce their NHV which can reach as low as 40,000 kJ/kg by comparison to the 42,500 kJ/kg for a conventional home-heating oil. This characteristic is not found in the specifications, but it is a main factor in price negotiations for fuels in terms of cost per ton. Therefore it is subject to frequent verification.

There are formulas for determining the NHV$_w$ of heavy fuels as function of their density and their sulfur content. The simplest is the following:

$$NHV_w = 55.5 - 14.4\,\rho - 0.32\,S$$

where the NHV$_w$ is expressed in MJ/kg, ρ is the density in kg/l at 15°C, and S is the sulfur content in wt %.

d. Conradson Carbon and Asphaltenes Content of Heavy Fuels

The Conradson Carbon of a heavy fuel can often reach 5 to 10%, sometimes even 20%. It is responsible for the combustion quality, mainly in rotating tip atomizing burners.

The asphaltene content is found either directly by precipitation using n-heptane (NF T 60-115 or ASTM D 32), or indirectly by correlation with the Conradson Carbon. It can vary from 4 or 5% to as much as 15 or 20% in extreme cases.

A high level of asphaltenes or high Conradson Carbon always causes problems in combustion (Feugier et al., 1985). In engines this tendency is counteracted by technological means such as increasing the injection pressure. In industrial burners, the quantities of unburned solids, called weighted indices and expressed in mg/m^3 of smoke emitted, are always correlated with the asphaltene content and the Conradson Carbon (Figure 5.16).

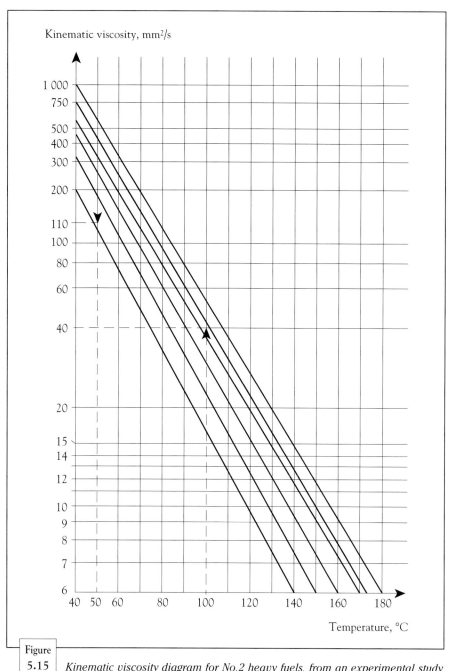

Figure 5.15 *Kinematic viscosity diagram for No.2 heavy fuels, from an experimental study of BNPét (Bureau de Normalisation du Pétrole, 1978).*

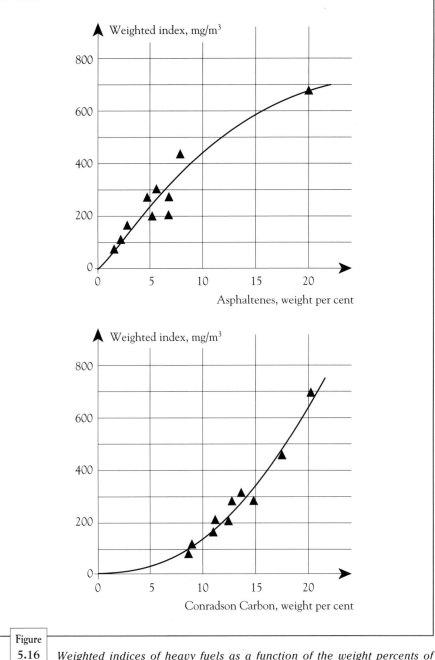

Figure
5.16 *Weighted indices of heavy fuels as a function of the weight percents of asphaltenes and Conradson Carbon.*
Measurements of weighted indices are conducted under normal conditions of temperature and pressure (25°C, atmospheric pressure).
The furnace output is 0.1 MW and excess air is 3 wt %.

e. Presence and Effect of Contaminants in Heavy Fuels

Emission problems of SO_2 and NO_x, linked to the presence of sulfur and nitrogen in heavy fuels will be examined later.

Other heavy fuel contaminants are metals (vanadium, nickel, sodium) coming from the crude oil itself or metallic salts (aluminum silicates) coming from catalysts in conversion steps. The aluminum silicates should not exceed 300 ppm (30 ppm of aluminum), for these materials exert a strong abrasive action on the engine cylinders and injection systems. They can however be eliminated partially by centrifuging and filtration.

Vanadium present in the crude oil is concentrated in the heavy fuels where levels of 200-300 ppm are possible. During combustion, this metal forms salt complexes with sodium, also present in trace quantities, which can be very corrosive when they deposit as liquids. To control this problem, it is sometimes necessary to design engine cooling systems that reduce the temperature of the exhaust valve seats, keeping it below the melting point of the vanadium and sodium salts. In industrial furnaces, combustion is frequently carried out with low excess air which leads to a higher vanadium salt melting point. That is usually adequate for avoiding corrosion.

f. Cetane Number of Heavy Fuels Used in Diesel Engines

The cetane number of heavy fuels is generally between 30 and 40, but its determination is inexact considering the difficulties encountered when testing such products in a CFR engine. One possible way to operate is to dilute the fuel with diesel fuel or home-heating oil and estimate a blending index. There are also formulas to predict the auto-ignition qualities of heavy fuels using their physical characteristics (Fiskaa et al., 1985). We cite as an example the relation giving the CCAI (Calculated Carbon Aromaticity Index):

$$CCAI = \rho_{15} - 140.7 \log \left[\log \left(\nu_{50} + 0.85 \right) \right] - 80.6$$

where ν_{50} represents the kinematic viscosity at 50°C, mm²/s, and ρ_{15} the density at 15°C in kg/l.

The auto-ignition delay of a heavy fuel measured in the engine increases linearly with the CCAI; it is therefore desirable that the latter value be as low as possible.

One other characteristic linked to the quality of combustion is called the CII (Calculated Ignition Index) and is expressed by the equation:

$$CII = 294.26 - 277.30 \, \rho_{15} + 13.263 \log \left[\log \left(\nu_{100} + 0.7 \right) \right]$$

where

ν_{100} = kinematic viscosity at 100°C, mm²/s

ρ_{15} = density at 15°C, kg/l

5.1.7.3 **Formulation of Heavy Fuels**

In the 1970's, heavy fuel came mainly from atmospheric distillation residue. Nowadays a very large proportion of this product is vacuum distilled and the distillate obtained is fed to conversion units such as catalytic cracking, visbreaking and cokers. These produce lighter products — gas and gasoline — but also very heavy components, that are viscous and have high contaminant levels, that are subsequently incorporated in the fuels.

Table 5.21 gives the characteristics of stocks used today to formulate heavy fuels. A few of these are strongly downgraded compared with their homologs of 1970s.

Characteristics	Atmospheric residue (Arabian Light)	Vacuum residue (VR)	Visbroken residue (on VR)	LCO (low sulfur)	HCO (low sulfur)
Kinematic viscosity at 100°C mm²/s	46	1500	7500	1	1.7
Density kg/l	0.972	1.026	1.040	0.899	0.942
Conradson carbon weight %	10.8	18.9	26.9	< 0.1	< 0.1
Asphaltenes weight %	3.2	6.4	16.6	0	0
Sulfur weight %	3.4	4.5	4.8	0.5	1.0
Nitrogen ppm	2500	3000	3500	370	1050
Metals ppm	65	215	257	0	0

Table 5.21 | *Typical characteristics of some refinery stocks used in the production of heavy fuels.*

For the refiner, the main problem is to meet the specifications for kinematic viscosity and sulfur content. Dilution by light streams such as home-heating oil and LCO, and selection of feedstocks coming from low-sulfur crude oils give him a measure of flexibility that will nevertheless lead gradually to future restrictions, most notably the new more severe anti-pollution rules imposing lower limits on sulfur and nitrogen contents.

In the future it will be difficult to avoid deterioration of certain characteristics such as viscosity, asphaltene and sediment contents, and cetane number. The users must employ more sophisticated technological means to obtain acceptable performance. Another approach could be to diversify the formulation of heavy fuel according to end use. Certain consuming plants require very high quality fuels while others can accept a lower quality.

5.2 Properties Related to Storage and Distribution of Petroleum Products

The properties linked to storage and distribution do not directly affect the performance of engines and burners, but they are important in avoiding upstream incidents that could sometimes be very serious. We will examine in turn the problems specific to gasoline, diesel fuel, jet fuel and heavy fuel.

5.2.1 Problems Related to the Storage and Distribution of Gasoline

5.2.1.1 Oxidation Tendency of Gasoline

In the presence of oxygen even at ambient temperature, hydrocarbons can undergo a process of deterioration by oxidation that can form viscous substances commonly called gums. This can cause a variety of undesirable incidents: blockage of engine fuel pump membranes, plugging of needle valves or injectors, sticking of the carburetor level control mechanism or even sticking (or gumming up) of piston rings in their grooves. Anti-oxidant additives of the alkyl-p-phenylenediamine or alkyl p-amino-phenol families are added to the gasoline in concentrations of 10 to 20 ppm as soon as it is produced in the refinery to avoid these problems. However, several methods exist to verify the quality of the finished products.

The "existing" gum content (NF M 07-004) is obtained by evaporating a 50 cm^3 fuel sample placed in a constant temperature bath at 160°C while subjected to a stream of air for 30 minutes. After evaporation, the residue is weighed giving the "unwashed" gum content. An extraction with heptane follows which leaves only the "existing" gums. The French specification concerns existing gums after washing and sets the upper limit at 10 mg/100 ml.

This procedure indicates the presence or lack of detergents in the fuel. In fact, when the fuel has additives, a large difference is observed between the weight of deposits before and after washing.

The method for determining potential gums (NF M 07-013) consists of artificially promoting the oxidation to attempt to simulate conditions of prolonged storage. The gums are then recovered by filtration after a storage period of 16 hours at 100°C and under an oxygen pressure of 7 bar. The potential gums content has no official specification; it is considered only as an indication. Moreover, its significance becomes very doubtful for fuels containing oxygenates such as alcohols and ethers.

Finally, the determination of the induction period (NF M 07-012) also reveals the potential of gum formation during storage. The fuel sample is contained in a bomb filled with oxygen at 100°C, under a pressure of 7 bar and the oxygen pressure is monitored with time. The time corresponding to the first drop in pressure is noted, symptomatic of incipient oxidation. If no further events take place, the test is stopped after 960 minutes. This time corresponds thus to the maximum induction period.

The European specification for unleaded gasoline has set a minimum induction period of 360 minutes.

5.2.1.2 **Fouling Tendency of Gasoline Engine Components**

Gasoline engine equipment such as carburetors, injectors, intake manifolds, valve systems and combustion chambers, are subject to fouling by the fuel itself, the gases recycled from the crankcase, or even dust and particulates arriving with poorly filtered air. Three types of problems then result:

- fuel system misadjustment resulting in unstable operation
- higher fuel consumption and pollutant emissions, particularly when idling
- Octane Requirement Increase (ORI).

To avoid these problems, refiners commonly use additives called "detergents" (Hall et al., 1976), (Bert et al., 1983). These are in reality surfactants made from molecules having hydrocarbon chains long enough to ensure their solubility in the fuel and a polar group that enables them to be absorbed on the walls and prevent deposits from sticking. The most effective chemical structures are succinimides, imides, and fatty acid amines. The required dosages are between 500 and 1000 ppm of active material.

The additives capable of controlling the octane requirement increase, have as far as they are concerned, a complex structure and are closely guarded industrial secrets.

There are no official specifications for obtaining a minimum level of engine cleanliness from a fuel. However, all additives in France are subject to approval by the *Direction des Carburants (DHYCA),* with the objective of having data that prove, first of all, the product to be harmless, and second, the product's effectiveness. Likewise, the automotive manufacturers, in establishing their specifications, set the minimum performance to be obtained by the fuel with regard to engine cleanliness.

Table 5.22 indicates the test references to be conducted on engines according to procedures established by the *Coordinating European Council (CEC).* Note that these are long-duration tests, some lasting dozens or even hundreds of hours and they are costly.

5.2.1.3 **Water Tolerance of Gasoline**

In a conventional gasoline containing hydrocarbons or even ethers, the presence of water is not a problem; in fact, water is totally soluble up to about 50 ppm at ambient temperature. Beyond this value water separates without affecting the hydrocarbon phase and the "water leg" can be withdrawn if necessary. On the other hand, in the presence of alcohols (ethanol and especially methanol), trace amounts of water can cause a separation of two phases: one is a mixture of water and alcohol, the other of hydrocarbons (Cox, 1979).

A. For Gasoline

Desired property	Test reference	Type of engine	Test duration, hours
Carburetor cleanliness	CEC F 02 T 81	Renault R5	12
Injector cleanliness	GFC	Peugeot 205	200
Admission valve cleanliness	CEC F 05 T 92	Mercedes M 102 E	60
Combustion chamber cleanliness	CEC PF 28	Renault F2N	400

B. For Gasoline Fuel

Desired property	Test reference	Type of engine	Test duration, hours
Injector cleanliness	CEC PF 26	Peugeot XU D9	6
Longevity	Peugeot	Peugeot XU7 DE	300

Table
5.22 | *Tests conducted on motors to verify the harmlessness and effectiveness of additives.*

The tendency to separate is expressed most often by the cloud point, the temperature at which the fuel-alcohol mixture loses its clarity, the first symptom of insolubility. Figure 5.17 gives an example of how the cloud-point temperature changes with the water content for different mixtures of gasoline and methanol. It appears that for a total water content of 500 ppm, that which can be easily observed considering the hydroscopic character of methanol, instability arrives when the temperature approaches $0°C$. This situation is unacceptable and is the reason that incorporating methanol in a fuel implies that it be accompanied by a cosolvent. One of the most effective in this domain is tertiary butyl alcohol, TBA. Thus a mixture of 3% methanol and 2% TBA has been used for several years in Germany without noticeable incident.

When ethanol is present, the risk of separation is much less than with methanol. Nevertheless, the ethanol should be relatively anhydrous (less than 3000 ppm water); moreover, if a fuel containing ethanol comes in contact with a water layer, a migration of ethanol toward the water is observed creating a fuel quality problem manifested by lower octane number and an environmental quality problem in that the water will need to be treated. Distribution of ethanol-based fuels requires extra precaution to ensure dryness in distribution systems.

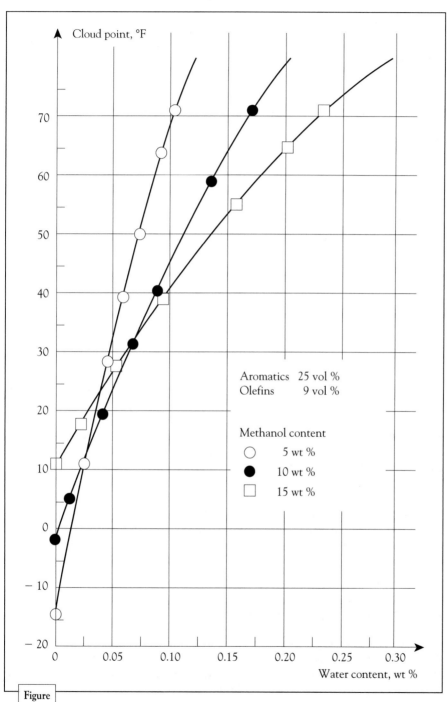

Figure
5.17 *Stability of a methanol-fuel mixture versus water content.*

5.2.1.4 **Gasoline Evaporative Losses**

Hydrocarbon losses through evaporation are inevitable in spite of all the preventive steps that are or will be employed. Vapor recovery systems are obligatory in all fuel storage operations and service station systems ("Stage 1"). These measures will soon extend to filling vehicle fuel tanks ("Stage 2"). Furthermore, new gasoline automobiles throughout Europe will be equipped with fuel tank vapor traps beginning the 1 January 1993. They are activated carbon canisters that trap and store the volatile hydrocarbons when the vehicle is stationary. When the vehicle is moving, the canisters are swept with air and the vapors are recovered as fuel. However this technique is not completely effective and needs to be complemented by very strict control of the fuel's vapor pressure; a study conducted in the United States shows that for vehicles equipped with canisters, a reduction of 1 psi (69 mbar) in the vapor pressure causes a 46% reduction in evaporation for stationary cold vehicles and a 9% reduction for vehicles still stationary but after a period of warm operation.

Among the compounds susceptible to evaporation, particular attention is focused on benzene. In the two conditions indicated above, for equal benzene contents in the fuel (1.5% volume), the benzene evaporative losses are reduced by 21% and 11%, respectively, when the vapor pressure decreases by 1 psi, that is, 69 mbar.

These data, given for illustrative purposes, show that even for the coming years the refiner will be pressed to reduce the gasoline vapor pressure as much as possible.

5.2.2 **Precautions to Observe for Diesel Fuel Use**

The most important point in the use of diesel fuel is its cold temperature behavior. The subject has been addressed previously because it directly affects the engine operation in winter conditions.

We will examine here other quality criteria having less impact yet still being important.

5.2.2.1 **Diesel Fuel Stability in Storage and Use**

Changes in diesel fuel between its formulation in the refinery and its distribution result from various chemical reactions leading to the formation of gums and sediment. This is particularly noticeable in diesel fuels containing stocks from conversion processes such as catalytic cracking, visbreaking, and coking with traces of nitrogen compounds that cause or favor the chemical degradation reactions.

Two criteria are used to characterize the behavior of diesel fuel in this area; these are color and resistance to oxidation.

a. Diesel Fuel Color

The tendency of the color to become darker with time is often indicative of chemical degradation. The test is conducted with the aid of a colorimeter (NF T 60-104 and ASTM D 1500) and by comparison with colored glass standards. The scale varies from 0.5 to 8. The French specifications stipulate that diesel fuel color should be less than 5, which corresponds to an orange-brown tint. Generally, commercial products are light yellow with indices from 1 to 2.

b. Resistance of Diesel Fuel to Oxidation

The procedure most commonly employed (NF M 07-047 or ASTM D 2274) is to age the diesel fuel for 16 hours while bubbling oxygen into it at 95°C. The gums and sediment obtained are recovered by filtration and weighed. There is no official French specification regarding oxidation stability; however, in their own specifications, manufacturers have set a maximum value of 1.5 mg/100 ml.

5.2.2.2 Injector Fouling Tendency in Diesel Engines

For passenger prechamber diesel engines, carbonaceous deposits can form in the pintle nozzle injectors. The deposits come from thermal cracking of the diesel fuel, either accompanied or not by slow oxidation (Montagne et al., 1987). This phenomenon is shown by removing the injectors and measuring the air flow likely to pass through as a function of needle lift as shown in Figure 5.18.

After operating for several hours, a flow deviation is still observed that can be considered as acceptable because the automotive manufacturer has taken it into account during the development of the engine. On the other hand, a more pronounced trend such as that shown in Figure 5.18 would be unacceptable because it will affect the engine noise, driving comfort and pollutant emissions.

The level of injector fouling is most often illustrated in terms of residual flow (RF) expressed as a percentage of the flow under new conditions for a given needle lift. An RF on the order of 20% for a lift of 0.1 mm is a good compromise. This level may not be achieved with certain aromatic or naphthenic diesel fuels. The best recourse is then detergent additive addition.

Figure 5.19 gives a normal increase in RF as a function of the additive content; it is then possible to adjust the dosage in accordance of the required quality criterion.

In order to meet French manufacturers' fuel quality specifications, a diesel fuel should contain a detergent additive whose effectiveness will have been demonstrated by the procedure described previously.

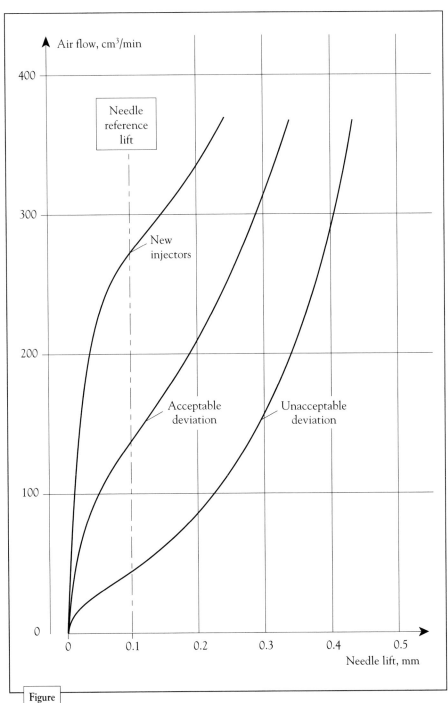

Figure 5.18   *Flow measurements for diesel injectors under a pressure drop of 0.6 bar.*

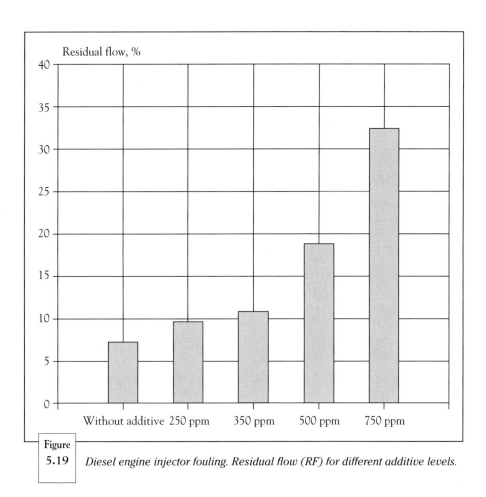

Figure 5.19   *Diesel engine injector fouling. Residual flow (RF) for different additive levels.*

5.2.2.3 **Diesel Fuel Safety as Characterized by its Flash Point**

The flash point of a petroleum liquid is the temperature to which it must be brought so that the vapor evolved burns spontaneously in the presence of a flame. For diesel fuel, the test is conducted according to a "closed cup" technique (NF T 60-103). The French specifications stipulate that the flash point should be between 55°C and 120°C. That constitutes a safety criterion during storage and distribution operations. Moreover, from an official viewpoint, petroleum products are classified in several groups according to their flash points which should never be exceeded.

The flash point depends closely on the distillation initial point. The following empirical relationship is often cited:

$$FP = IP - 100$$

where the flash point (FP) and the initial point (IP) of the diesel fuel are in °C.

It should be noted finally that adding gasoline to diesel fuel which was sometimes recommended in the past to improve cold behavior conflicts with the flash point specifications and presents a serious safety problem owing to the presence of a flammable mixture in the fuel tank airspace. Adding a kerosene that begins to boil at 150°C does not have the same disadvantage from this point of view.

5.2.3 **Problems Related to Storage and Distribution of Jet Fuel**

Jet fuel is subject to particular attention in all operations that precede and accompany its use in accordance with the draconian air transport safety regulations (Anon., 1983).

5.2.3.1 **Elimination of Water Traces in Jet Fuel**

Leaving the refinery, jet fuel has generally no free water and contains only a small quantity of dissolved water. But humidity from the air and tank breathing result in continuous intrusion of water that must be then removed by decanting and filtration. This is why jet fuel needs to be tested for its ability to separate the contained water.

A first approach to testing, ASTM D 1094, is to create, using a potassium phosphate reagent, a separation between two layers, hydrocarbon and aqueous. The degree of separation of the two phases is estimated by attributing a grade from 1 to 3 and the appearance of the interface by five levels of observation: 1, 1b, 2, 3, and 4. The specifications establish both the quality of separation (2 is the maximum) and the appearance of the interface (1b maximum).

Another method is available for measuring a jet fuel's ability to disengage water is to pass it through a coalescing medium. The method is called the Water Separation Index Modified (WSIM) described in ASTM D 2550. A water and jet fuel emulsion is sent through a glass fiber coalescer. The turbidity of the solution is measured at the outlet by the amount of light transmitted. The WSIM is a number between 0 and 100, corresponding to an increasing tendency to reject water. The specifications for Jet A1 establish a minimum WSIM of 85. A similar method exists, ASTM D 3948, which employs a portable instrument enabling measurements to be taken in the field. The index obtained is called Micro Separometer Surfactants (MSEP); it is related to the WSIM by the empirical relation:

$$MSEP = 0.6\,WSIM + 40$$

For jet fuels, the elimination of free water using filters and coalescers by purging during storage, and the limit of 5 ppm dissolved water are sufficient to avoid incidents potentially attributable to water contamination: formation of micro-crystals of ice at low temperature, increased risk of corrosion, growth of micro-organisms.

However, of all the petroleum products, jet fuel is the one receiving the most careful scrutiny.

5.2.3.2 **Corrosion Protection during Utilization of Jet Fuel**

Corrosion protection is indispensable, especially concerning certain vulnerable parts of the aircraft such as the combustion chamber and turbine. The potential hazards are linked to the presence of sulfur in various forms: mercaptans, hydrogen sulfide, free sulfur, and sulfides.

Until 1992, the total sulfur content of jet fuel was limited to 0.2 wt. %. Starting in 1993, a reduction to 0.1% was instituted apparently without major incident since for commercial products, lower levels (to 500 ppm) had been observed very often.

It is noteworthy, however, that traces of sulfur can have beneficial effects on the anti-wear resistance of fuel injection pumps. It is thus undesirable to reduce the sulfur content to extremely low values unless additives having lubricating qualities are added. Independently from total sulfur content, the presence of mercaptans that are particularly aggressive towards certain metal or synthetic parts is strictly controlled. The mercaptan content is thereby limited to 0.002% (20 ppm) maximum. The analysis is performed chemically in accordance to the NF M 07-022 or ASTM D 3227 procedures.

The Doctor Test, NF M 07-029 or ASTM D 484, is conducted in qualitative tests for free sulfur, free H_2S, and mercaptans. The specifications require a negative Doctor Test for these three types of contaminants.

Finally, other tests to control jet fuel corrosivity towards certain metals (copper and silver) are used in aviation. The corrosion test known as the "copper strip" (NF M 07-015) is conducted by immersion in a thermostatic bath at 100°C, under 7 bar pressure for two hours. The coloration should not exceed level 1 (light yellow) on a scale of reference. There is also "the silver strip" corrosion test (IP 227) required by British specifications (e.g., Rolls Royce) in conjunction with the use of special materials. The value obtained should be less than 1 after immersion at 50°C for four hours.

5.2.3.3 **Electrical Conductivity of Jet Fuel**

An fuel-air mixture explosion can be initiated by a sudden discharge of static electricity. Yet, while flowing in systems, a fluid develops an electrical charge which will take as long to dissipate as the fluid is a poor conductor. The natural electrical conductivity of jet fuel is very low, on the order of a few picosiemens per meter, and it decreases further at low temperature.

It is believed that to avoid any risk of explosion, the electrical conductivity of jet fuel should fall between 50 and 450 pS/m. This level is attained using anti-static additives which are metallic salts (chromium, calcium) added at very low levels on the order of 1 ppm.

5.2.4 **Problems Related to Storage of Heavy Fuels**

One of the most important problems in using heavy fuels is that of the incompatibility during mixing operations of products coming from different sources. This results in the more or less rapid precipitation and agglomeration of certain components resembling muds that can block flow channels and filters. A standard test, ASTM D 27, is designed to estimate the degree of compatibility of a heavy fuel. The sample is diluted with diesel fuel and a drop of the mixture is allowed to fall on a filter paper. Examination of the circular stain, more specifically the concentric development of the stain around its axis has resulted in the establishment of a diagnostic tool and an acceptable threshold limit. In the refinery, the petroleum company can resort to this test to prepare products that are sufficiently compatible.

5.3 **Motor Fuels, Heating Fuels and Environmental Protection**

The idea of "clean" motor and heating fuels, that is, those having an improved impact on the environment, has been developing since the beginning of the 1980s, first in the United States. It has since then appeared in Europe and will most certainly have its impact in the rest of the world beyond the year 2000.

5.3.1 **Justification for Deep Desulfurization**

In the past, reducing the sulfur content was mainly concerned with the heaviest products, most particularly the fuel oils. This development is explained by a legitimate concern to reduce SO_2 emissions, notably in areas around large population centers. This is how low sulfur heavy fuels —having a maximum of 2% sulfur— and very low sulfur (1% sulfur) came into being. Currently the whole range of petroleum products, particularly motor fuels, should be strongly desulfurized for reasons we will explain hereafter.

5.3.1.1 **Sulfur Content in Gasolines**

The French specification for sulfur in all types of gasolines —regular, premium, with or without lead— is 0.1% maximum, that is, 1000 ppm. This value is easily achieved because in the majority of commercial products, the content is less than 500 ppm.

However, such a level can still be considered too high for vehicles having 3-way catalytic converters. In fact, results observed in the United States (Benson et al., 1991) and given in Figure 5.20 show that exhaust pollutant emissions, carbon monoxide, hydrocarbons and nitrogen oxides, increase from 10 to 15% when the sulfur level passes from 50 ppm to about 450 ppm. This is explained by an inhibiting action of sulfur on the catalyst; though

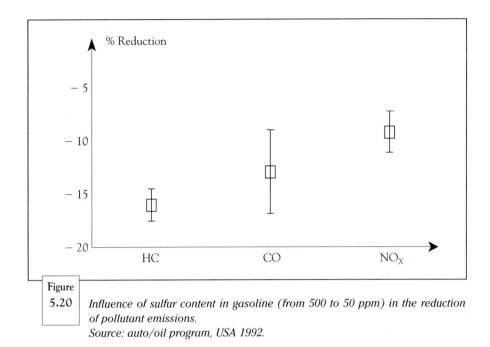

Figure
5.20

Influence of sulfur content in gasoline (from 500 to 50 ppm) in the reduction of pollutant emissions.
Source: auto/oil program, USA 1992.

reversible, i.e., it is capable of disappearing when a sulfur-free gasoline is used again, this deterioration should be avoided. For this reason the automotive industry insists on deep desulfurization of the gasoline pool in Europe as well as in the United States. The French automotive manufacturers have set, in their "quality" specifications a maximum sulfur content of 300 ppm, with an average of ten samples below 200 ppm. Eventually, sulfur levels from 50 to 100 ppm in the gasolines could be required by car manufacturers.

There is another problem linked to the presence of sulfur in gasoline: the potential emission of trace amounts of hydrogen sulfide (H_2S) in the exhaust of vehicles having catalytic converters and operating under the transitory conditions of a rich mixture, for example, during cold starting. Sulfur stored on the catalyst is re-emitted in the form of H_2S. This phenomenon is undesirable more because of the bad odor than the danger it presents. However, it should be avoided for better driving comfort.

5.3.1.2 **Sulfur Content in Diesel Fuel**

We have previously stated that the sulfur content of diesel fuel will be limited in Europe to 0.2% as of 1 October 1994 and to 0.05% as of 1 October 1996.

These developments will reduce the total emissions of SO_2 but the effect will remain limited. The reduction in sulfur levels in diesel fuel from 0.2 to 0.05% is admitted to reduce yearly emissions of SO_2 in France to only 10 to 12%.

The main justification for diesel fuel desulfurization is related to particulate emissions which are subject to very strict rules. Part of the sulfur is transformed first into SO_3, then into hydrated sulfuric acid on the filter designed to collect the particulates. Figure 5.21 gives an estimate of the variation of the particulate weights as a function of sulfur content of diesel fuel for heavy vehicles. The effect is greater when the test cycle contains more high temperature operating phases which favor the transformation of SO_2 to SO_3. This is particularly noticeable in the standard cycle used in Europe (ECE R49).

In any and all cases, desulfurization of diesel fuel is a necessary condition for attaining very low particulate levels such as will be dictated by future regulations (Girard et al., 1993).

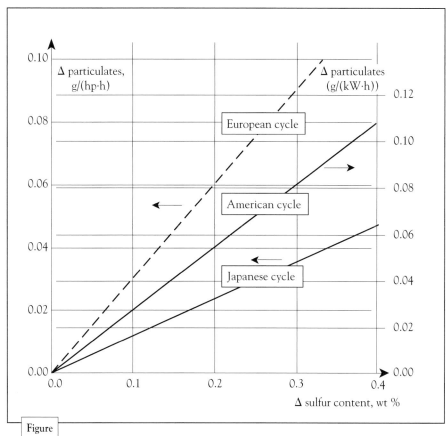

Figure
5.21

Effect of sulfur content of diesel fuel on particulate emissions according to different tests.

Desulfurization will become mandatory when oxidizing catalysts are installed on the exhaust systems of diesel engines. At high temperatures this catalyst accelerates the oxidation of SO_2 to SO_3 and causes an increase in the weight of particulate emissions if the diesel fuel has not been desulfurized. As an illustrative example, Figure 5.22 shows that starting from a catalyst temperature of 400°C, the quantity of particulates increases very rapidly with the sulfur content.

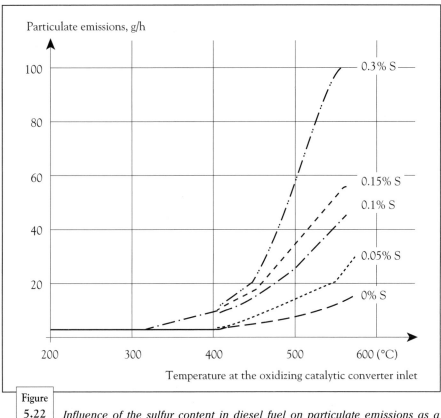

Figure
5.22   *Influence of the sulfur content in diesel fuel on particulate emissions as a function of the catalytic converter inlet temperature.*
Source: Peugeot.

Finally, sulfur has a negative effect on the performance of the catalyst itself. One sees for example in Figure 5.23 that the initiation temperature increases with the sulfur level in the diesel fuel, even between 0.01% and 0.05%. Yet, in the diesel engine, characterized by relatively low exhaust temperatures, the operation of the catalyst is a determining factor. One can thus predict an ultimate diesel fuel desulfurization to levels lower than 0.05%.

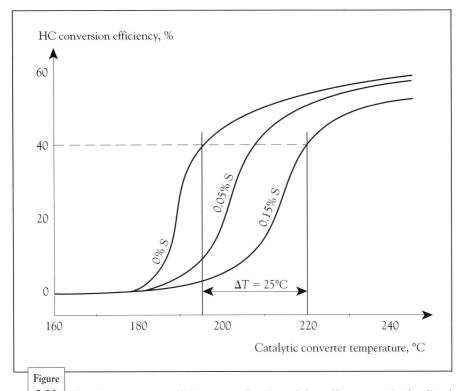

Figure
5.23 *Catalytic converter efficiency as a function of the sulfur content in the diesel fuel and the temperature.*

5.3.1.3 **Sulfur Content of Heavy Fuel**

The European regulations have set SO_2 emission limits for industrial combustion systems. They range from 1700 mg/Nm^3 for power generation systems of less than 300 MW and to 400 mg/Nm^3 for those exceeding 500 MW; between 300 and 500 MW, the requirements are a linear interpolation (Figure 5.24). To give an idea how difficult it is to meet these requirements, recall that for a fuel having 4% sulfur, the SO_2 emissions in a conventional boiler are about 6900 mg/Nm^3; this means that a desulfurization level of 75% will be necessary to attain the SO_2 content of 1700 mg/Nm^3 and a level of 94% to reach 400 mg/Nm^3.

There are, however, technological means available to burn incompletely desulfurized fuels at the same time minimizing SO_2 emissions. In the auto-desulfurizing AUDE boiler developed by IFP, the effluent is treated in place by an absorbent based on lime and limestone: calcium sulfate is obtained. This system enables a gas desulfurization of 80%; it requires nevertheless a relatively large amount of solid material, on the order of 200 kg per ton of fuel.

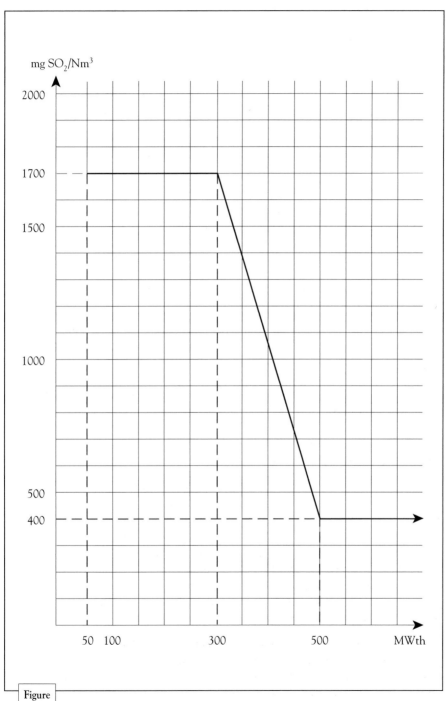

Figure
5.24   *Emission limits of sulfur dioxide for heating oils (heavy fuels essentially).*
JOCE N° L336/9 of 7.12.88.

5.3.2 Influence of the Chemical Composition of Motor Fuels and Heating Oils on the Environment

There always is a relation between fuel composition and that of hydrocarbon emissions to the atmosphere, whether it concerns hydrocarbon emissions from evaporative losses from the fuel system, or from exhaust gases. This is the reason that environmental protection regulations include monitoring the composition of motor and heating fuels. We will describe here the regulations already in existence and the work currently underway in this area with its possible effects on refining.

5.3.2.1 Benzene Content in Gasoline

A European Directive, 85/210/EEC, limits benzene content to 5% by volume in all gasolines: regular, premium, with or without lead. This level is easily achieved, since the average value in 1993 was less than 3%. In France, for example, average benzene concentrations of 1.7% and 2.6% were reported for leaded and unleaded premium fuels, respectively, in 1993.

This rule is justified by the need to limit the benzene emissions from evaporation (Tims, 1983); Figure 5.25 shows that emissions increase linearly with the benzene content of the fuel. It is noteworthy that current legislation limits the measured evaporation to 2 g per test conducted in accordance with a standard procedure (Sealed Housing for Evaporative Determination, or SHED). Yet for a fuel containing 5% benzene, an evaporation of 0.7 g benzene /test is observed.

Around 2000, the regulations should become more severe. In this area, a European limit of benzene of 3% appears very probable; certain countries such as Germany are even looking at 1%. In Italy, it was decided towards the end of 1991, to limit benzene to 2.5% for leaded and unleaded fuels in the seven largest cities characterized by having heavy atmospheric pollution; concurrently, in these same cities, the overall aromatic contents of gasolines should not exceed 33%.

For the refiner, the reduction in benzene concentration to 3% is not a major problem; it is achieved by adjusting the initial point of the feed to the catalytic reformers and thereby limiting the amount of benzene precursors such as cyclohexane and C_6 paraffins. Further than 3% benzene, the constraints become very severe and can even imply using specific processes: alkylation of benzene to substituted aromatics, separation, etc.

5.3.2.2 Relations between Gasoline Composition and Pollutant Emissions

The implementation of very effective devices on vehicles such as catalytic converters makes extremely low exhaust emissions possible as long as the temperatures are sufficient to initiate and carry out the catalytic reactions; however, there are numerous operating conditions such as cold starting and

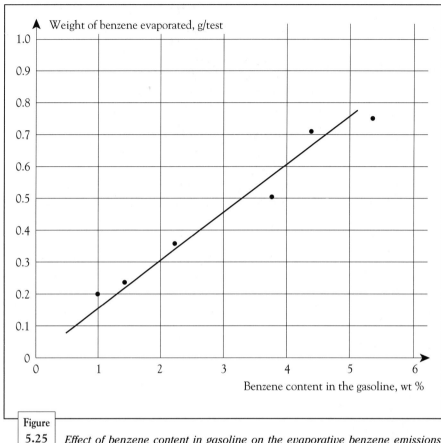

Figure
5.25 *Effect of benzene content in gasoline on the evaporative benzene emissions during the SHED standardized test.*

acceleration for which the catalyst is not wholly effective. It is then necessary to find the fuel characteristics that can minimize the emissions; we will consider here three themes that are currently under serious study.

a. "Conventional" Pollutants

These are carbon monoxide, CO, unburned hydrocarbons (HC), and the nitrogen oxides, NO_x. In the U.S.A., a program called Auto/Oil (Burns et al., 1992), conducted by automotive manufacturers and petroleum companies, examined the effect of overall parameters of fuel composition on evaporative emissions and in the exhaust gases. The variables examined were the aromatics content between 20 and 45%, the olefins content between 5 and 20%, the MTBE content between 0 and 15% and finally the distillation end point between 138 and 182°C (more exactly, the 95% distilled point).

Table 5.23 gives the results obtained on the American automotive fleet. The pollutant emissions attributable to one or another of the parameters stated above does not generally exceed 10 to 15%. However, certain tendencies merit attention; for example, the presence in the fuel of an oxygen compound like MTBE contributes to reducing the CO and hydrocarbon emissions; a reduction in the aromatics content goes equally in the same direction. This work has led to the concept of "reformulated fuel" in the United States, that is presenting physical-chemical characteristics adapted to minimizing the pollutant emissions. We will go more deeply into the idea of reformulated fuel in the following pages.

Parameter studied		Action on the level of pollutants, %		
		CO	HC	NO$_x$
Aromatics	(45 → 20%)	−13.6%	−6.3%	2.1%
Olefins	(20 → 5%)	1.5%	6.1%	−6.0%
MTBE	(0 → 15%)	−11.2%	−5.1%	1.4%
90% distillation point	(182 → 138°C)	0.8%	−21.6%	5.0%

Table 5.23

Influence of the chemical composition of the fuel on pollutant emissions from vehicles in the US (auto/oil program).

b. Specific Pollutants

Outside of carbon monoxide for which the toxicity is already well-known, five types of organic chemical compounds capable of being emitted by vehicles will be the focus of our particular attention; these are benzene, 1-3 butadiene, formaldehyde, acetaldehyde and polynuclear aromatic hydrocarbons, PNA, taken as a whole. Among the latter, two, like benzo [a] pyrene, are viewed as carcinogens. Benzene is considered here not as a motor fuel component emitted by evaporation, but because of its presence in exhaust gas (see Figure 5.25).

Table 5.24 shows that these specific pollutants are present only in small proportions (about 8%) of the total organic compounds emitted by the motor, but they are particularly feared because of their incontestable toxicity. Prominent among them is benzene.

The action taken to reduce the level of these toxic products is to modify the formulation of the fuel. Table 5.25 gives some general trends brought to light during the Auto/Oil American program: the reduction already cited in the CO emissions by employing oxygenated fuels, decreasing benzene emissions with low aromatic fuels or by having a low distillation end point, the relationship with olefins in the fuel and the formation of butadiene in the

Type of pollutant	Weight % of volatile organic compounds contained in exhaust gas	Relative risk factor
Benzene	6.2%	1.000
Butadiene	0.5%	5.290
Formaldehyde	0.9%	0.243
Acetaldehyde	0.5%	0.043

Table
5.24   *Evaluation of the concentrations of four toxic pollutants in exhaust gas (order of magnitude).*

Parameter studied	Recorded effect, %				
	Benzene	Butadiene	Formaldehyde	Acetaldehyde	CO
Aromatics (45 → 20 %)	−42	≈0	+24	+21	−13
Olefins (20 → 5 %)	≈0	−30	≈0	≈0	≈0
MTBE (0 → 15 %)	≈0	≈0	+26	≈0	−11
90% Distillation Point (182 → 138 °C)	−10	−37	−27	−23	≈0

Table
5.25   *Influence of the chemical composition of fuels on emissions of toxic materials (auto/oil program).*

exhaust, coupling of MTBE–formaldehyde. Reducing the distillation end point always acted favorably on the emissions of each of the toxic products identified above.

c. Formation of Tropospheric Ozone

Ozone, known for its beneficial role as a protective screen against ultra-violet radiation in the stratosphere, is a major pollutant at low altitudes (from 0 to 2000 m) affecting plants, animals and human beings. Ozone can be formed by a succession of photochemical reactions that preferentially involve hydrocarbons and nitrogen oxides emitted by the different combustion systems such as engines and furnaces.

More precisely, the rate of ozone formation depends closely on the chemical nature of the hydrocarbons present in the atmosphere. A reactivity scale has been proposed by Lowi and Carter (1990) and is largely utilized today in ozone prediction models. Thus the values indicated in Table 5.26 express the potential ozone formation as O_3 formed per gram of organic material initially present. The most reactive compounds are light olefins, cycloparaffins, substituted aromatic hydrocarbons notably the xylenes, formaldehyde and acetaldehyde. Inversely, normal or substituted paraffins,

Compound	Absolute reactivity (g O_3/g compound)	Relative reactivity
Methane	0.0102	1
Ethane	0.147	14
Propane	0.33	32
n-Butane	0.64	63
i-Butane	0.85	83
Ethylene	5.3	519
Propylene	6.6	647
1-Butene	6.1	598
Isobutene	4.2	412
Acetylene	0.37	36
Butadiene	7.7	755
Benzene	0.28	27
Toluene	1.9	186
Ethyl benzene	1.8	176
Xylenes	5.2 to 6	509 to 588
Other dialkylbenzenes	3.9 to 5.3	382 to 519
Trialkylbenzenes	5.6 to 7.5	549 to 735
Formaldehyde	6.2	608
Acetaldehyde	3.8	372
Methanol	0.40	39
Ethanol	0.79	77

Table 5.26   *Reactivities compared for selected organic compounds with respect to ozone formation.*

benzene, alcohols and ethers have very little reactivity as far as ozone formation is concerned.

To estimate the effect of automobile traffic and motor fuels on ozone formation, it is necessary to know the composition of exhaust gas in detail. Figure 5.26 gives an example of a gas phase chromatographic analysis of a conventional unleaded motor fuel.

For each type of component, its relative reactivity in ozone formation was taken into account which makes it possible to characterize by weighting the behavior of the overall motor fuel under the given experimental conditions. The overall reactivity is in fact governed by a limited number of substances: ethylene, isobutene, butadiene, toluene, xylenes, formaldehyde, and acetaldehyde. The fuels of most interest for reducing ozone formation are those which contribute towards minimizing emissions of the above substances.

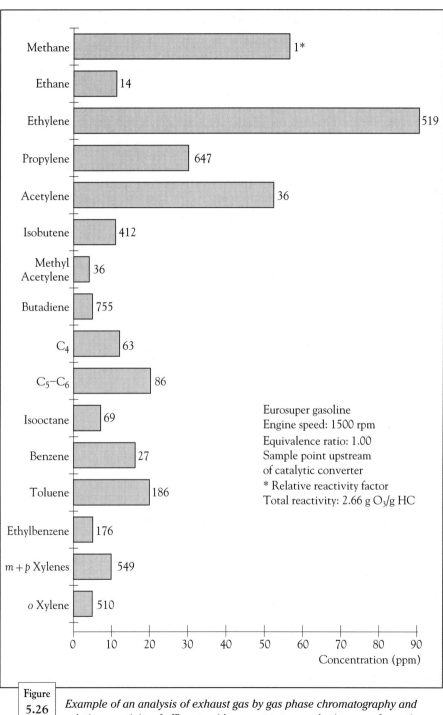

**Figure
5.26**

Example of an analysis of exhaust gas by gas phase chromatography and relative reactivity of effluents with respect to tropospheric ozone formation.

d. "Reformulated" Gasolines

Gasolines said to be reformulated are designed with all aspects of environmental protection being considered: reducing evaporative losses and conventional exhaust system pollutants, extremely low emissions of toxic substances, the lowest reactivity regarding ozone formation. The general action paths are known: reduction of volatility, lowering the levels of aromatics, olefins, sulfur, reducing the distillation end point, addition of oxygenates. Table 5.27 gives an example of a reformulated gasoline's characteristics suggested in 1992 by the Arco Company in the United States. Claims for the pollution improvements are also noted. This is an extreme example of that which would be expected as a result of drastic modification of motor fuel. However, in the United States, local pollution problems observed in a number of urban population centers have already launched safeguarding measures applicable to fuel compositions. These include:

- obligatory addition of oxygenates to a total oxygen concentration level of 2.7% for "non-attainment" regions (those having excessively high CO levels)

- reduction of aromatics to 25%, accompanied by other moves to reduce volatile hydrocarbon emissions and the resulting formation of tropospheric ozone.

Type of fuel	Average	Reformulated
Reid vapor pressure, psi*	8.6	6.7
Aromatics, volume %	34.4	21.6
Olefins, volume %	9.7	5.5
Oxygen, weight %	0	2.7
90% distillation temperature, °C	162	145
Sulfur, ppm	349	41
Reduction claimed CO THC** NMHC** NO_x Ozone	– – – – –	−26% −31% −36% −26% −39%
Total toxic compounds	–	−46%
Variation in consumption	–	+4.7%

Table 5.27   *Characteristics of an "extremely" reformulated gasoline and its impact on the environment. Source: Arco.*

* 1 psi = 69 mbar.
** THC = total hydrocarbons   NMHC = non-methane hydrocarbons.

In Europe and elsewhere in the world, the trend towards reformulated gasoline has scarcely begun; it is very likely, however, that it will be felt around the beginning of 2000, with more or less profound impact on the refining industry.

5.3.2.3 **Relations between Diesel Fuel Composition and Pollutant Emissions**

The study of the relations between diesel fuel composition and pollution caused by the diesel engine is the focus of considerable attention, particularly in Europe where this line of thought has been rapidly developing in recent years.

Numerous works have been directed towards studying the influence of diesel fuel hydrotreatment on emissions.

Table 5.28 gives the modifications in physical/chemical characteristics resulting from deeper and deeper hydrotreatment (Martin et al., 1992). The sulfur contents could thus be reduced to first as low as a few hundred ppm, then to a few ppm. The level of aromatics in the selected example drops from 39% to 7% while the cetane number increases from 49 to 60. Note here that such a treatment, possible through experimental means, does not correspond to current industrial practice because of its high cost and its very high hydrogen consumption.

Product designation	A	A$^+$	A^{++}	A^{+++}	A^{++++}
Density, kg/l at 15°C	0.862	0.850	0.849	0.838	0.827
Viscosity at 20°C, mm2/s	5.55	5.34	5.22	5.12	4.90
Sulfur content, ppm	11,600	640	230	22	4
Nitrogen content, ppm	216	150	135	17	0.2
Cetane number	49.0	50.4	49.0	53.9	60.2
Composition, weight %					
Paraffins	36.5	36.2	36.8	37.0	41.4
Naphthenes	24.3	24.5	36.5	37.7	51.8
Monoaromatics	14.2	23.1	21.9	20.2	6.0
Diaromatics	15.4	12.8	12.6	4.5	0.8
Triaromatics	1.8	1.0	0.9	0.4	0.0
Thiophenes	7.7	2.4	1.4	0.3	0.0
Total aromatics	**39.1**	**39.3**	**36.8**	**25.4**	**6.8**

Table
5.28   *Effect of hydrotreatment on the characteristics of gas oil.*

Tables 5.29 and 5.30 show an example of the effects of hydrotreated diesel fuels on a diesel passenger car already having a low level of pollution owing to technical modifications such as sophisticated injection and optimized combustion. In the standard European driving cycle (ECE + EUDC), between

Type of diesel fuel*	Particulate emissions			
	ECE 15.04 (cold) (g/test)	EUDC (g/test)	ECE 15.04 (hot) (g/test)	ECE + EUDC (g/km)
A	0.68	1.17	0.48	0.168
A$^+$	0.56	1.06	0.37	0.147
A^{++}	0.52	1.14	0.40	0.151
A^{+++}	0.39	1.00	0.34	0.126
A^{++++}	0.30	0.93	0.29	0.112

Table 5.29   *Influence of hydrotreating a diesel fuel on particulate emissions.*

Fuel treatment		Effects on pollutants	
Aromatics	26 → 4%	CO	−30%
Naphthenes	33 → 50%	HC	−40%
Sulfur	3000 → 5 ppm	NO$_x$	0
		Particulates	−25%

Table 5.30   *Influence of deep hydrotreatment on the emissions of a diesel vehicle passenger car – ECE and EUDC cycle.*

the diesel fuel extremes, particulate emission reductions of 25% are observed; the CO and hydrocarbon emission reductions of 30 and 40% respectively are also very significant. On the other hand, the diesel fuel composition exerts very little effect on nitrogen oxide emissions. Figure 5.27 gives a satisfactory correlation between emissions of CO, HC and particulates, and the cetane number.

In the future, European and worldwide refining should evolve toward the production of relatively high cetane number diesel fuels either by more or less deeper hydrotreating or by judicious choice of base stocks. However, it is not planned to achieve levels of 60 for the near future as sometimes required by the automotive manufacturers.

Finally it is likely that attention will be focused on emissions of polynuclear aromatics (PNA) in diesel fuels. Currently the analytical techniques for these materials in exhaust systems are not very accurate and will need appreciable improvement. In conventional diesel fuels, emissions of PNA thought to be carcinogenic do not exceed however, a few micrograms per km, that is a car will have to be driven for several years and cover at least 100,000 km to emit one gram of benzopyrene for example! These already very low levels can be divided by four if deeply hydrotreated diesel fuels are used.

---

* Refer to characteristics in Table 5.28.

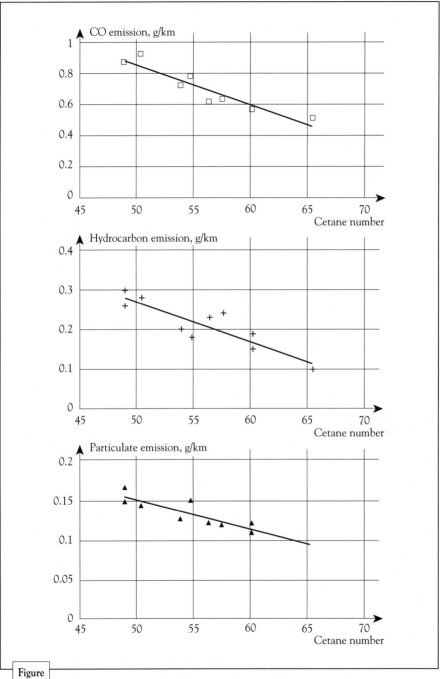

Figure 5.27 *Effect of cetane number on the emissions of carbon monoxide, unburned hydrocarbons and particulates for a diesel passenger car.*
ECE + EUDC standard cycle.

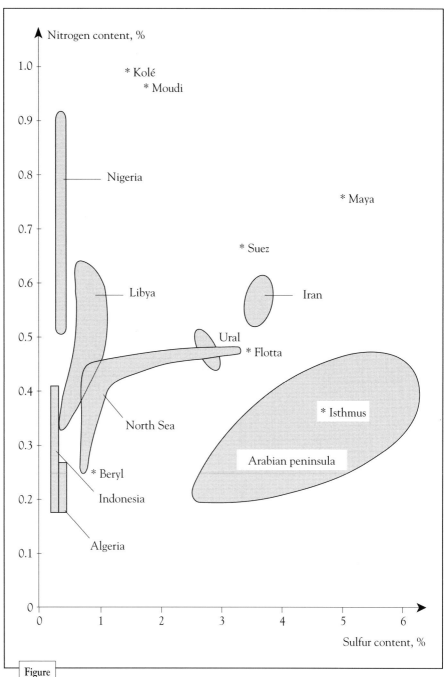

Figure
5.28

Sulfur and nitrogen contents in 550°C⁺ vacuum residues according to crude oil origin. Source: Total.

5.3.2.4 **Influence of the Nitrogen Content of Heavy Fuels on Nitrogen Oxide Emissions**

Besides containing sulfur which is directly responsible for SO_2 emissions, heavy fuels have significant amounts of nitrogen combined in complex heterocyclic structures. Figure 5.28 shows sulfur and nitrogen concentration ranges in vacuum residues ($550°C^+$) as a function of the crude oil origin. Note that nitrogen contents between 3000 and 5000 ppm are common. Yet the European standards for NO_x emissions for combustion in industrial installations, set a maximum smoke threshold of 450 mg/Nm^3. In burners, the nitrogen oxides come from both nitrogen in the air and that contained in the fuel.

$$NO_{x\ total} = NO_{x\ air} + NO_{x\ fuel}$$

The quantity coming from air is practically invariant and corresponds to a level approaching 130 mg/Nm^3. Nitrogen present in the fuel is distributed as about 40% in the form of NO_x and 60% as N_2. With 0.3% total nitrogen in the fuel, one would have, according to stoichiometry, 850 mg/Nm^3 of NO_x in the exhaust vapors. Using the above hypothesis, the quantity of NO_x produced would be:

$$850 \times 0.4 + 130 = 470 \text{ mg } NO_x/\text{Nm}^3,$$

a value slightly higher than the standard.

A conclusion is that meeting the regulations for NO_x emissions in industrial combustion practically implies a limit in the nitrogen content of fuel of 3000 ppm.

This justifies all the work undertaken to arrive at fuel denitrification which, as is well known, is difficult and costly. Moreover, technological improvements can bring considerable progress to this field. That is the case with "low NO_x" burners developed at IFP. These consist of producing separated flame jets that enable lower combustion temperatures, local oxygen concentrations to be less high and a lowered fuel's nitrogen contribution to NO_x formation. In a well defined industrial installation, the burner said to be of the "low NO_x" type can attain a level of 350 mg/Nm^3, instead of the 600 mg/Nm^3 with a conventional burner.

Finally, it is by means of synergy between the refining processes and the combustion techniques that the emissions of NO_x due to industrial installations can be minimized.

6

Characteristics of Non-Fuel Petroleum Products

Bernard Thiault

Demand for non-fuel petroleum products increases from year to year. For example, in France, their share of the total petroleum market rose from 9% to 15.7% between 1973 and 1992.

Non-fuel petroleum products cover an extremely wide range and are distinguished as much by their nature and physical aspects as by their types of application.

We will examine the most important of these products in this chapter:

- solvents   (6.1)
- naphthas   (6.2)
- lubricants   (6.3)
- waxes and paraffins   (6.4)
- asphalts (bitumen)   (6.5)
- other products   (6.6)

6.1 Characteristics of Petroleum Solvents

Petroleum solvents are relatively light petroleum cuts, in the C_4 to C_{14} range, and have numerous applications in industry and agriculture. Their use is often related to their tendency to evaporate; consequently, they are classified as a function of their boiling points.

6.1.1 **Nomenclature and Applications**

Petroleum solvents, or solvent naphthas, are grouped in four categories:

a. *Special Boiling Point Spirits*

The special boiling point spirits (SBP's) have boiling ranges from 30 to 205°C and are grouped in subdivisions according to Table 6.1.

Classification	Density kg/l @ 15°C (approximate)	Boiling range, °C	End use
Spirit A	0.675	40 – 100	Rubber-based adhesives cleaning fluids, degreasing fluids
Spirit B	0.675	60 – 80	Fat extraction, vegetable oil mills, tallow manufacture
Spirit C	0.700	70 – 100	Fat extraction, vegetable oil mills, rubber industry, heating
Spirit D	0.710	95 – 103	Dehydration of alcohol
Spirit E	0.730	100 – 130	Rubber industry, cleaning fluids, degreasing fluids
Spirit F	0.740	100 – 160	Rubber industry, cleaning fluids, degreasing fluids
Spirit G (petroleum ether)	0.645	30 – 75 (approx.)	Petroleum gas equipment, perfume extraction
Spirit H	≤ 0.765	≥ 10% at 70 max. 205	Blowtorches

Table 6.1   *Special boiling point spirits.*

b. *White-Spirits*

White-spirits are solvents that are slightly heavier than SBP's and have boiling ranges between 135 and 205°C. A "dearomatized" grade exists. These solvents are used essentially as paint thinners although their low aromatic content makes them unsuitable for lacquers, cellulosic paints and resins.

c. *Lamp Oils*

Dearomatized or not, lamp oils correspond to petroleum cuts between C_{10} and C_{14}. Their distillation curves (less than 90% at 210°C, 65% or more at 250°C, 80% or more at 285°C) give them relatively heavy solvent properties. They are used particularly for lighting or for emergency signal lamps. These materials are similar to *"kerosene solvents"*, whose distillation curves are between 160 and 300°C and which include solvents for printing inks.

d. Pure Aromatics – Benzene, Toluene and Xylenes

Benzene, toluene and xylenes are used either as solvents or as basic intermediates for the chemical and petrochemical industries.

Independently from the uses reviewed here, a few other applications of petroleum solvents are given below:

- solvents for glues and adhesives
- vehicles for insecticides, fungicides and pesticides
- components for reaction media such as those for polymerization.

6.1.2 **Desired Properties of Petroleum Solvents**

The essential properties of the various types of solvents are related to the following characteristics:

- volatility
- solvent properties
- degree of purity
- odor
- toxicity.

a. Volatility

Volatility is one of the most important properties of a hydrocarbon solvent. Volatility has a direct relation to the time it takes to evaporate the solvent and, therefore, to the drying time for the dissolved product. The desired value of volatility varies greatly with the nature of the dissolved product and its application temperature. Therefore, whether it be an ink that needs to dry at ambient temperature, sometimes very fast, or whether it be an extraction solvent, the volatility needs are not the same.

Volatility is generally characterized by a distillation curve (the quantity distilled as a function of temperature). Often, only the initial and final boiling points are taken into account along with, possibly, a few intermediary points.

Another measurement characterizing volatility is that of vapor pressure.

b. Solvent Power

Solvent power characterizes the miscibility of solute and solvent. This concept covers two types of uses: dissolving a solid or reducing the viscosity of a liquid. The solvent power should be as high as possible. However, a solvent used as an extractant should also be selective, i.e., extract certain substances preferentially from the feed being treated.

There are several criteria used to define solvent power. Chemical analysis is ideal because it can indicate the proportion of hydrocarbons known to be good solvents: in particular, the aromatics.

Nevertheless, this type of analysis, usually done by chromatography, is not always justified when taking into account the operator's time. Other quicker analyses are used such as "FIA" *(Fluorescent Indicator Analysis)* (see paragraph 3.3.5), which give approximate but usually acceptable proportions of saturated, olefinic, and aromatic hydrocarbons. Another way to characterize the aromatic content is to use the solvent's "aniline point" the lowest temperature at which equal volumes of the solvent and pure aniline are miscible.

c. Degree of Purity

The required degree of purity varies with the application but the requirements in this domain are sometimes very important. Several tests are employed in the petroleum industry.

If sulfur is a contaminant, its content can be measured, but it may suffice to characterize its effects by the copper strip corrosion test, or by the *"doctor test"*.

The tendency for a solvent to form deposits by polymerization of impurities such as olefins is measured by the test for "potential gums". Olefin content can also be represented by the "bromine index", which is a measure of the degree of unsaturation (see paragraph 3.4.1).

The solvent can contain traces of acidic or basic compounds which are measured by titration.

Finally, color is always examined since the solvent should be colorless when pure. Traces of heavy aromatics give the solvent a yellow tint.

d. Odor

Odor is of prime importance because a petroleum solvent is often used in closed rooms; moreover, the idea of odor is tied instinctively in the public image to toxicity. Odor is a function of the solvent's composition and volatility. Generally, the paraffin hydrocarbons are less odorous while the aromatics are more so.

There is no reliable standard method to characterize odor and specifications often indicate merely "a not-unpleasant odor".

e. Safety and Toxicity

Petroleum solvents are very flammable and can cause an explosion in the presence of air. For this reason, their flash points, directly related to volatility, are always specified.

Another danger, intoxication by inhalation, is related to the benzene content. A maximum limit is often set for this compound.

6.2 **Characteristics of Naphthas**

Naphthas constitute a special category of petroleum solvents whose boiling points correspond to the class of white-spirits (see paragraph 6.1).

They are classified apart in this text because their use differs from that of petroleum solvents; they are used as raw materials for petrochemicals, particularly as feeds to steam crackers. Naphthas are thus industrial intermediates and not consumer products. Consequently, naphthas are not subject to governmental specifications, but only to commercial specifications that are re-negotiated for each contract. Nevertheless, naphthas are in a relatively homogeneous class and represent a large enough tonnage so that the best known properties to be highlighted here.

Two types of specifications are written into supply contracts for naphthas; they concern the composition and the level of contaminants.

Composition is normally expressed by a distillation curve, and can be supplemented by compositional analyses such as those for aromatics content. Some physical properties such as density or vapor pressure are often added. The degree of purity is indicated by color or other appropriate test (copper strip corrosion, for example).

Sometimes analyses are required for particular compounds such as sulfur, chlorine and lead, or for specific components such as mercaptans, hydrogen sulfide, ethers and alcohols.

6.3 **Characteristics of Lubricants, Industrial Oils and Related Products**

This heading covers such a large number of products and applications that it is difficult to give a complete inventory. For this reason the standards organizations, starting with *ISO (International Organization for Standardization),* have published a series of standards to classify these products.

6.3.1 **Nomenclature and Applications**

a. *Classification*

The ISO 8681 standard, which treats all the petroleum products, groups lubricants, industrial oils and related products in the L Class. The international standard ISO 6743/0, accepted as the French standard NF T 60-162, subdivides the L Class into 18 families or categories.

Table 6.2 summarizes the principal product classes.

Furthermore, each sub-category given in Table 6.2 can be divided according to product viscosities, which are classified in the international standard ISO 3448 (French standard NF ISO 3448, index T 60-141).

ISO Standard	Category	Number of sub-categories
ISO 6743-1	Total loss systems	3
ISO 6743-2	Spindle bearings, bearings and associated clutches	2
ISO 6743-3A	Compressors	12
ISO 6743-3B	Gas and refrigeration compressors	9
ISO 6743-4	Hydraulic fluids	17
ISO 6743-5	Turbines	12
ISO 6743-6	Gears	11
ISO 6743-7	Metalworking	17
ISO 6743-8	Temporary protection against corrosion	18
ISO 6743-9	Greases	5670
ISO 6743-10	Miscellaneous	22
ISO 6743-11	Pneumatic tools	9
ISO 6743-12	Heat transfer fluids	5
ISO 6743-13	Slideways	1
ISO 6743-14	Heat treatment	26
Being studied	Motor oils	

Table 6.2   *Classification of lubricants, industrial oils and related products.*

The classification of motor oils has not been completed in the ISO standard because the technical differences between motors in different parts of the world, particularly Europe and the United States, make the implementation of a single system of classification and specifications very difficult. In practice, different systems coming from national or international organizations are used. The best known is the SAE *viscosity classification* from the Society of Automotive Engineers, developed in the United States.

The SAE classification has existed since 1911 and has undergone several revisions. The latest version is designated by the symbol SAE J 300 followed by the date of the latest revision.

The classification defines the "viscosity grades" for which the characteristics correspond either to winter climatic conditions (grades aW where W designates "Winter"), or to summer conditions (b-type grades). Thus, an oil designated by a type aWb number is a multigrade oil, capable of maintaining its defined viscosity qualities in winter as in summer. The six W grades are defined by a maximum cold viscosity (from $-30°C$ to $-5°C$ according to the grade, measured by rotating viscometer, "CCS", for Cold

Cranking Simulator), by a pumpability temperature limit measured by a rotating mini viscometer, and by the minimum kinematic viscosity at 100°C. The five summer grades are defined by bracketing kinematic viscosities at 100°C.

The 15W40 or 15W50 oils are the most widespread in temperate climates (Western Europe), while the 20W40 or 20W50 oils are used in relatively warm climates (Mediterranean countries, Middle East, South America). The 5W or 10W grades are used in countries having severe winters such as Scandinavia and Canada.

A chart by J. Groff relates the SAE grades, the kinematic viscosity, and the viscosity indices. This correlation is given in Figure 6.1 in the 1994 edition.

At the beginning of each year, the new version of the SAE Handbook is published for which Volume 3 is entitled:

"Engines, Fuels, Lubricants, Emissions and Noise
Cooperative Engineering Program"

edited by

Society of Automotive Engineers, Inc.
400 Commonwealth Drive
Warrendale, Pa 15087 – 0001 USA

b. Composition of Lubricating Oils

Each lubricating oil is composed from a main "base" stock, into which additives are mixed to give the lubricant the properties required for a given application.

Lubricant bases can be mineral (petroleum origin) or synthetic.

• The conventional mineral bases result from the refining of vacuum distillation cuts and deasphalted atmospheric residues. According to the crude oil origin and the type of refining they undergo, the structures of these bases can be essentially paraffinic, isoparaffinic, or naphthenic. The conventional scheme for lubricating oil production involves the following steps: selection of distillates having appropriate viscosities, elimination of aromatics by solvent extraction in order to improve their VI (viscosity index), extraction of high freezing point paraffins by dewaxing and finally light hydrogen purification treatment (see Figure 10.13).

Different treatments provide lubricant bases having accentuated isoparaffinic structures: these are the bases from hydrorefining, hydrocracking and hydroisomerization (see paragraph 10.3.2.2.c.2).

• The many varieties of synthetic bases:

– Olefin polymers: alpha-olefin polymers (PAO), polybutenes and alkylaromatics, in particular the dialkylbenzenes (DAB). This class of compounds is the most widespread and accounted for 44% of the synthetic base market in France in 1992.

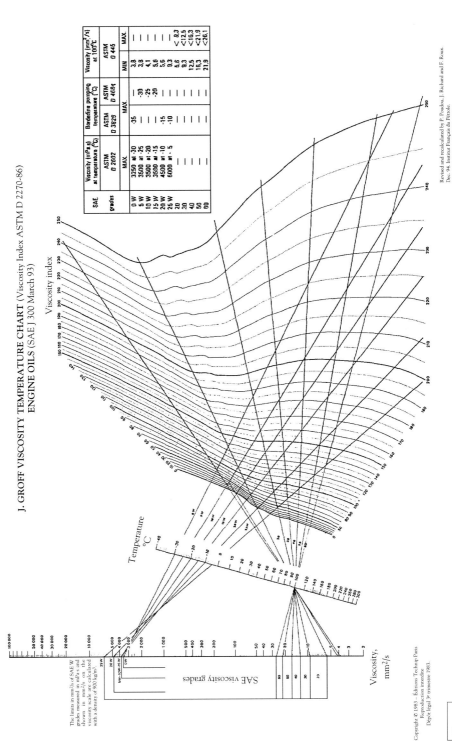

Figure 6.1   *Viscosity/temperature chart of J. Groff (ASTM D 2270-86 viscosity index). New SAE classification.*

- Organic polyesters, obtained either from a diacid and a mono-alcohol, or from poly-alcohols and a monoacid, or from di-alcohols and a diacid. This class represented 29% of the synthetic base market in France in 1992.
- Polyalkylene glycols, particularly polypropylene glycols, represent 17% of French market in 1992.
- Phosphoric esters, not common as bases, but used as additives.
- Special products: silicones, silicates, polyphenylethers, used rarely and for very specialized applications because they are expensive materials.

The properties of bases are modified by additives to meet the specifications. The principal classes of additives are (see chapter 9) as follows:

- VI additives to improve the viscosity index: polymethacrylates, polyacrylates, olefin polymers.
- Additives to lower the pour point: polyacrylates.
- Detergent and dispersant additives: sulfonates, thiophosphonates, phenates, salicylates more or less overbased, succinimides.
- Anti-wear and extreme pressure additives: phosphoric esters, dithiophosphates, sulfur-containing products such as fatty esters and sulfided terpenes or chlorinated products such as chlorinated paraffins.
- Additives for lubrication under extreme conditions: fatty esters, fatty acids, etc.
- Antioxidant and deactivation additives: substituted phenols, dithiophosphates, dithiocarbamates, alkylated aromatic amines.
- Corrosion inhibitors: partial esters of succinic acid, fatty acids, sulfonates, phenates, amine phosphates.
- Anti-foaming agents: polydimethylsiloxanes, fluorosilicones, acrylates.

During the formulation of an oil, blending of all these components gives an extremely wide variety of products described in the classification. Nevertheless, the lubricating greases make up a special product category among them.

c. Lubricating Greases

A grease is by definition a semi-solid material obtained by dispersing a gelling agent in a liquid lubricant. A multitude of gelling agents blended with a large variety of liquid lubricants (mineral or synthetic) can be used; this accounts for the large number of greases.

The selection of gelling agents and liquid lubricants is conducted according to the requirements for a specific use:

- high or low temperature properties
- mechanical stability (resistance to shearing and centrifugation)
- resistance to water.

These properties are mainly associated with the gelling agent and liquid lubricant.

Other properties such as:

• oxidation resistance

• corrosion protection

• wear and seizing protection

are mainly associated with specialized additives.

The most common liquid lubricants are: mineral oils (usually naphthenic), esters (either diesters or complex esters), polyalpha olefins and polyalkylene glycols.

Using one or another type depends on the desired properties for the grease, in particular the high or low temperature properties.

There are numerous possible gelling agents:

• inorganic gelling agents

• organic gelling agents (soaps or others).

The great number of possible associations between the types of liquid lubricants and gelling agents results in many different categories of greases:

Greases Having Inorganic Gelling Agents

These are mainly the silicon-based greases (hydrophobic calcined silica) or grafted bentones. Silica and bentones can be associated with practically all types of liquid lubricants. The greases based on these two gelling agents have high dropping points, and are thereby utilized for high temperature service. Their mechanical stability is generally average to mediocre as is their resistance to water. The silicon-based greases are translucent (in the absence of inorganic compounds); the bentone greases are opaque.

Greases Having Organic Gelling Agents

Within this category, the greases are divided into those based on simple soaps and those based on complex soaps. The latter generally have better high temperature and structural stability properties under high mechanical shear; they also have higher resistance to water than their simple soap-based counterparts.

The simple soaps are obtained by the reaction of fatty acids of animal or vegetable origin, hydrogenated or not (or their triglycerides, i.e., their glycerol esters) with a base. The bases most commonly used are lithium hydroxide, calcium hydroxide, and sodium hydroxide, to make the "lithium" greases (or lithium soap), the "calcium" greases (or calcium soap), and the "sodium" greases (or sodium soap). The aluminum soap greases are most often prepared by reacting aluminum isopropylate with fatty acids. The barium-based greases are practically no longer in use. The most common fatty acids are tallow-derived acids, stearic acid, 12-hydroxy-stearic acid. The most commonly used triglycerides are tallow and hydrogenated castor oil.

The most common combinations are as follows:

- Lithium hydroxide with 12-hydroxy-stearic acid (or hydrogenated castor oil); they form the family of lithium greases very commonly used for general lubrication and bearing lubrication.
- Lime with tallow-derived fatty acids: they are the so-called calcium greases that are often used as subframe greases and water-resistant greases.
- Sodium hydroxide with stearic acid: they constitute the sodium greases, used in the lubrication of bearings under dry conditions and gear trains.

The "complex" greases are obtained by the reaction of bases with mixtures of organic and/or inorganic acids. The three groups of complex greases are:

- Calcium complex soap greases, obtained by the reaction of lime and a mixture of fatty acids and acetic acid. These greases offer good high temperature and anti-wear/extreme pressure properties related to the presence, in the soap, of calcium acetate that acts as solid lubricant; they have good mechanical stability.
- Lithium complex soap greases, obtained by the reaction of lithium hydroxide and a mixture of 12-hydroxy-stearic acid with an organic diacid (generally azelaic acid). These greases have very good high temperature properties, remarkable mechanical stability, a good resistance to water, and replace conventional lithium soap greases whenever the latter's properties become marginal for the intended application.
- Aluminum complex greases, obtained by the reaction of aluminum isopropylate with a mixture of benzoic acid and fatty acids. These greases have a remarkable resistance to water, very good adhesion to metallic surfaces, good mechanical stability properties and resistance to temperature. They are less common than the first two types.

There are also greases with organic gelling agents other than soaps. These are polyurea-based greases obtained by the reaction of a di-isocyanate with an amine (or mixture of amines). These greases possess excellent high temperature properties and good mechanical stability.

6.3.2 **Properties Desired in Lubricants, Industrial Oils and Related Products**

The properties sought for these products depend on the type of application; it is useful to distinguish motor oil from industrial lubricants.

6.3.2.1 **Properties of Motor Oils**

For every use constraint there is a series of corresponding characteristics that the oil should have. The situation is summarized in Table 6.3.

Constraint	Required oil properties
Motor performance	Lubrication properties Constant viscosity (viscosity index) Fluidity at low temperature Proper viscosity at high temperature
Maintaining motor cleanliness	Detergent and dispersant power
Corrosion and anti-wear protection	Anti-corrosion and anti-wear power High viscosity at high shear rates
High temperature operation	Thermal stability
Oil change interval	Oxidation stability
Low consumption	High viscosity, low volatility
Gasket compatibility	Adapted composition, low aggressivity
Energy economy	Low viscosity, reduced friction
Control of emissions	Low consumption Low volatility, constant viscosity
Environment	Absence of toxic compounds such as polychlorinated biphenyls (PCB's)

Table
6.3   *Characteristics of Motor Oils.*

6.3.2.2 **Properties of Industrial Oils**

Owing to the large number of types of industrial lubricants, the number of constraints, and therefore the number of desired properties, is very large. The main industrial oils are summarized in Tables 6.4 and 6.5, the first giving the constraints common to all applications, and the second addressing the more specific requirements. A few essential properties appear from these tables:

a. Viscosity

This is the essential characteristic for every lubricant. The kinematic viscosity is most often measured by recording the time needed for the oil to flow down a calibrated capillary tube. The viscosity varies with the pressure but the influence of temperature is much greater; it decreases rapidly with an increase in temperature and there is abundant literature concerning the equations and graphs relating these two parameters. One can cite in particular the ASTM D 341 standard.

b. Viscosity Index (VI)

The VI is a number that results from a calculation involving the viscosities at 40°C and 100°C. It characterizes the capacity of the lubricant to maintain a constant viscosity through a large range in temperature. This property can be improved by additives.

c. Pour Point

The pour point is the lowest temperature at which an oil can still pour while it is cooled, without agitation, under standardized conditions. The pour point of paraffinic bases is linked to the crystallization of n-paraffins. The pour point of naphthenic bases is related to a significant viscosity increase at low temperatures. This property can be improved by additives.

d. Aniline Point

This value characterizes the level of aromatics in non-formulated oils; the aniline point is higher for low aromatic contents.

e. Volatility

Volatility can be characterized either indirectly, by measurement of the flash point (the temperature to which the oil must be heated for inflammation of its vapor to become possible) or by direct measurement, following the Noack method.

f. Conradson Carbon Residue

This measurement characterizes the tendency of an oil to form carbonaceous deposits upon undergoing carbonization.

Constraint	Required oil properties
Large range of service temperatures	Constant viscosity (viscosity index)
	Pour point, thermal stability
Protection of lubricated members	Viscosity selection Anti-wear power
	Anti-corrosion power
Maintaining cleanliness	Detergent and dispersive powers
	Filterability
Life span	Resistance to oxidation
Volume reduction in service	Thermal stability
	Resistance to oxidation, deaeration
Controlling emissions of gas and fog	Low volatility
Shop environment	Odor
Skin toxicity	Low polynuclear aromatic (PNA) content
Gasket compatibility	Adapted composition, low aggressivity
Environment	Absence of toxic compounds such as
	polychlorobiphenyls (PCB's)

Table 6.4   *Industrial oils. General requirements.*

Type of lubricant	Constraints	Desired properties
For air compressors	Operating safety	Thermal stability, Volatility Resistance to oxidation Extreme pressure and anti-wear (compressors) properties Low coking tendency (hot reciprocating compressors)
	Environment	Biodegradability
For gas compressors	Danger of gas/oil reaction	Adapted composition
For refrigeration compressors	Miscibility with the refrigeration fluids (including compressors with the new refrigeration fluids)	Adapted composition
For combined cycle turbines	Single lubrication system	Low pour point Extreme pressure and anti-wear properties Hydrolysis stability Water separation
For gear trains	Protection from seizing and rapid wear	Extreme-pressure and anti-wear properties Resistance to oxidation Thermal stability High viscosity Low pour point Anti-foaming properties Anti-corrosion properties
Hydraulic fluids	Environment	Biodegradability
Greases	Mechanical and rheological behavior and its persistence	Consistency and viscosity Mechanical stability Oxidation resistance
	Protection	Anti-corrosion power
	Lubricating performance	Extreme pressure and anti-wear properties Oil separation
	Imperviousness	Consistency
	Systems under pressure	Pumpability

Table
6.5   *Industrial oils. Special complementary requirements.*

g. Mechanical Properties

There are numerous tests for characterizing the mechanical properties of lubricants: cone penetration of greases, extreme pressure tests (as in the four-ball test), etc.

h. Anti-Oxidant Properties

There are available standard accelerated oxidation tests that consist of passing air or oxygen through an oil at elevated temperature. The test is conducted with or without the presence of catalysts or water.

i. Performance Properties

Performance can be illustrated for example by the time necessary for deaeration or de-emulsification of oils, anti-rust properties, copper strip corrosion test, the flash point in closed or open cup, the cloud and pour points, the foaming characteristics, etc.

6.4 Characteristics of Waxes and Paraffins

During the production of mineral oils from vacuum distillates, one of the process steps, "dewaxing", removes the high melting point materials in order to improve the oil's pour point. Dewaxing produces paraffins and waxes, the first coming from light distillates, and the second from medium or heavy distillates.

Paraffins consist mainly of straight chain alkanes, with a very small proportion of isoalkanes and cycloalkanes. Their freezing point is generally between 30°C and 70°C, the average molecular weight being around 350. When present, aromatics appear only in trace quantities.

Waxes are less well defined aliphatic mixtures of n-alkanes, isoalkanes and cycloalkanes in various proportions. Their average molecular weights are higher than those of the paraffins: from 600 to 800.

Applications for these products cover a wide range. When completely dearomatized, paraffins have markets in the food industry, especially in food packaging. Generally containing polymer additives, paraffins are very useful for impregnating paper or cardboard imparting water resistance. Paraffins or waxes are also found in particle board. This non-exhaustive list of applications for waxes and paraffins can not be ended without mentioning the manufacture of candles, polishes, cosmetics and coatings.

6.4.1 Desired Properties of Waxes and Paraffins

The characteristic properties of waxes and paraffins can be grouped in three classes

a. Physical Properties

Independently of crystallization properties, it is useful to characterize the behavior of a product with respect to the following criteria:

- freezing and congealing temperatures
- hardness and consistency
- oil content, which if too high, causes harmful side effects. Note, however, that it is not a true oil but, by convention and by reference to the manufacturing process, it is considered as the fraction of the material that is soluble at $-32°C$ in methylethylketone
- viscosity, particularly important for paper coating
- color, measured by comparison to colored glass standards
- odor, measured by a "panel" of at least five observers.

b. Mechanical Properties

The mechanical properties of waxes and solid paraffins are of considerable importance for most applications and numerous tests have been developed for characterizing the hardness, the brittleness, and resistance to rupture.

The most widespread test is the penetration of a needle or cone.

c. Properties of Food-Grade Paraffins

These properties concern paraffins that are part of food packaging materials. Their potential toxicity could be attributable to aromatic residues. The latter are thereby characterized directly or indirectly by:

- analyzing the ultraviolet absorption spectra in the spectral zone corresponding to aromatics
- determining the level of "carbonizable materials"; that is, observing the paraffin's change in color by treating it with concentrated sulfuric acid.

6.5 Characteristics of Asphalts (Bitumen)

The words *asphalt* and *bitumen* take on different meanings depending on type of industry or national custom. In the petroleum industry and in the United States, asphalt is the term most widely used for products derived from refining operations, typically atmospheric or vacuum distillation residues, containing cementitious, high molecular weight polar materials called asphaltenes. In Europe, although the terms bitumen and asphalt are often used interchangeably for the same products, bitumen is considered to be more correct. In France, there is a technical distinction between the two terms, and because of this, the word bitumen will replace asphalt in most cases in this chapter.

In France, bitumen belong to a category of products called "hydrocarbon binders". They are defined and classified in the French Standard, NF T 65000. The hydrocarbon binders comprise:

- bitumen that are semi-solid or solid materials extracted from petroleum
- bitumen emulsions that result from a dispersion of bitumen in a receiving phase, usually aqueous
- tars, which are produced by from coal coking at high temperatures.

The term bitumen is used in France to designate petroleum products, as in Great Britain and Germany. In the United States on the other hand, the equivalent material is designated by the expression "asphalt-cement". In France, asphalt is a mastic, a mixture of bitumen and powdered minerals, poured in place. This mixture can be either natural or reconstituted by an industrial process. Asphalt (French meaning) is utilized on roads, particularly in urban centers as well as for sidewalk surfacing.

6.5.1 **Classification of Bitumen**

The term "hydrocarbon binder" covers:

- bitumen coming directly from blended petroleum refining asphalts
- cutbacks (bitumen which are mixed with a more or less volatile petroleum solvent, generally a non-commercial grade kerosene. The viscosity of these products is thus lowered enabling a lower working temperature to be used
- fluxed bitumen, which are mixtures of bitumen with a low viscosity oil. These binders are often more viscous than cutbacks. In France, the fluxing material is generally an oil derived from coal, but can be an oil of petroleum origin. In many other countries, oils of petroleum origin are the only fluxing materials used
- bitumen emulsions.

In addition to the more important categories, there are:

- mixed fluxed bitumen, where the dilution oil is a mixture of products from both petroleum and coal origins
- composite bitumen which are mixtures of bitumen and tar, or of bitumen and coal-tar pitch in which bitumen has the largest portion
- modified bitumen, which are bitumen supplemented by materials from various origin, generally polymers that modify certain properties.

Within these categories, there is further classification:

- The standard NF T 65-001 gives a classification for bitumen as a function of their hardness. This is measured using a "needle penetrability" test, which measures the penetration depth of a weighted needle into the bitumen. Five grades have been defined.
- The standard NF T 65-002 defines five grades of cutbacks according to their pseudo-viscosities.

- The standard NF T 65-003 defines in the same manner three grades of fluxed bitumen.

- The standard NF T 65-004 classifies the types of composite bitumen: it distinguishes three grades of bitumen-tars by their pseudo-viscosities and two grades of bitumen-coal tar pitch by their penetrabilities.

- The standard NF T 65-011 distinguishes the bitumen emulsions by their ionic nature (anionic or cationic), their stability with respect to agglomerates and weight content of base binder. There are 20 grades of emulsions.

6.5.2 **Bitumen Manufacture**

There are several processes to manufacture bitumen from crude oil:

- The distillation of crudes chosen for their yield in heavy fractions is the most common means. Bitumen is extracted from the residue from a vacuum distillation column (a few dozen mm of mercury), the latter being fed by atmospheric distillation residue. Unlike the practice of a decade ago, it is now possible to obtain all categories of bitumen, including the hard grades.

- Solvent deasphalting. This is an extraction of the heaviest fractions of a vacuum residue or heavy distillate. The extract is used to produce the bitumen. The separation is based on the precipitation of asphaltenes and the dissolution of the oil in an alkane solvent. The solvents employed are butane or propane or a butane-propane mixture. By selecting the proper feedstock and by controlling the deasphalting parameters, notably temperature and pressure, it is possible to obtain different grades of bitumen by this process.

- Air blowing consists of circulating air countercurrently in a bitumen in order to oxidize it. This operation forms molecules of high molecular weight and of different structure from the initial material. It is thus possible to produce hard "grades" of very high softening points.

6.5.3 **Bitumen Applications**

There are two main categories of bitumen applications:

- Road paving. This includes bitumen, cutbacks and fluxed bitumen as well as emulsions. Each of these products is subject to very special application techniques. This list is completed by the use of poured asphalt, even though this product is better suited to smaller surfaces: sidewalks, courts, etc., than to pavements. Since the middle of the 1980's, air-blown bitumen is no longer used for road construction.

- The industrial applications, for which blown bitumen are very often used. Some of the industrial applications are given below:

- Waterproofing, whether it has to do with protecting civil engineering structures or roofs or terraces. Poured asphalt, often placed in layers with kraft paper, oxidized bitumen or modified bitumen can be used, generally with copolymer. The modified bitumen are used for the making prefabricated multi-layer waterproofing composites.
- Filler for cracks and fissures, particularly in highways. Mixtures of bitumen, heavy oils, polymer or sulfur are used.
- Soundproofing in buildings, automobiles, home appliances, for example.
- Electrical insulation with oxidized bitumen: electrical cabling, condensers, batteries.
- Bitumen paints and varnishes, which are mixtures of hard bitumen, usually oxidized, and a light or very light solvent.

6.5.4 **Desired Bitumen Properties**

The principal characteristics of bitumen are its softening point and its needle penetrability. In France the latter has always been the basis for bitumen classification and class designation. Yet, the former is more representative of a bitumen's capacity to deform when the service temperature increases. The other properties have more or less importance depending on the application.

a. *Needle Penetrability*

Penetrability is the depth, expressed in tenths of a millimeter, a standard steel needle penetrates into a bitumen sample at 25°C. The needle carries a weight of 100 g and the test is applied for five seconds. The corresponding test method is relatively difficult to carry out and is defined in France by the standard NF T 66-004, and in the USA by the method ASTM D 583. Penetration is related to the viscosity.

b. *Softening Point*

The softening point is the temperature at which a bitumen becomes soft under standard conditions. The measurement uses the "ring and ball" method which is standardized in France (NF T 66-008) and in the USA (ASTM D 36). A steel ball of precisely defined dimensions and weight is placed on the bitumen sample that is surrounded by a metal ring also precisely defined. The assembly is heated gradually. When the bitumen sample becomes soft enough for the ball to pass through and travel a vertical distance of 2,5 cm, the corresponding temperature is called the softening point. This measurement is also related to the viscosity.

c. *Density*

Density is measured by a pycnometer (NF T 66-007, ASTM D 70).

d. Fraass Brittle Point

This test attempts to characterize the brittleness of bitumen at low temperatures. It consists of measuring the temperature at which fissures appear on a bitumen film spread on a blade as it is repeatedly flexed. This test is delicate and of questionable reliability, but it is currently the only one that allows the elastic behavior of bitumen on decreasing temperature to be characterized. It is standardized in France (T 66-026).

e. Resistance to Hardening (RTFOT)

A bitumen sample is oxidized at high temperature under well defined conditions and its physical characteristics are measured before and after this artificial ageing process. The method is defined in France as AFNOR T 66-032 and in the USA by ASTM D 2872 (Rolling Thin-Film Oven Test).

f. Solubility

This measurement provides a definition of the bitumen content in bitumen materials as the portion soluble in carbon disulfide (in France, in trichloroethylene, carbon tetrachloride or tetrachloroethylene). The method is defined by AFNOR NF T 66-012 or IP 47, or ASTM D 4 (the latter is not equivalent to the others).

g. Ductility

Ductility is the elongation, at the moment of failure, of a standard bitumen briquette that is stretched at a predetermined speed and temperature. References are the NF T 66-006, ASTM D 113, IP 32 methods.

h. Volatility

Volatility can be characterized in different ways:
- Flash point (cutbacks, fluxed bitumen). Standards NF T 66-009 and IP 113
- Loss of weight on heating. NF T 66-011 and ASTM D 6. The loss of weight on heating can also be measured during the RTFOT test.

6.6 Other Products

In this section we will discuss:
- white oils
- aromatic extracts
- coke.

6.6.1 White Oils

This term refers to highly refined lubricating oils for which dearomatization, in particular, has been pushed to the extreme. These products are sometimes designated by the expression "liquid petrolatum".

Historically, the white oils were manufactured from light mineral oils (spindle oil) that had undergone severe oleum treatment. This process, which has the disadvantage of leaving an acid sludge residue, is being increasingly replaced by hydrotreatment (catalytic hydrogenation under severe conditions) of light distillates.

There are two categories of white oils: technical white oils and medicinal white oils. The technical white oils, which are already highly dearomatized, are used for specialized lubricants, particularly in the textile industry, and also as components in cosmetics, as plasticizers in the rubber or plastics industries, or as emulsion bases for certain pulverized agriculture products. The medicinal white oils, whose dearomatization is pushed further still, are used in pharmaceuticals, or in the food industry, wherever residual oils might be in contact with food.

These white oils are subject to specifications from various organizations: *"Codex"* in France, *"British Pharmacopoeia"* (BP) in the United Kingdom, and *"National Formulary"* (NF) in the USA.

White oils can be characterized by their physical properties as base oils: density, viscosity, flash point, etc.

For medicinal or food grade white oils, a very high purity is required. This is controlled by tests analogous to food-grade paraffins:

• analysis for traces of aromatics by UV absorption spectrometry
• test for carbonizable matter.

6.6.2 **Aromatic Extracts**

In the manufacture of base oils, one of the refining operations is to extract with the aid of an appropriate solvent (furfural most often) the most aromatic fractions and the polar components. When free of solvent, the extracted aromatic fraction can eventually be refined, particularly to remove color or to thicken it, or still further, to fractionate it. The term, "aromatic extract" is used in every case.

The aromatic extracts are black materials, composed essentially of condensed polynuclear aromatics and of heterocyclic nitrogen and/or sulfur compounds. Because of this highly aromatic structure, the extracts have good solvent power.

The aromatic extracts have been used in the paint industry to partially replace linseed oil. They are still used for producing printer's ink. In addition, they are finding a variety of applications as plasticizers in the rubber industry or for the manufacture of plastics such as PVC.

The aromatic extracts are sought mainly for their solvent power. They are characterized particularly by componential analyses such as the separation according to hydrocarbon family by liquid phase chromatography.

6.6.3 **Coke**

All modern refineries have conversion units, designed to transform black effluent streams into lighter products: gas, gasoline, diesel fuel. Among these conversion units, "coking" processes take place by pyrolysis and push the cracking reaction so far that the residue from the operation is very heavy; it is called "coke".

There are few coking units in the world, and the majority of them is found in the United States, such that coke production is marginal. Different coking processes have been described, but only two have survived (see Chapter 10):

- *"Delayed Coking"*, is a semi-continuous process, developed at the end of the 1930's. The reaction is conducted at 450–500°C under relatively low pressure, four atmospheres, maximum.
- *"Fluid Coking"*, developed in 1953. The reaction proceeds at atmospheric pressure, at about 500–550°C, in a reactor whose feed is mixed in a fluidized bed of hot coke which maintains the desired temperature.

Petroleum coke is in reality a hydrocarbon whose C/H ratio is very high; it is usually higher than 20 and can attain 1000 after calcination. It is not, therefore, elementary carbon.

Coking units are operated to optimize the light products produced, coke being considered as a by-product. Its quality is not too important. Generally speaking, the quality of coke produced varies widely according to the feed, the operating conditions, and the process.

- **Influence of the feed:** coke produced from distillation residue is less structured, less crystalline than that from a cracking residue. If the residue feeding the unit is highly contaminated with sulfur and metals, it is still coke, but is disqualified for certain applications.
- **Influence of operating conditions:** This concerns the temperature, the pressure and the residence time. The more severe the conditions are, the harder is the coke produced.
- **Influence of the type of process:** *"fluid coking"*, compared to *"delayed coking"* makes a harder coke that contains less volatile matter and forms finer grains.

It is possible to modify the quality of the coke by calcination at high temperatures (1200–1400°C); this has the effect of reducing the volatile material and to increase the density.

Petroleum coke is an excellent fuel, and that is its main use, especially for the coke from *"fluid coking"*. There are some other markets that have to do with calcined coke: electrodes for aluminum production or for all other electrolytic cells, carbons for electro-mechanical equipment, graphite, and pigments.

7

Standards and Specifications of Petroleum Products

Bernard Thiault

The terms *standards* and *specifications* are constantly confused and interchanged in everyday use. The general opinion is that they are synonymous; yet these two terms cover different concepts. Therefore, they —as well as the organizations that are responsible for their development— need to be defined.

7.1 Definitions of the Terms Specification and Standard

Specifications represent, as indicated in dictionaries, the definition of the characteristics that a construction, a material, a product, etc., must have. The specifications for industrial products, such as petroleum products, are thus lists of terms and conditions that the products must meet. There are many types of specifications for petroleum products:

- Governmental specifications. In France, they are published through inter-ministerial directives and are prepared at the Ministry of Industry by the "DHYCA" (Direction des Hydrocarbures et des Carburants). They govern the characteristics that products must adhere to in all French territories.

- Customs specifications, that have jurisdiction over the characteristics related to fiscal matters.

- For countries in the European Union, the specifications issued by European directives that replace, when they exist, the national specifications.

In everyday language, the term "specifications" often means "governmental specifications".

Standards also give definitions for the characteristics of a material or product, or they provide the means and methods to implement quality tests for them. The difference lies in their method of preparation, therefore, in their legal status. A standard is the result of a consensus between all parties concerned. These parties represent the manufacturers of the product or material, the consumers who are the industries or user services or, ultimately, consumer associations, as well as, finally, governments.

Standards are generally not made into law and therefore are not enforced but depend on voluntary compliance. Their only strength lies in the consensus obtained during their preparation. There are, nevertheless, a few exceptions; it can happen that a decree or directive gives a standard an obligatory nature.

There are two types of standards concerning petroleum products:

- The standards for classifications and characteristics. They look completely like specifications but they are not enforced by law. Instead of speaking about standards for characteristics, it is more common to talk about standards for specifications. In spite of its very general usage, this expression is unfortunate because it continues to foster confusion between standards and specifications.

- The standards for tests, which are cited in reference to the standards for characteristics. In order for a client and his supplier to compare their results during a commercial transaction, it is important that their laboratories rigorously follow the same operating procedures. The procedures are thus precisely defined in the standards.

Moreover, it is useful to distinguish between the standards prepared by the *official standards organizations* and the *professional* standards. The former's mission is to ensure that the conditions of the consensus of the widest assemblage of interested parties are followed. The professional standards are prepared by recognized professional organizations but limit the consensus to only the participating organizations.

7.2 Organizations for Standardization

In the same way that standards either can be limited to a consensus between professionals or they can be official, the standards organizations can be either professional or official.

7.2.1 Recognized Professional Organizations

There are many professional organizations; their influence is more or less in proportion to the number of members and they can be national or international. Among those dealing with petroleum products, the following examples are cited:

- In France:
  - *CNOMO (Comité de Normalisation des Moyens de Production)* which prepares for the two national automobile manufacturers the texts that serve as the basis for supplier contracts
  - *GFC (Groupement Français de Coordination pour le développement des essais de performances des lubrifiants et des combustibles pour moteurs)* the membership of which includes petroleum companies, additive manufacturers, automobile manufacturers and a few consumers. The *GFC* is interested mainly in mechanical testing.
- In the United Kingdom, the *IP (Institute of Petroleum)*, for which most of the texts, but not all, are taken as British standards with the reference, BS (British Standard).
- In the USA:
  - *ASTM (American Society for Testing and Materials)*
  - *API (American Petroleum Institute)*
  - *SAE (Society of Automotive Engineers)*.
  In spite of their authority and international prestige, these institutes are not the official standards organizations, and participation in their work is restricted to those who have paid the membership fees.
- In Europe, the *CEC (Coordinating European Council for the development of performance tests for lubricants and engine fuels)* federalizes the *GFC* and its counterparts in other European countries.

7.2.2 **Official Standards Organizations**

The accreditation of a standard is an official act (signed by the Ministry of Industry in France). To prepare standards, governments have mandated private organizations which are responsible for continuously following the rules to reach a maximum consensus. There is only one such organization per country. They are, moreover, grouped at the European and international levels.

At the international level, the official organization is the *ISO (International Organization for Standardization)* and its counterpart for the electrical industries, the *IEC (International Electrotechnical Commission)*.

At the European level, the national organizations from both the *EU (European Union,* formerly the *EEC)* and the *EFTA (European Free Trade Association)* are grouped in the *CEN (European Committee for Standardization)* and its counterpart for the electrical industries: *CENELEC.* The standards published by these two organizations have authority in all *EU* countries as well as the *EFTA* countries who have voted their approval, which means that these standards have replaced their corresponding national standards.

Among the official standards organizations are: in France, *AFNOR (Association Française de Normalisation)*; in the United Kingdom, *BSI (British*

Standards Institution); in Germany, *DIN (Deutsches Institut für Normung)*, in the United States, *ANSI (American National Standards Institute)*.

The national organizations are often relayed into each profession by a body created and financed by this profession and which undertakes all or part of the work in preparing the standards. In the petroleum industry, this role is carried out in France by the *BNPét (Bureau de Normalisation du Pétrole)* and in Germany by the *FAM (Fachausschuss Mineralöl-und Brennstoffnormung)*, in the United Kingdom by the *IP (Institute of Petroleum)*, and in the USA by the *ASTM (American Society for Testing and Materials)*. In the first two cases, the standards are published only by the national organizations (*AFNOR* and *DIN* respectively), while the *IP* and the *ASTM* also publish their own documents, only some of which are adopted by the *BSI* and *ANSI,* respectively.

All these organizations have developed numerous working procedures, with very little difference between each other. These procedures seem at first heavy and cumbersome, but following them allows a consensus to be reached. Thus, for example, free access for all to the standardization commissions' work is guaranteed, and the existence of *"lobbies"* is avoided.

7.3 **Evolution of the Standards and Specifications**

The standards and the specifications are in constant change and should, of course, be continuously kept up-to-date for several reasons:

- Changes in the consumer needs which are manifested by the appearance of new types of products, new manufacturing processes and new uses for products, each requiring specific qualities.
- Changes in technology, particularly when it concerns:
    - New analytical methods. From this point of view, the development of instrument technology is emphasized that provides either gains in productivity or quality, or more advanced analyses.
    - More powerful computers. Associated with analytical instruments, computers have provided the latter with many new possibilities.
    - Calculational methods. Associating the analysis, the knowledge of the property-structure relationships, and the calculation methods has made possible the replacement of costly and arduous test methods by quicker tests whose results are linked by calculations to the characteristic under study. Some examples are the cetane number, in some cases, the octane number, or the characteristics of LPG (refer to Chapter 3).
- Changes in regulations. Today changes in regulations concern essentially fiscal aspects, environmental protection, and the campaign against toxic materials. The new regulations in France are of course linked to those in Europe.

7.4 **Specifications for Petroleum Products in France**

For each category of petroleum products, either governmental or customs specifications or standards for characteristics generally exist, but sometimes there may be nothing but those conditions usually required in commercial contracts.

As an example, the following tables show the main specifications and characteristics that are in effect in France as of December 1993.

These specifications and characteristics are defined with references to standard test methods which the different parties to a contract should conduct for quality control. The tables that follow show specifically the standards that are applicable in France, but a more general table in Appendix 2 shows the main test methods commonly referenced in specifications.

Collections of specifications are kept up-to-date and are available either at the *Chambre Syndicale du Raffinage*[1], or at the *CPDP (Centre Professionnel du Pétrole)*[2], or in documentation available at *AFNOR*[3] and for which the references are given in the tables hereafter.

---

[1] CSR, UFIP, 4 avenue Hoche - 75008 PARIS - Tel. (1) 40537000.
[2] CPDP, Tour Corosa, BP 282 - 92505 RUEIL MALMAISON CEDEX - Tel. (1) 47089404.
[3] AFNOR, Tour Europe, cedex 7 - 92049 PARIS LA DÉFENSE - Tel. (1) 42915555.

Product / Characteristics	Commercial butane	Commercial propane	LPG motor fuel (from NF EN 589) (see AFNOR document M 40-003)
Définition (in vol %)	Mainly butanes/butenes < 19% propane/propylene	90% approx. propane/propylene 10% ethane/ethylene/butanes/butenes	Mainly propane, butanes and propylene/butenes/pentanes/pentenes
Odor	Characteristic	Characteristic	Characteristic and detectable at conc. > 20% lower explosive limit (NF EN 589, Appendix A)
Density (NF M 41-008)	≥ 0.559 kg/l at 15°C (≥ 0.513 kg/l at 5°C)	≥ 0.502 kg/l at 15°C (≥ 0.443 kg/l at 50°C)	
Gauge vapor pressure (NF M 41-010)	≤ 690 kPa at 50°C	≥ 830 kPa at 37.8°C (≥1150 kPa at 50°C) ≤1440 kPa at 37.8°C (≤1930 kPa at 50°C)	
Absolute vapor pressure (NF M 41-010/ISO 4256)			≥ 250 kPa at −10°C ≤ 1550 kPa at 40°C (ISO 4256)
Sulfur content		≤ 0.005 % weight (NF M 41-009)	After odorization (NF EN 24260) ≤ 200 mg/kg
Sulfur compounds	No reaction at sodium plumbite test (NF M 41-006)		Pass hydrogen sulfide test (ISO 8819, future NF EN 28819)
Copper strip corrosion (NF M 41-007)	≤ 1 b	≤ 1 b	≤ 1 (ISO 6251)
Water content	No decanted water	Non-detectable by cobalt bromide test (NF M 41-004)	Pass valve freezing test (ASTM D 2713, ISO 13758)
Evaporation (NF M 41-012)	≤ 1°C Final boiling point (so-called 95% point method) ≤ −15°C		
Evaporation residue (NF M 41-015)			≤ 100 mg/kg
Motor octane number (MON) (NF EN 589 Appendix B)			≥ 89
Diene content as equivalent to 1,3 butadiene (NF EN 27941)			≤ 0.5% mole

Other frequently required characteristics

Chromatographic analysis — M 41-013
Calculation of characteristics (density, vapor pressure, heating value) — M 41-014
Hydrogen sulfide in propane — NF M 41-009
Evaporation residue in butane and propane — NF M 41-015

Table 7.1   *Specifications and test methods for LPG (Liquefied Petroleum Gas).*

Characteristics	Specifications
Definition	Mixture of hydrocarbons of mineral or synthetic origin
Color	Red (4 mg/l of scarlet red)
Density at 15°C (NF T 60-101)	0.700 kg/l ≤ density ≤ 0.750 kg/l
Distillation (NF M 07-002) (vol % including losses)	between 10% and 47% at 70°C between 40% and 70% at 100°C ≥ 85% at 180 °C ≥ 9% at 210°C
Difference between 5% − point 90% (including losses) End point Residue	> 60°C ≤ 215°C ≤ 2% vol
Vapor pressure at 37.8°C (RVP) (NF EN 12) Volatility index (VLI) (VLI = RVP + 7 E70) (where E70 = % evapored at 70°C)	From 20/6 to 9/9 45 kPa ≤ RVP ≤ 79 kPa VLI ≤ 900 From 10/4 to 19/6 and from 10/9 to 31/10 50 kPa ≤ RVP ≤ 86 kPa VLI ≤ 1000 From 1/11 to 9/4 55 kPa ≤ RVP ≤ 99 kPa VLI ≤1150
Sulfur content (NF EN 24260)	≤ 0.20 weight %
Copper strip corrosion (NF M 07-015)	3 h at 50°C: ≤ 1 b
Existent gum content (NF EN 26246)	≤ 10 mg/100 cm^3
Research octane number (RON) (NF EN 25164)	89 ≤ RON ≤ 92
Lead content (NF EN 23830)	< 0.15 g of metallic lead per liter of gasoline (Pb (Et)$_4$ or Pb (Me)$_4$ or their mixture)
Benzene content (NF M 07-062)	≤ 5 vol %
Additives	Authorized by the DHYCA (Ministry of Industry)

Other characteristic often required	
Motor octane number (MON) (MON ≤ 95 in customs specification)	NF EN 25163

Table 7.2	*Specifications and test methods for gasoline (see AFNOR information document M 15-001).*

Characteristics	Specifications
Definition	Mixture of hydrocarbons of mineral or synthetic origin and, possibly, oxygenates
Color	Pale yellow
Density at 15°C (NF T 60-101)	$720 \text{ kg/m}^3 \leq \text{density} \leq 770 \text{ kg/m}^3$
Distillation (NF M 07-002) (vol % including losses)	between 10% and 47% at 70°C between 40% and 70% at 100°C $\geq 85\%$ at 180°C $\geq 90\%$ at 210°C
Difference point 5% − point 90% (including losses) End point Residue	$> 60°C$ $\leq 215°C$ ≤ 2 vol %
Vapor pressure at 37.8°C (RVP) (NF EN 12) Volatility index (VLI) (VLI = RVP + 7 E70) (where E70 = % evaporated at 70°C)	From 20/6 to 9/9 $45 \text{ kPa} \leq \text{RVP} \leq 79 \text{ kPa}$ $\text{VLI} \leq 900$ From 10/4 to 19/6 and from 10/9 to 31/10 $50 \text{ kPa} \leq \text{RVP} \leq 86 \text{ kPa}$ $\text{VLI} \leq 1000$ From 1/11 to 9/4 $55 \text{ kPa} \leq \text{RVP} \leq 99 \text{ kPa}$ $\text{VLI} \leq 1150$
Sulfur content (NF EN 24260)	≤ 0.15 weight %
Copper strip corrosion (NF M 07-015)	3 h at 50°C: ≤ 1 b
Existent gum content (NF EN 26246)	$\leq 10 \text{ mg/100 cm}^3$
Research octane number (RON) and Motor octane number (MON) (NF EN 25164, NF EN 25163)	$97 \leq \text{RON} \leq 99$ $\text{MON} \geq 86$
Lead content (NF EN 23830)	≤ 0.15 g of metallic lead/l of premium gasoline ($Pb \text{ (Et)}_4$ or $Pb \text{ (Me)}_4$ or their mixture)
Benzene content (NF M 07-062)	≤ 5 vol %
Additives	Authorized by the DHYCA (Ministry of Industry)

Other characteristics often required

Stability to oxidation (induction period method)	NF M 07-012
Oxygenates content	NF M 07-054

Table 7.3 *Specifications and test methods for premium gasoline (see AFNOR information document M 15-005).*

Characteristics	Specifications		
Definition	Mixture of hydrocarbons of mineral or synthetic origin and, possibly, oxygenates		
Aspect	Clear and bright		
Color	green (2 mg/l of blue + 2 mg/l of yellow)		
Density (NF T 60-101/NF T 60-172)	725 kg/m^3 ≤ density ≤ 780 kg/m^3 at 15 °C		
Sulfur content (NF EN 24260/M 07-053)	≤ 0.10 weight % (≤ 0.05 weight % as of 1 Jan. 95)		
Copper strip corrosion (NF M 07-015/ISO 2160)	3 h at 50°C: class 1		
Existent gum content after washing (NF EN 26246)	≤ 5 mg/100 ml		
Research octane number (RON) (NF EN 25164) Motor octane number (MON) (NF EN 25163)	RON ≥ 95 MON ≥ 85		
Lead content (EN 237/M 07-061)	≤ 0.013 g/l		
Benzene content (EN 238/ASTM D 2267)	≤ 5 vol %		
Stability to oxidation (NF M 07-012/ISO 7536)	≥ 360 min		
Phosphorus content	No phosphorus compound should be present		
Water tolerance	No separated water		
Volatility	from 20/6 to 9/9 classe 1 from 10/4 to 19/6 and from 10/9 to 31/10 classe 3 from 1/11 to 9/4 classe 6		
	Classe 1	**Classe 3**	**Classe 6**
Vapor pressure RVP (kPa) min (NF EN 12) max	35 70	45 80	55 90
% evaporated at 70°C E70, vol % min (NF M 07-002/ISO 3405) max	15 45	15 45	15 47
Volatility index VLI VLI = 10 RVP + 7 (E70) (RVP en kPa)	≤ 900	≤ 1000	≤ 1150
% evaporated at 100°C, vol % min (NF M 07-002/ISO 3405) max	40 65	40 65	43 70
% evaporated at 180°C, vol % (NF M 07-002/ISO 3405)	≥ 85	≥ 85	≥ 85
Final boiling point (NF M 07-002/ISO 3405)	≤ 215°C	≤ 215°C	≤ 215°C
Distillation residue % vol (NF M 07-002/ ISO 3405)	≤ 2%	≤ 2%	≤ 2%
Ethanol acidity (if present) (ISO 1388/2)	≤ 0,007 % weight % as acetic acid		

Table 7.4	*Specifications and test methods for unleaded premium gasoline (from the standard NF EN 228; see AFNOR information document M 15-023).*

Characteristics	Specifications
Definition	Mixture of hydrocarbons of mineral or synthetic origin
Density at 15°C (NF T 60-101/ISO 3675)	820 kg/m^3 $\leq$ density $\leq$ 860 kg/m^3
Distillation (NF M 07-002/ISO 3405) in vol % (including losses)	< 65% at 250°C $\geq$ 85% at 350°C $\geq$ 95% at 370°C
Viscosity at 40°C (NF T 60-100/ISO 3104)	2 mm^2/s $\leq$ ν < 4.5 mm^2/s
Sulfur content (NF EN 24260/M 07-053)	$\leq$ 0.3 weight % ($\leq$ 0.2% from 1/10/94 to 30/9/96, $\leq$ 0.05% from 1/10/96 onward)
Water content (NF T 60-154/ASTM D 1744)	$\leq$ 200 mg/kg
Ash content (NF EN 26245/M 07-045)	$\leq$ 0.01 weight %
Sediment content (DIN 51419)	$\leq$ 24 mg/kg
Copper strip corrosion (NF M 07-015/ISO 2160)	3 h at 50°C: classe 1
Stability to oxidation (NF M 07-047/ASTM D 2274)	$\leq$ 25 g/m^3
Cetane number (NF M 07-035/ISO 5165)	$\geq$ 49
Cetane index (ISO 4264)	$\geq$ 46
Conradson carbon (on the residue of 10 vol % distillation) (NF T 60-116/ISO 10,370)	$\leq$ 0.30 weight %
Flash point (NF EN 22719)	> 55°C
Additives	Authorized by the DHYCA (Ministry of Industry)
Cold filter plugging point (CFPP) (NF EN 116/M 07-042)	from 1/11 to 30/04 : $\leq$ −15°C from 1/05 to 31/10 : 0°C Severe cold : $\leq$ −20°C

Other characteristics often required	
Cloud point	NF EN 23015
Pour point	NF T 60-105
Neutralization index	NF T 60-112
Sediment content	NF M 07-020

Table 7.5	*Specifications and test methods for diesel fuel (normal and severe cold grades) (from the standard NF EN 590; see AFNOR information document M 15-007 and M 15-022).*

Characteristics	Specifications
Aspect	Clear and bright
Total acidity (ASTM D 3242/IP 354)	≤ 0.015 mg KOH/g
Aromatics (NF M 07-024/ASTM D 1319/IP 156)	≤ 20 vol %
Olefins (NF M 07-024/ASTM D 1319/IP 156)	≤ 5 vol %
Total sulfur (ASTM D 1266/ASTM D 2622)	≤ 0.30 weight %
Mercaptan sulfur (NF M 07-022/ASTM D 3227/IP 342) or "Doctor Test" (NF M 07-029/ASTM D 235/IP 30)	≤ 0.002 weight % Negative
Distillation (NF M 07-002/ASTM D 86/IP 123)	$\leq 204°C$ at 10 vol % $\leq 300°C$ at end point residue ≤ 1.5 vol % losses ≤ 1.5 vol %
Flash point (NF M 07-011/IP 170) (ASTM D 3828/IP 303)	$\geq 38°C$
Density at 15°C (NF T 60-101/ASTM D 1298/IP 360) (NF T 60-172/ASTM D 4052/IP 365)	775 kg/m^3 $\leq$ density ≤ 840 kg/m^3
Freezing point (NF M 07-048/ASTM D 2386/IP 16)	$\leq -47°C$
Viscosity at $-20°C$ (NF T 60-100/ASTM D 445/IP 71)	≤ 8.0 mm^2/s
Net heating value (NF M 07-030/ASTM D 2382/IP 12 or ASTM D 1405/IP 193) or "Aniline Gravity Product" (NF M 07-021/ASTM D 611/ IP 2 and standards for density)	≥ 42.8 MJ/kg ≥ 4800
Smoke point (NF M 07-028/ASTM D 1322) or Luminometer index (ASTM D 1740) or {Smoke point {and naphthalenes ASTM D 1840)	≥ 25 mm ≥ 45 ≥ 20 mm ≤ 3.0 vol %
Copper strip corrosion (NF M 07-015/ASTM D 130/IP 154)	2 h at 100°C : ≤ 1
Silver Corrosion (IP 227)	4 h at 50°C : ≤ 1
Thermal stability (JFTOT) (NF M 07-051/ASTM D 3241/IP 323) ΔP of filter Tube grading (visual) Color deposit	$\leq 25,0$ mmHg ≤ 3 Nil
Existent gum content (NF EN 26246/ASTM D 381/IP 131)	≤ 7 mg/100 ml
Water tolerance (ASTM D1094/IP 289) Interface grading Separation grading	≤ 1 b ≤ 2
WSIM *(Water Separation Index Modified)* (M 07-050/ASTM D 2550/ASTM D 3948)	≥ 85 or ≥ 70 with anti-static additive
Conductivity (ASTM D 2624/IP 274)	50 pS/m $\leq \lambda \leq 450$ pS/m

Table 7.6 *Specifications and test methods for jet fuel. The specifications of jet fuels are set at the international level and are written into the "Aviation Fuel Quality Requirements for Jointly Operated Systems".*

Characteristics	Specifications
Definition	Mixture of hydrocarbons of mineral or synthetic origin
Color	Red
Viscosity at 20°C (NF T 60-100)	≤ 9.5 mm^2/s
Sulfur content (NF EN 24260)	≤ 0.3 weight %
Distillation (NF M 07-002) (in vol % including losses)	$< 65\%$ at 250°C
Flash point (NF T 60-103)	≥ 55°C

Table 7.7   *Specifications and test methods for home-heating oil (in France, FOD) (see AFNOR information document M 15-008).*

Table 7.8, which gathers the French government specifications and test methods concerning hydrocarbon solvents, is divided into three parts:

- Table 7.8a: Special Boiling Point Spirits
- Table 7.8b: White Spirits
- Table 7.8c: Lamp Oils

Characteristics	Specifications		
Definition	Mixture of hydrocarbons of mineral or synthetic origin		
Color	Colorless		
Density at 15°C (NF T 60-101)	SBP A	:	675 kg/m^3 approx.
	SBP B	:	675 kg/m^3 approx.
	SBP C	:	700 kg/m^3 approx.
	SBP D	:	710 kg/m^3 approx.
	SBP E	:	730 kg/m^3 approx.
	SBP F	:	740 kg/m^3 approx.
	SBP G	:	645 kg/m^3 approx.
	SBP H	: ≤ 765 kg/m^3	
Distillation (NF M 07-002) vol % including losses	SBP's A through G : ≥ 10% at 70°C		
		≥ 50% at 140°C	
		≥ 95% at 195°C	
	SBP H	: ≥ 90% at 210°C	
Difference 5% − 90% point (including losses)	SBP's A through G :	≤ 60°C	
	SBP H	: ≥ 60°C	
End Point	SBP H	: ≤ 205°C	
Residue	SBP H	: < 2.5 vol %	

Characteristics	Specifications
Vapor pressure at 37.8°C (NF EN 12)	SBP H : ≤ 70 kPa
Flash point (Abel – Pensky) (NF M 07-036)	SBP's A through G : ≤ 21°C
Sulfur content (NF EN 24260)	SBP H : ≤ 0.20 weight %
Copper strip corrosion (NF M 07-015) 3 h at 50°C	SBP H : ≤ 1 b
Existent gum content (NF EN 26246)	SBP H : ≤ 12 mg/100 cm^3
Lead content (NF EN 23830)	The addition of lead (tetraethyl or tetra methyl) is forbidden

Table 7.8a	*Special Boiling Point Spirits (SBP's) (see AFNOR information document M 15-002 for SBP H).*

Characteristics	Specifications
Definition	Mixture of hydrocarbons of mineral or synthetic origin
Color (NF M 07-003)	≥ 22
Odor	Non unpleasant

Distillation (NF M 07-002)
  Difference point 5% − point 90%
  (including losses)
  Initial Point
  End Point
  Residue
  Losses
	
	$\leq\ 60°C$
	$\geq 135°C$
	$\leq 205°C$
	< 1.5 vol %
	≤ 1 vol %

Sulfur content (NF EN 24260)	≤ 0.05 weight %
Sulfur compounds (NF M 07-029)	No reaction
Copper strip corrosion (NF M 07-015)	3 h at 100°C: $\leq 1a$
Abel flash point (NF M 07-011)	$\geq 30°C$
Aromatics content (NF M 07-024)	Dearomatized white-spirit: ≤ 5 vol %

Table
7.8b   *White-spirits (see AFNOR information documents M 15-006 and M 15-014).*

Characteristics	Specifications
Definition	Mixture of hydrocarbons of mineral or synthetic origin
Aspect	Limpid
Color (NF M 07-003)	≥ 21

Distillation (NF M 07-002)
    vol % including losses
	$< 90\%$ at 210°C
	$\geq 65\%$ at 250°C
	$\geq 80\%$ at 285°C
	Dearomatized: $\leq 90\%$ at 210°C
	    $\geq 65\%$ at 250°C

Initial Point Difference Initial point − End point	Dearomatized: ≥ 180°C Dearomatized: ≤ 65°C
Sulfur content (NF EN 24260)	≤ 0.13 weight %
Copper strip corrosion (NF M 07-015)	3 h at 50°C: ≤ 1 b
Total acidity (NF T 60-112)	≤ 3 mg KOH/100 cm^3
Flash point (NF M 07-011)	≥ 38°C Dearomatized: ≥ 45°C
Smoke point (NF M 07-028)	≥ 21 mm
Aromatics content (NF M 07-024)	≤ 5 vol %

Other characteristics often required	
Benzene content	NF M 07-044, NF M 07-062, ASTM D 2267, ASTM D 3606
Aniline point	NF M 07-012
Bromine number	ASTM D 2710

Table
7.8c   *Lamp oil (see AFNOR information documents M 15-003 and M 15-004).*

Characteristics	Test method
Saybolt color	NF M 07-003
Density	NF T 60-101, NF T 60-172
Distillation	NF T 67-101, NF M 07-002 ISO 3405, ASTM D 1078
Aromatics content	NF M 07-024
Sulfur content	NF T 60-142, NF M 07-059
Vapor pressure	NF EN 12, M 07-079
Copper strip corrosion	NF M 07-015
Mercaptan sulfur	NF M 07-022
H_2S content	UOP 163
Chlorine content	UOP 588/UOP 779
Lead content	Colorimetry (IP 224) or atomic absorption
Ethers content	Gas chromatography
Alcohols content	Gas chromatography

Table
7.9 *Specifications and test methods for naphthas.*
These products are industrial intermediates and are not subject to governmental specifications. The characteristics that are often required in commercial contracts are given below.

Characteristics	Specifications	
Products	FOL N° 1 (M 15-010)	FOL N°2 (M 15-011, BTS: M 15-012, TBTS: M 15-013)
Definition	Mixture of hydrocarbons of mineral or synthetic origin	
Viscosity (NF T 60-100) at 50°C { at 100°C	$>$ 15 mm^2/s $\leq$ 110 mm^2/s	$>$ 110 mm^2/s $<$ 40 mm^2/s
Sulfur content (NF M 07-025)	$\leq$ 2 weight %	N° 2 : $\leq$ 4 weight % N° 2 BTS : $\leq$ 2 weight % N° 2 TBTS : $\leq$ 1 weight %
Distillation (NF M 07-002) vol. % (including losses)	$\leq$ 65% at 250°C $\leq$ 85% at 350°C	
Flash point Luchaire (NF T 60-103) Pensky-Martens (NF EN 22719)	$\geq$ 70°C $\geq$ 60°C (bunker and fishing boats)	
Water content (NF T 60-113)	$\leq$ 0.75 weight %	$\leq$ 1.5 weight %

Other characteristics often required	
Insoluble matter content	NF M 07-063
Conradson Carbon	NF T 60-116
Density	NF T 60-101, NF T 60-172
Ash content	NF M 07-045/EN 7
Pour point	NF T 60-105
Asphaltenes	NF T 60-115
Metals:	
Vanadium	NF M 07-027 and spectrometric methods
Nickel	Spectrometric methods
Sodium	NF M 07-038
Abrasive particles (Si, Al)	Spectrometric methods
Distillation	NF M 07-002
Gross heating value	NF M 07-030
Hydrogen content	Microanalysis

Table 7.10	*Specifications and test methods for heavy fuel oil (in France, FOL). The French specifications distinguish two grades: FOL No. 1 and the heavier FOL No. 2 which can require supplementary specifications such as BTS (Low Sulfur content) and TBTS (Very Low Sulfur content). See AFNOR information documents M 15-010, M 15-011, M 15-012, M 15-013.*

Distillation (NF M 07-002) vol. % (including losses)	≤ 65% at 250°C ≤ 85% at 350°C

Other characteristics often required	
Kinematic viscosities at 40°C and 100°C	NF T 60-100
Viscosity index	NF T 60-136
Pour point	NF T 60-105
Flash point	NF EN 22592
Conradson carbon	NF T 60-116
Neutralization index	NF T 60-112
Density	NF T 60-101, NF T 60-172
Color	NF T 60-104 or NF M 07-003
Foaming tendency	NF T 60-129
Demulsibility	NF T 60-125
Sulfur content	All standardized methods
Hydrocarbon families	NF T 60-155
Polynuclear aromatics content	IP 346

Supplementary characteristics (formulated lubricating oils)	
Analysis of organic functional groups	Infra-Red Absorption
Elemental Analysis	Atomic absorption spectrometry
	X-Ray fluorescence spectrometry
	Plasma emission spectrometry
Water content	Karl Fisher (NF T 60-154) or
	Azeotropic distillation (NF T 60113)
Copper strip corrosion	NF M 07-015
Anti-rust power	NF T 60-151
CCS viscosity	ASTM D 2602
Brookfield viscosity	NF T 60-152
Deaeration	NF T 60-149
Noack volatility	NF T 60-161

Characteristics specified for lubricating greases	
Cone penetrability	NF T 60-132/NF T 60-140
Dropping point	NF T 60-102
Bleeding tendency	IP 121/ASTM D 1742
Apparent viscosity	NF T 60-139
Oxidation	ASTM D 942
Water resistance	ASTM D 1264
Copper strip corrosion	ASTM D 4048
Anti-rust properties	NF T 60-135/ASTM D 1743
Sulfated ash content	NF T 60-144

Complementary characteristics for insulating oils	
Corrosive sulfur	NF T 60-131
Stability to oxidation	CEI 74
Breakdown voltage	CEI 156
Dielectric dissipation factor	CEI 247
Anti-oxidant additives	CEI 666
Leconte de Nouy interfacial tension	ISO 6295

Table 7.11 *Lubricants, industrial oils and related products. There are no French specifications for these products, but there are customs specifications.*

Customs specifications

Characteristics	Specifications			
	Paraffin	Paraffinic residue	Crude wax	Refined wax
Congealing temperature (NF T 60-128)	$\geq 30°C$	$\geq 30°C$	$\geq 30°C$	$\geq 30°C$
Density at 70°C (NF T 60-101)	< 942 kg/m^3	< 942 kg/m^3	< 942 kg/m^3	< 942 kg/m^3
Cone penetrability at 25°C worked (NF T 60-132) non-worked (NF T 60-119)	< 350 < 80	< 350 < 80	< 350 < 80	< 350 < 80
Oil content in weight % (NF T 60-120)	< 3.5	≥ 3.5		
Color			> 3	≤ 3
Viscosity at 100°C (or at congealing temperature $+10°C$)	< 9 mm^2/s	< 9 mm^2/s	$9 \leq v \leq 46$	$9 \leq v \leq 46$

Other characteristics often specified	
Melting point	NF T 60-114/NF T 60-121
Odor	NF T 60-158
Softening point	NF T 60-147
Needle penetration	NF T 60-123

Characteristics for food-grade paraffins	
Codex test	French pharmacopoeia
UV absorption	FDA test "Subpart 172-886" *(Federal Drug Administration)*
Carbonizable material	NF T 60-134

Table 7.12	*Specifications and test methods for paraffins and waxes.* *There are no French specifications for these products, but only the customs specifications.*

Characteristics of bitumen

The characteristics specified in the French standard NF T 65-001 pertain to the following tests:

Ring and ball softening point (RB)	NF T 66-008
Penetrability at 25°C (100 g during 5 s)	NF T 66-004
Relative density (specific gravity) at 25°C	NF T 66-007
Weight loss on heating	NF T 66-011
Hardening per RTFOT *(Rolling thin-film oven test)*	NF T 66-032/ASTM D 2872
Flash point	NF EN 22592
Ductility at 25°C	NF T 66-006
Solubility	NF T 66-012
Paraffin content	NF T 66-015
Fraass brittle point	NF T 66-026

Other characteristic of interest for bitumen

Asphaltene content	NF T 60-115

Characteristics of cutbacks

The characteristics specified in the French standard NF T 65-002 pertain to the following tests:

Pseudo-viscosity	NF T 66-005
Relative density (specific gravity) at 25°C	NF T 66-007
Distillation	NF T 66-003
Penetrability at 25°C on distillation residue	NF T 66-004
Flash point	NF T 66-009

Characteristics of bitumen emulsions

The characteristics specified in the French standard NF T 65-011 pertain to the following tests:

Water content	NF T 66-023
Pseudo-viscosity	NF T 66-020
Homogeneity to screening	NF T 66-016
Particle polarity	NF T 66-021
Rupture index	NF T 66-017, NF T 66-019
Storage stability	NF T 66-022
Passive adhesiveness of a cationic emulsion	NF T 66-018

Table
7.13 *Bitumen.*
There are no specifications, but there are two French standards grouping the characteristics of bitumen and cutbacks.

Characteristics of food-grade white oils

Density at 15°C	NF T 60-101, NF T 60-172
Kinematic viscosity	NF T 60-100
Pour point	NF T 60-105
Flash point, Cleveland	NF EN 22592
Neutralization index	NF T 60-112
Saybolt color	NF M 07-003
Aromatics	UV absorption
	(French pharmacopoeia or ASTM D 2269)
Carbonizable material	French pharmacopoeia

Table
7.14   *Food-grade white oils.*

Evaluation of Crude Oils

Sami G. Chatila

Crude oils appear as liquids of varying viscosities. Their color can range from green (crude from Moonie, Australia) to dark brown (crude from Ghawar, Saudi Arabia). They can have an odor of hydrogen sulfide, turpentine or simply hydrocarbon.

Their chemical compositions are very complex and depend essentially on their age, that is, the phase of development of the kerogene, regardless of the origin of the crude (Speight, 1991) (see Chapter 1).

Knowledge of a crude oil's overall physical and chemical characteristics will determine what kind of initial treatment —associated gas separation and stabilization at the field of production— transport, storage, and of course, price.

A detailed study of the properties of the potential products is of prime technical and economic importance, because it allows the refiner to have a choice in selecting feedstocks for his different units for separation, transformation and conversion, to set their operating conditions, in order to satisfy the needs of the marketplace in the best ways possible.

8.1 Overall Physical and Chemical Properties of Crude Oils Related to Transport, Storage And Price

8.1.1 Specific Gravity of Crude Oils

Specific gravity is important commercially because the crude oil price depends partly on this property. The specific gravity is expressed most often in degrees API (see Chapters 1 and 4).

During loading and unloading of crude oil tankers, the specific gravity of the crude is measured to confirm it meets the specifications for the case where payment is made on a barrel basis, or when the volume is converted into weight if the transaction is based on a price per ton (Hayward et al., 1980).

Within the same geographical region, the crude specific gravity varies from one reservoir to another. In Saudi Arabia, for example, the crude from the Ghawar field has an average standard specific gravity on the order of 0.850 (34° API) while the specific gravity of the crude from the nearby Safaniyah field is 0.893 (27° API).

Within the same reservoir, we also observe variations of specific gravity from one well to another: for example, 0.848 (38.4° API) and 0.861 (32.8° API) in the Ghawar field.

The specific gravities of crudes fall generally between 0.800 and 1.000 as shown in Table 8.1, even though crudes having specific gravities outside this range exist: 0.787 (48.2° API) for crudes in Barrow South, Alaska, USA and Santa Rosa, Venezuela and 1.028 (6° API) for the crude from Bradley Canyon, California, USA.

Generally, the crude oils are classed according to specific gravity in four main categories:

- light crudes $\qquad$ sp. gr. $d_4^{15} < 0.825$
- medium crudes $\qquad$ $0.825 < $ sp. gr. $d_4^{15} < 0.875$
- heavy crudes $\qquad$ $0.875 < $ sp. gr. $d_4^{15} < 1.000$
- extra-heavy crudes $\qquad$ sp. gr. $d_4^{15} > 1.000$

Table 8.1 gives the average specific gravities of some typical crude oils.

Crude oil name	Country of origin	Specific gravity $\left(d_4^{15}\right)$
Hassi Messaoud	Algeria	0.804
Bu-Attifel	Libya	0.822
Arjuna	Indonesia	0.836
Bonny Light	Nigeria	0.837
Kirkuk	Iraq	0.845
Ekofisk	North Sea (Norway)	0.846
Minas	Indonesia	0.845
Arabian Light	Saudi Arabia	0.858
Kuwait	Kuwait	0.870
Cyrus	Iran	0.940
Boscan	Venezuela	1.000

Table 8.1   *Specific gravities of typical crude oils.*

8.1.2 **Crude Oil Pour Point**

When crude petroleum is cooled, there is no distinct change from liquid to solid as is the case for pure substances. First there is a more or less noticeable change in viscosity, then, if the temperature is lowered sufficiently, the crude oil ceases to be fluid, and approaches the solid state by thickening. This happens because crude oil is a complex mixture in which the majority of components do not generally crystallize; their transition to the solid state does not therefore occur at constant temperature, but rather along a temperature range, for which the parameters are a function of the crude oil's previous treatment. Knowledge of the crude's previous history is very important. Preheating to 45-65°C lowers the temperature of the pour point because the crude petroleum contains seeds of paraffinic crystals, and these are destroyed during preheating. If the crude is preheated to a higher temperature (about 100°C), an increase in pour point is observed which is due to the vaporization of light hydrocarbons; the crude has become heavier.

The pour point of crude oils is measured to give an approximate indication as to their "pumpability". In fact, the agitation of the fluid brought on by pumping can stop, slow down or destroy the formation of crystals, conferring on the crude additional fluidity beyond that of the measured pour point temperature.

The measurement of this temperature is defined by the standard NF T 60-105 and by the standard ASTM D 97.

Both test stipulate moreover, preheating the sample to 45 to 48°C.

Crude oil pour points usually are between −60°C and +30°C (Table 8.2).

Crude oil name	Country of origin	Pour point, °C
Hassi Messaoud	Algeria	−60
Zarzaïtine	Algeria	−24
Dahra	Libya	−1
Ozouri	Gabon	−16
Abqaiq	Saudi Arabia	−24
Kuwait	Kuwait	−42
Gash Saran	Iran	−12
Bachaquero	Venezuela	+15
Boscan	Venezuela	+15

Table 8.2   *Pour points for selected crudes.*

8.1.3 **Viscosity of Crude Oils**

The measurement of a crude oil's viscosity at different temperatures is particularly important for the calculation of pressure drop in pipelines and refinery piping systems, as well as for the specification of pumps and exchangers.

The change in viscosity with temperature is not the same for all crudes.

The viscosity of a paraffinic crude increases rapidly with decreasing temperature; on the other hand, for the naphthenic crudes, the increase in viscosity is more gradual.

The viscosity is determined by measuring the time it takes for a crude to flow through a capillary tube of a given length at a precise temperature. This is called the kinematic viscosity, expressed in mm^2/s. It is defined by the standards, NF T 60-100 or ASTM D 445. Viscosity can also be determined by measuring the time it takes for the oil to flow through a calibrated orifice: standard ASTM D 88. It is expressed in Saybolt seconds (SSU).

Some conversion tables for the different units are used and standardized (ASTM D 2161).

Certain calibrated orifice instruments (Engler-type) provide viscosity measurements at temperature lower than pour point. This is possible because the apparatus agitates the material to the point where large crystals are prevented from forming; whereas in other methods, the sample pour point is measured without agitation.

This is, for example, the case for crude from Dahra (Libya) which, with a pour point of $-1°C$, gives a viscosity of $2.4°E$ or $16 \ mm^2/s$ at $0°C$, or the crude from Coulomnes (France) whose viscosity is close to $20°E$ at $0°C$ whereas its pour point is $+12°C$.

Table 8.3 gives the viscosity of some crude oils at $20°C$.

Crude oil name	Country of origin	Viscosity, mm^2/s
Zarzaïtine	Algeria	5
Nigerian	Nigeria	9
Dahra	Libya	6
Safaniyah	Saudi Arabia	48
Bachaquero	Venezuela	5500
Tia Juana	Venezuela	70

Table 8.3 *Viscosity of selected crude oils at 20°C.*

8.1.4 **Vapor Pressure and Flash Point of Crude Oils**

The measurement of the vapor pressure and flash point of crude oils enables the light hydrocarbon content to be estimated.

The vapor pressure of a crude oil at the wellhead can reach 20 bar. If it were necessary to store and transport it under these conditions, heavy walled equipment would be required. For that, the pressure is reduced (< 1 bar) by separating the high vapor pressure components using a series of pressure reductions (from one to four "flash" stages) in equipment called "separators", which are in fact simple vessels that allow the separation of the two liquid and vapor phases formed downstream of the pressure reduction point. The different components distribute themselves in the two phases in accordance with equilibrium relationships.

The resulting vapor phase is called "associated gas" and the liquid phase is said to be the crude oil. The production of gas is generally considered to be unavoidable because only a small portion is economically recoverable for sale, and yet the quantity produced is relatively high. The reservoirs in the Middle East are estimated to produce 0.14 ton of associated gas per ton of crude.

Safety standards govern the manipulation and storage of crude oil and petroleum products with regard to their flash points which are directly linked to vapor pressure.

One generally observes that crude oils having a vapor pressure greater than 0.2 bar at 37.8°C (100°F), have a flash point less than 20°C.

During the course of operations such as filling and draining tanks and vessels, light hydrocarbons are lost. These losses are expressed as volume per cent of liquid. According to Nelson (1958), the losses can be evaluated by the equation:

$$\text{Losses (volume \%)} = \frac{\text{RVP} - 1}{6}$$

the Reid vapor pressure being expressed in psi, (pounds per square inch)*.

To reduce these losses, the crude oils are stored in floating roof tanks.

The measurement of vapor pressure is defined by the standards NF M 07-007 and ASTM D 323, flash points by the standards NF M 07-011 and ASTM D 56. (see Chapter 7).

Table 8.4 gives the vapor pressures and the flash points of some crude oils.

---

* 1 psi = 6.9 kPa.

Crude oil name	Country of origin	RVP (bar)	Flash point (°C)
Hassi Messaoud	Algeria	0.75	< 20
Nigerian	Nigeria	0.26	< 20
Kirkuk	Iraq	0.29	< 20
Qatar	Qatar	0.50	< 20
Kuwait	Kuwait	0.51	< 20
Bachaquero	Venezuela	0.06	46

Table 8.4   *Reid vapor pressures and flash points of selected crude oils.*

8.1.5 **Sulfur Content of Crude Oils**

Crude oils contain organic sulfur compounds, dissolved hydrogen sulfide, and sometimes even suspended sulfur. Generally speaking, the total sulfur content of a crude is between 0.05 and 5% by weight (Table 8.5), ratios that are in accordance with organic materials thought to be the origin of crude oil*.

Note, however, the particular cases of sulfur in crude oil from Rozel Point (Utah, USA): 13.95% in crudes from Etzel (Germany): 9.6% or crude from Gela

Crude oil name	Country of origin	Weight % sulfur
Bu Attifel	Libya	0.10
Arjuna	Indonesia	0.12
Bonny Light	Nigeria	0.13
Hassi Messaoud	Algeria	0.14
Ekofisk	North Sea (Norway)	0.18
Arabian Light	Saudi Arabia	1.80
Kirkuk	Iraq	1.95
Kuwait	Kuwait	2.50
Cyrus	Iran	3.48
Boscan	Venezuela	5.40

Table 8.5   *Sulfur content of selected crude oils.*

---

* For purposes of information one can compare these levels with those of plants (0.1 to 0.4%), mollusks (0.4%) and the human body (0.14%), but the highest levels are observed in marine algae (13% in Macrocytis pyrifera), and in the tissues of certain bacteria which can even contain elemental sulfur (25% in *Beggiatoa Albea* tissues).

(Sicily): 7.8%, which are too high to be able to say with certainty that sulfur comes only from organic material alone.

8.1.5.1 **Origin of Sulfur** (Rall et al., 1972)

Sulfur comes mainly from the decomposition of organic matter, and one observes that with the passage of time and of gradual settling of material into strata, the crude oils lose their sulfur in the form of H_2S that appears in the associated gas, a small portion stays with the liquid. Another possible origin of H_2S is the reduction of sulfates by hydrogen by bacterial action of the type desulforibrio desulfuricans (Equation 8.1):

$$4H_2 + SO_4^= \xrightarrow{\text{(bacteria)}} H_2S + 2OH^- + 2H_2O \tag{8.1}$$

Hydrogen comes from the crude and the sulfate ions are held in the reservoir rock.

The H_2S formed can react with the sulfates or rock to form sulfur (Equation 8.2) that remains in suspension as in the case of crude from Goldsmith, Texas, USA, or that, under the conditions of pressure, temperature and period of formation of the reservoir, can react with the hydrocarbons to give sulfur compounds:

$$3H_2S + SO_4^= \longrightarrow 4S + 2OH^- + 2H_2O \tag{8.2}$$

H_2S reacts in another way with the olefinic hydrocarbons producing thiols and sulfur compounds (Equation 8.3 and 8.4):

$$CH_3 - CH = CH_2 + H_2S \longrightarrow \underset{\underset{S-H}{|}}{CH_3 - CH - CH_3} \tag{8.3}$$

$$\underset{\underset{S-H}{|}}{CH_3 - CH - CH_3} + CH_3 - CH = CH_2 \longrightarrow CH_3 - \underset{\underset{CH_3}{|}}{CH} - S - \underset{\underset{CH_3}{|}}{CH} - CH_3 \tag{8.4}$$

These reactions can explain the absence of olefins in crude oil, their presence being detected only in the crudes of low sulfur content. The sulfur content in crude from Bradford which is the one of the rare crudes containing olefins is about 0.4%.

Knowledge of the nature and quantity of sulfur compounds contained in crudes and petroleum cuts is of prime importance to the refiner, because in constitutes a constraint in the establishment of refinery flow sheets and the preparation of finished products. In fact a few of these products contain or entrain corrosive materials which, during refinery operations, reduce the service life of certain catalysts such as reforming catalysts, degrade the quality of finished products by changing their color and by giving them an unpleasant odor, reduce the service life of lubricating oils, without mentioning atmospheric pollution from formation of SO_2 and SO_3 during the combustion of petroleum fuels, and fires caused by contact between iron sulfide on the piping and air.

8.1.5.2 **Nature of Sulfur Compounds
Contained in Crude Oil**

Practically, one measures the quantity of total sulfur (in all its forms) contained in crude oil by analyzing the quantity of SO_2 formed by the combustion of a sample of crude, and the result is taken into account when evaluating the crude oil price. When they are present, elementary sulfur and dissolved H_2S can also be analyzed.

The sulfur compounds are classed in six chemical groups.

a. Free Elemental Sulfur S

Free sulfur is rarely present in crude oils, but it can be found in suspension or dissolved in the liquid. The crude from Goldsmith (Texas, USA.) is richest in free sulfur (1% by weight for a total sulfur content of 2.17%). It could be produced by compounds in the reservoir rock by sulfate reduction (reaction 8.2).

b. Hydrogen Sulfide H_2S

H_2S is found with the reservoir gas and dissolved in the crude (< 50 ppm by weight), but it is formed during refining operations such as catalytic cracking, hydrodesulfurization, and thermal cracking or by thermal decomposition of sulfur-containing hydrocarbons during distillation.

In the 1950's, crude oils were either corrosive (sour), or non-corrosive (sweet). Crudes containing more than 6 ppm of dissolved H_2S were classed as sour because, beyond this limit, corrosion was observed on the walls of storage tanks by formation of scales of pyrophoric iron sulfides.

At this point in time, the total sulfur content of crudes was not taken into consideration, since most of them were produced and refined in the United-States and contained less than 1%, and only the gasoline coming from corrosive crudes needed sweetening (elimination of thiols) for them to meet the specifications then in force. Today all crudes containing more than one per cent sulfur are said to be "corrosive".

c. The Thiols

Of the general formula, $R - S - H$, where R represents an aliphatic or cyclic radical, the thiols — also known as mercaptans — are acidic in behavior owing to their $S - H$ functional group; they are corrosive and malodorous. Their concentration in crude oils is very low if not zero, but they are created from other sulfur compounds during refining operations and show up in the light cuts, as illustrated in Table 8.6.

Table 8.7 gives some of the mercaptans identified in crude oils.

Nature of cut (temperature interval, °C)		Mercaptan sulfur, %	Total sulfur, %	% mercaptan sulfur / total sulfur
Crude petroleum		0.0110	1.8	0.6
Butane		0.0228	0.0228	100
Light gasoline	(20-70°C)	0.0196	0.0240	82
Heavy gasoline	(70-150°C)	0.0162	0.026	62
Naphtha	(150-190°C)	0.0084	0.059	14
Kerosene	(190-250°C)	0.0015	0.17	0.9
Gas oil	(250-370°C)	0.0010	1.40	< 0.1
Residue	(370⁺°C)	0	3.17	0

Table 8.6 *Distribution of mercaptan sulfur among the different cuts of Arabian Light crude oil.*

Name	Chemical formula	Boiling point, °C	Cut
Methanethiol	$CH_3 - SH$	6	Butane Gasoline
Ethanethiol	$CH_3 - CH_2 - SH$	34	Gasoline
2 methylpropanethiol	$CH_3 - CH - CH_2 - SH$ $\quad\quad\mid$ $\quad\quad CH_3$	85	Gasoline
2 methylheptanethiol	$CH_3 - CH - (CH_2)_5 - SH$ $\quad\quad\mid$ $\quad\quad CH_3$	186	Kerosene
Cyclohexanethiol	CH_2 $CH_2 \quad CH_2$ $CH_2 \quad CH_2$ $\quad CH$ $\quad\mid$ $\quad S-H$	159	Gasoline

Table 8.7 *Mercaptans identified in crude oils.*

d. The Sulfides

The sulfides are chemically neutral; they can have a linear or ring structure. For molecules of equal carbon number, their boiling points are higher than those of mercaptans; they constitute the majority of sulfur containing hydrocarbons in the middle distillates (kerosene and gas oil).

Table 8.8 gives some examples of sulfides identified in crude oils.

Name	Chemical formula	Boiling point, °C	Cut
3 Thiapentane	$CH_3 - CH_2 - S - CH_2 - CH_3$	92	Gasoline
2 Methyl – 3 thiapentane	$CH_3 - CH - S - CH_2 - CH_3$ $\quad\quad\;\; \mid$ $\quad\quad CH_3$	108	Gasoline
Thiacyclohexane		141.8	Gasoline
2 Methylthiacyclo-pentane		133	Gasoline
Thiaindane		235.6	Kerosene
Thiabicyclooctane		194.5	Kerosene and gas oil

Table 8.8   *Sulfides identified in the crude oils.*

e. The Disulfides

These compounds are difficult to separate and, consequently, few have been identified:

- Dimethyl disulfide   (2,3 dithiobutane)

  $CH_3 - S - S - CH_3$

- Diethyl disulfide   (2,3 dithiohexane)

  $CH_3 - CH_2 - S - S - CH_2 - CH_3$

It is only recently that the more complex substances like trinaphtheno-diphenyldisulfide have been able to be identified.

f. Thiophene and Derivatives

The presence of thiophene and its derivatives in crude oils was detected in 1899, but until 1953, the date at which the methyl-thiophenes were identified in kerosene from Agha Jari, Iran crude oil, it was believed that they came from the degradation of sulfides during refining operations. Finally, their presence was no longer doubted after the identification of benzothiophenes and their derivatives (Table 8.9), and lately of naphthenobenzothiophenes in heavy cuts.

Name	Chemical formula	Boiling point °C	Cut
Thiophene	$\begin{smallmatrix} C - C \\ \parallel \quad \parallel \\ C \diagdown_S\diagup C \end{smallmatrix}$	84	Gasoline
Dimethylthiophene	$\begin{smallmatrix} C - C \diagup^{CH_3} \\ \parallel \quad \parallel \\ C \diagdown_S \diagup C \diagdown_{CH_3} \end{smallmatrix}$	141.6	Gasoline and Kerosene
Benzothiophene	(benzothiophene ring structure)	219,9	Kerosene
Dibenzothiophene	(dibenzothiophene ring structure)	300	Gas oil

Table 8.9	Thiophene derivatives identified in crude oils.

The major part of the sulfur contained in crude petroleum is distributed between the heavy cuts and residues (Table 8.10) in the form of sulfur compounds of the naphthenophenanthrene or naphthenoanthracene type, or in the form of benzothiophenes, that is, molecules having one or several naphthenic and aromatic rings that usually contain a single sulfur atom.

Note that the total sulfur levels are different from those appearing in Table 8.6 as a result of having different distillation ranges.

Cut	Light gasoline	Heavy gasoline	Kerosene	Gas oil	Residue	Crude
Temperature interval, °C	20-70	70-180	180-260	260-370	370$^+$	
Specific gravity d_4^{15}	0.648	0.741	0.801	0.856	0.957	
Average molecular weight	75	117	175	255	400	
Total sulfur, weight %	0.024	0.032	0.202	1.436	3.167	1.80
$\dfrac{\text{Number of moles of sulfides}}{\text{Total number of moles}}$	$\dfrac{1}{1800}$	$\dfrac{1}{855}$	$\dfrac{1}{90}$	$\dfrac{1}{9}$	$\dfrac{1}{2.5}$	

Table 8.10	Distribution of total sulfur in the different cuts of crude Arabian Light (as in Table 8.6).

8.1.6 **Nitrogen Content of Crude Oils**

Crude oils contain nitrogen compounds in the form of basic substances such as quinoline, isoquinoline, and pyridine, or neutral materials such as pyrrole, indole, and carbazole.

These compounds can be malodorous as in the case of quinoline, or they can have a pleasant odor as does indole. They decompose on heating to give organic bases or ammonia that reduce the acidity of refining catalysts in conversion units such as reformers or crackers, and initiate gum formation in distillates (kerosene, gas oil).

Table 8.11 gives some typical nitrogen and sulfur values.

Crude	% S	% N	N/S × 100
Kirkuk	2.0	0.10	5
Kuwait	2.5	0.15	6
Gash Saran	1.6	0.23	14

Table 8.11 *Nitrogen and sulfur contents and their N/S ratios in selected crudes.*

8.1.7 **Water, Sediment, and Salt Contents in Crude Oils**

Crude oils contain, in very small quantities, water, sediments and mineral salts most of which are dissolved in the water, the remainder found as very fine crystals.

These materials can damage equipment by means of corrosion, erosion, deposits, plugging, catalyst poisoning, etc.

8.1.7.1 **Water Content of Crude Oils**

In the crude, water is found partly in solution and partly in the form of a more-or-less stable emulsion; this stability is due to the presence of asphaltenes or certain surfactant agents such as mercaptans or naphthenic acids.

The water content of crude oils is determined by a standardized method* whose procedure is to cause the water to form an azeotrope with an aromatic (generally industrial xylene). Brought to ambient temperature, this azeotrope separates into two phases: water and xylene. The volume of water is then measured and compared with the total volume of treated crude.

---

* NF T 60-113 and ASTM D 95.

The water content of crude oils at the wellhead is usually small as shown in Table 8.12; it generally increases during transport and storage and can attain 3%.

Crude oil name	Country of origin	Water content in volume %
Dahra	Libya	Traces
Safaniyah	Saudi Arabia	Traces
Arabian Light	Saudi Arabia	Traces
Zarzaïtine	Algeria	0.05
Mandgi	Gabon	0.6
Bachaquero	Venezuela	1.8

Table 8.12	*Water content of selected crudes.*

8.1.7.2 **Sediments (Bottom Sediments)**

Solids materials that are insoluble in hydrocarbon or water can be entrained in the crude. These are called "bottom sediments" and comprise fine particles of sand, drilling mud, rock such as feldspar and gypsum, metals in the form of minerals or in their free state such as iron, copper, lead, nickel, and vanadium. The latter can come from pipeline erosion, storage tanks, valves and piping systems, etc.: whatever comes in contact with the crude oil.

The presence of such substances in crude oil is highly undesirable because they can plug piping and contaminate the products.

During storage, sediments decant with the water phase and deposit along with paraffins and asphalts in the bottoms of storage tanks as thick sludges or slurries (BS&W). The interface between the water-sediment and the crude must be well monitored in order to avoid pumping the slurry into the refinery's operating units where it can cause serious upsets.

To lessen the risk of pumping sludges or slurries into a unit, the practice is to leave a safety margin of 50 cm (heel) below the outlet nozzle or install a strainer on the pump suction line. The deposits accumulate with time and the tanks are periodically emptied and cleaned.

The water and sediment contents of crude oils is measured according to the standard methods NF M 07-020, ASTM D 96 and D 1796, which determine the volume of water and sediments separated from the crude by centrifuging in the presence of a solvent (toluene) and of a demulsifying agent; Table 8.13 gives the bottom sediment and water content of a few crude oils.

Crude oil name	Water and sediment contents (BS&W) in volume %
Nigerian	0.1
Arabian Light	0.1
Dahra	0.6
Mandgi	0.8
Bachaquero	2.0

Table
8.13   *Bottom sediments and water content of some crude oils.*

8.1.7.3 Salt Content of Crude Oils

Regardless of their presence in very small amounts, on the order of a few dozen ppm (Table 8.14), mineral salts cause serious problems during crude oil treatment.

Chlorides of sodium, magnesium and calcium are almost always the prevailing compounds, along with gypsum and calcium carbonate.

The measurement of chlorides is standardized (NF M 07-023, ASTM 3230); the result of two measurements is expressed in mg of NaCl/kg of crude. Table 8.14 gives the contents of some crude oils; these values come from measurements taken in a refinery and thereby include the salts brought in by contamination.

The presence of salts in crude oils has several disadvantages:

• During production; sodium chloride can deposit in layers on tubing walls after partial vaporization of the water due to the pressure drop between bottomhole and wellhead; when these deposits become important large enough, the diameter of the well tubing is reduced,

Crude oil name	Country of origin	NaCl mg/kg (ppm weight)
Arabian Light	Saudi Arabia	25
Agha Jari	Iran	25
Hassi Messaoud	Algeria	30
Kuwait	Kuwait	35
Boscan	Venezuela	60
Bonny	Nigeria	135
Brega	Libya	155
Safaniyah	Saudi Arabia	280
Sarir	Libya	345

Table
8.14   *Salt contents of various crudes.*

which causes production loss. In order to reduce the impact of such incidents, freshwater is injected.

- In the refinery; the salts deposit in the tubes of exchangers and reduce heat transfer, while in heater tubes, hot spots are created favoring coke formation.

The major portion of salt is found in residues; as these streams serve as the bases for fuels, or as feeds for asphalt and petroleum coke production, the presence of salt in these products causes fouling of burners, the alteration of asphalt emulsions, and the deterioration of coke quality. Furthermore, calcium and magnesium chlorides begin to hydrolyze at 120°C. This hydrolysis occurs rapidly as the temperature increases (Figure 8.1) according to the reaction:

$$MgCl_2 + 2H_2O \longrightarrow Mg(OH)_2 + 2HCl$$

Hydrogen chloride released dissolves in water during condensation in the crude oil distillation column overhead or in the condenser, which cause corrosion of materials at these locations. The action of hydrochloric acid is favored and accelerated by the presence of hydrogen sulfide which results in the decomposition of sulfur-containing hydrocarbons; this forces the refiner to inject a basic material like ammonia at the point where water condenses in the atmospheric distillation column.

In addition, salts deactivate reforming and catalytic cracking catalysts.

These hazards are reduced drastically by desalting crude oils, a process which consists of coalescing and decanting the fine water droplets in a vessel by using an electric field of 0.7 to 1 kV/cm.

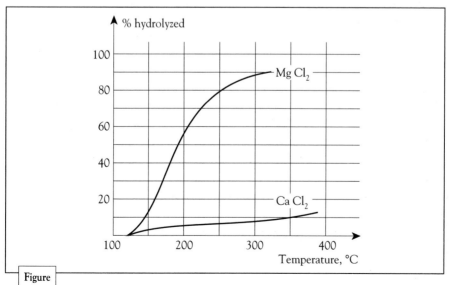

Figure 8.1   *Hydrolysis of chlorides contained in the crude as a function of temperature (salt content = 35 ppm weight − test duration = constant).*

8.1.7.4 **Crude Oil Acid Number**

Crude oils contain carboxylic acids. These are analyzed by titration with potassium hydroxide and the result of the analysis is expressed in mg of KOH/g crude.

Table 8.15 gives the acid content of some crudes.

Crude oil name	Country of origin	mg KOH/g
Hassi Messaoud	Algeria	0
Arabian Light	Saudi Arabia	0.07
Nigerian	Nigeria	0.15
Agha Jari	Iran	0.22
Bachaquero	Venezuela	2.9

Table **8.15**   *Acid numbers of certain crudes in mg KOH/g.*

It is worthwhile to mention that the distribution of naphthenic acids is not uniform in a crude oil since a maximum value is observed in the fractions distilled between 400 and 450°C and whose average specific gravity is 0.950 (Figure 8.2).

In the light or medium cuts, the acids are linear as in valeric acid $CH_3 - (CH_2)_3 - COOH$ or stearic acid $CH_3 - (CH_2)_{16} - COOH$.

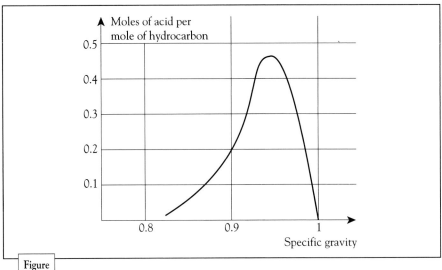

Figure **8.2**   *Distribution of the acid content as a function of the specific gravity of successive distillation fractions.*

The majority of acids contained in the diesel cuts are cyclic and come from cyclopentane or cyclohexane. They are better known as naphthenic acids.

p. Methylcyclohexane carboxylic acid	Methyl - 3 cyclopentylacetic acid

The presence of these acids in crude oils and petroleum cuts causes problems for the refiner because they form stable emulsions with caustic solutions during desalting or in lubricating oil production; very corrosive at high temperatures (350-400°C), they attack ordinary carbon steel, which necessitates the use of alloy piping materials.

8.2 TBP Crude Oil Distillation – Analysis of Fractions

The TBP (True Boiling Point) distillation gives an almost exact picture of a crude petroleum by measuring the boiling points of the components making up the crude; whence its name.

Crude petroleum is fractionated into around fifty cuts having a very narrow distillation intervals which allows them to be considered as ficticious pure hydrocarbons whose boiling points are equal to the arithmetic average of the initial and final boiling points, $T_m = (T_i + T_f)/2$, the other physical characteristics being average properties measured for each cut.

The different cuts obtained are collected; their initial and final distillation temperatures are recorded along with their weights and specific gravities. Other physical characteristics are measured: for the light fractions octane number, vapor pressure, molecular weight, PONA, weight per cent sulfur, etc., and, for the heavy fractions, the aniline point, specific gravity, viscosity, sulfur content, and asphaltene content, etc.

The determination of properties for each cut enables curves to be obtained for yields and properties as well as curves for iso-properties that are useful in the economic analyses of crude oils.

It is possible to calculate the properties of wider cuts given the characteristics of the smaller fractions when these properties are additive in volume, weight or moles. Only the specific gravity, vapor pressure, sulfur content, and aromatics content give this advantage. All others, such as viscosity, flash point, pour point, need to be measured. In this case it is preferable to proceed with a TBP distillation of the wider cuts that correspond with those in an actual refinery whose properties have been measured.

The most commonly referenced so-called wide cuts are the following:

Gas	$C_3 - C_4$
Light debutanized gasoline	$\left\{ \begin{array}{l} C_5 - 70°C \\ C_5 - 80°C \\ C_5 - 100°C \end{array} \right.$
Heavy gasoline	$\left\{ \begin{array}{l} 70 - 140°C \\ 80 - 180°C \\ 100 - 180°C \end{array} \right.$
Kerosene	$\left\{ \begin{array}{l} 160 - 260°C \\ 180 - 260°C \end{array} \right.$
Gas oil	$\left\{ \begin{array}{l} 260 - 325°C \\ 260 - 360°C \\ 260 - 370°C \\ 160 - 360°C \end{array} \right.$
Residue	$\left\{ \begin{array}{l} T > 325°C \\ T > 360°C \\ T > 370°C \end{array} \right.$

All the analytical results are represented as curves that enable easy and rational utilization.

In certain cases, the curves, $T = f$ (% distilled) or specific gravity = f (% distilled) curves show irregularities in light cuts that are rich in aromatics (Figure 8.3 – Saharan Crude). Moving toward the heavy cuts, the curves become smoother because the number of isomers becomes very large and their boiling points and specific gravities are very close in value.

Certain curves, $T = f$ (% distilled), level off at high temperatures due to the change in pressure and to the utilization of charts for converting temperatures under reduced pressure to equivalent temperatures under atmospheric pressure.

Currently the charts used most often for this purpose are those published by the API. (Maxwell and Bonnel charts) from the Technical Data Book (see Chapter 4).

8.3 Graphical Representation of Analyses and Utilization of the Results

8.3.1 Graphical Representation

The analytical results are represented as tables or curves and are usually used with a computer and an appropriate program.

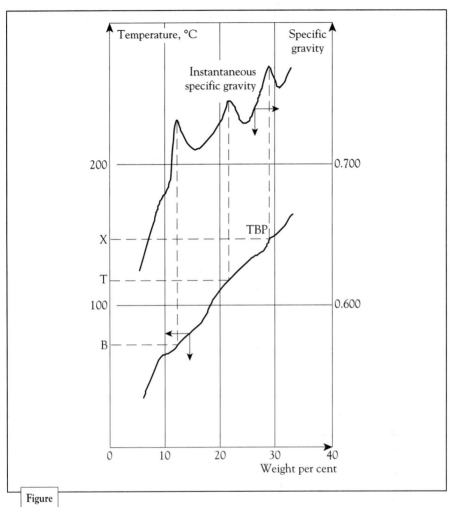

Figure
8.3 | *Initial portion of the TBP curve of a Saharan crude oil (Note the discontinuities due to the presence of aromatics: benzene B, toluene T, xylenes X).*

Showing the results as curves enables manual calculations to be made which are often useful in rough estimates.

8.3.1.1 Mid-Per Cent or Instantaneous Property Curves

The most important curve is the TBP distillation, properly defined as $T = f$ (% volume or weight). Figure 8.4 shows the distillation curves for an Arabian Light crude. The chart is used to obtain yields for the different cuts as a function of the selected distillation range.

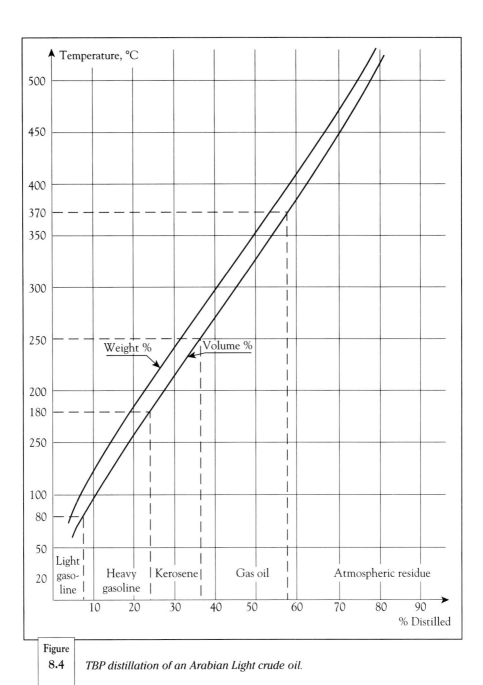

Figure 8.4 *TBP distillation of an Arabian Light crude oil.*

The graph gives the yields that the refiner would obtain at the outlet of the atmospheric distillation unit allowing him to set the unit's operating conditions in accordance with the desired production objectives.

8.3.1.2 **Property – Yield Curves**

These curves are drawn for all properties having for the ordinate axis an appropriate scale of the property, and for the abscissa the yield in volume or weight.

Generally, these curves give the properties of "gasolines" and residues.

In order to draw the property-yield curves for "gasolines", it suffices to choose the initial point, which could be C_5 or 20°C, the end point being variable and situated between the end point of the heaviest gasoline cut which can be produced (200-220°C) and about 350°C.

In order to draw the property-yield curves for residues, the end point is set, the initial point is variable and can be as low as 300°C.

It is advisable to provide re-sectioning between these two types of curves so that the properties of any cut can be deduced.

These curves are given in Figures 8.5 and 8.6 for gasolines, 8.7, 8.8, and 8.9 for residues.

Each point corresponding to the ordinate axis is the value of the cumulative property of the cut. The C_5-EP properties of "gasoline" cuts or IP-EP properties of "residue" cuts are obtained directly from the curves, while properties of other cuts are calculated either directly for the properties that are additive by volume, weight or moles, or by using blending indices.

8.3.1.3 **Iso-Property Curves**

This type of curve can be utilized for intermediate cuts between 250 and 400°C. They show the value of the property of a cut as a function of its initial point and its end point.

Unlike the property-yield curves, calculations are not necessary for determining the properties of a cut.

Figure 8.10 shows the aniline points and the pour points for intermediate cuts from an Arabian Light crude.

8.3.2 **Using the Curves**

Once the distillation intervals of cuts coming from atmospheric distillation and vacuum distillation are specified, the preceding curves give the properties of the selected cuts.

The comparison between the qualities and quantities obtained and those for marketable products, the refiner can estimate the capacities and the operating conditions of the various treatment units

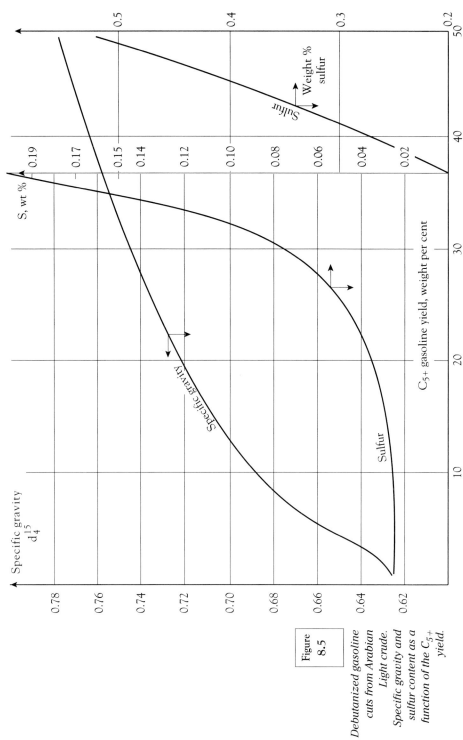

Figure
8.5

Debutanized gasoline cuts from Arabian Light crude. Specific gravity and sulfur content as a function of the C_{5+} yield.

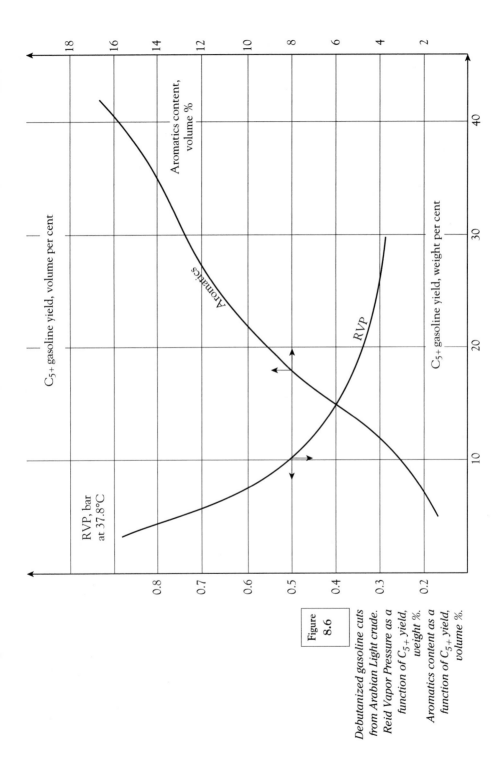

Figure 8.6

Debutanized gasoline cuts from Arabian Light crude. Reid Vapor Pressure as a function of C_{5+} yield, weight %. Aromatics content as a function of C_{5+} yield, volume %.

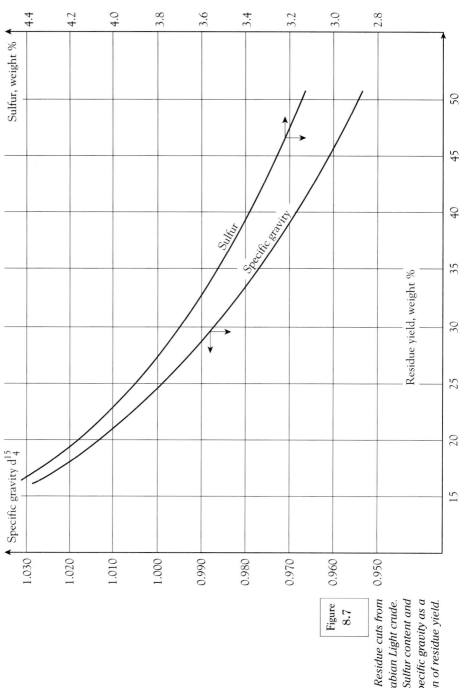

Figure 8.7

Residue cuts from Arabian Light crude. Sulfur content and specific gravity as a function of residue yield.

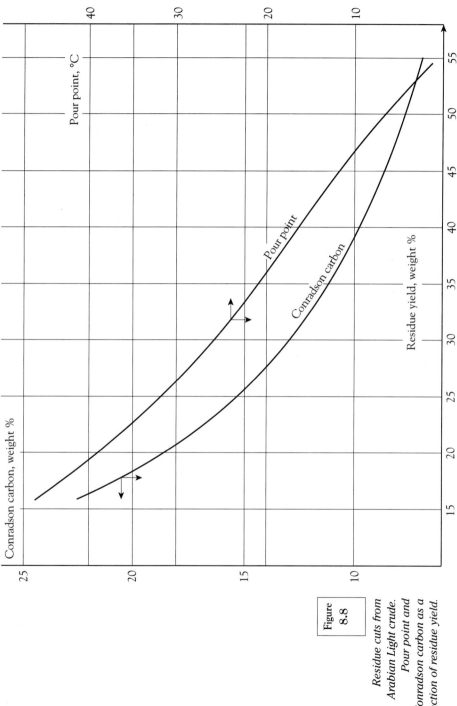

Figure 8.8

Residue cuts from
Arabian Light crude.
Pour point and
Conradson carbon as a
function of residue yield.

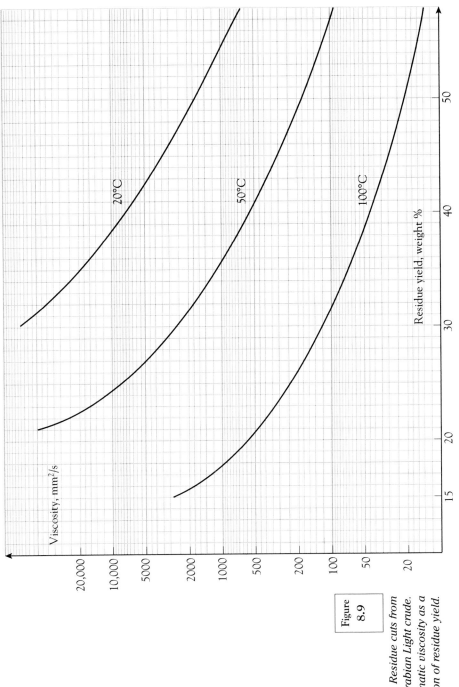

Figure
8.9

*Residue cuts from
Arabian Light crude.
Kinematic viscosity as a
function of residue yield.*

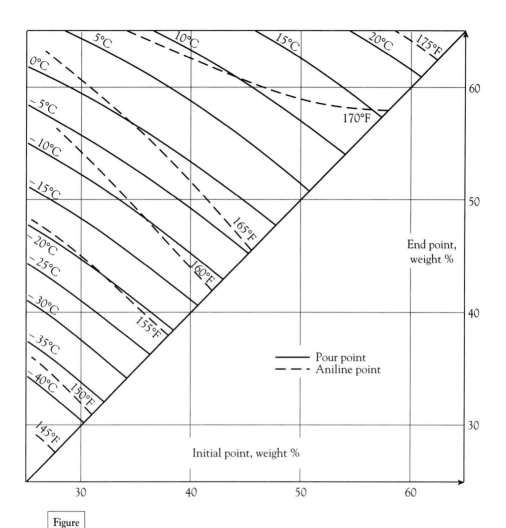

Figure
8.10

Iso-property curves for Arabian Light.
Pour point, °C and Aniline point, °F as a function of initial and end points of
the cut.

	Gas	Light gasoline	Heavy gasoline	Kerosene	Gas oil	Atmos. resid.
IP – EP, °C		20 – 80	80 – 180	180 – 250	250 – 370	370$^+$
Volume %	0 – 1.67	1.67 – 8.01	8.01 – 24.97	24.97 – 36.70	36.70 – 58.60	58.60 – 100
Yield, vol. %	1.67	6.34	16.96	11.73	21.9	41.4
Weight %	0 – 1.1	1.1 – 5.99	5.99 – 20.81	20.81 – 31.78	31.78 – 53.67	53.67 – 100
Yield, wt. %	1.1	4.89	14.82	10.97	21.89	46.33
Specific gravity d_4^{15}		0.659	0.747	0.798	0.854	0.956
Weight % S		0.024	0.036	0.157	1.39	3.18
RON (clear)		60.6	23.5			
Aromatics, vol. %		1.54	14.3	20.68		
Molecular weight		84.3	121.1	169		
Visc. in mm^2/s at 20°C				1.98	7	
Visc. in mm^2/s at 100°C				0.7	1.5	28.5
Pour point, °C				– 48	– 6	10

Table 8.16 *Characteristics of petroleum cuts from Arabian Light crude oil.*

For example, in the case of light Arabian crude (Table 8.16), the sulfur content of the heavy gasoline, a potential feedstock for a catalytic reforming unit, is of 0.036 weight per cent while the maximum permissible sulfur content for maintaining catalyst service life is 1 ppm. It is therefore necessary to plan for a desulfurization pretreatment unit. Likewise, the sulfur content of the gas oil cut is 1.39% while the finished diesel motor fuel specification has been set for a maximum limit of 0.2% and 0.05% in 1996 (French specifications).

From these data, it is possible either to size a desulfurization unit, or to set the operating conditions for an existing unit.

This type of study, applied over all the cuts, enables the refinery flow scheme to be defined in order to satisfy a given set of market conditions starting from one or more crude oil feedstocks.

A comparison of overall treatment costs (purchase and refining) for several crude oils enables the refiner to establish his feedstock requirements and to satisfy the market needs under the most economical conditions.

9

Additives for Motor Fuels and Lubricants

Bernard Sillion

The purpose of this chapter is to present a special class of compounds, additives, that plays an important role in the formulation of fuels as well as lubricants.

Indeed, it is not possible for refining products to meet from the start all the specifications that could be required; as a matter of fact, it often seems to be easier and less costly to correct a defect or to improve a property by the addition of a small quantity of additive rather than to proceed with a complete product reformulation.

We will reserve the word *additive* for substances, or mixtures of substances, capable of noticeably improving at least one property of the product in question, without altering the other intrinsic properties.

In the case of motor fuels, when the content of the material added is relatively high, we will use the term *components* of the mixture. Thus the ethers, used in the formulation of motor fuels in significant proportions, will not be included in our definition.

Furthermore, the formulators having shown plenty of imagination, it would seem illusory to provide an exhaustive list of additives used today. We will mention only the families of additives, protected of course by patents, but well established and widely commercialized.

This chapter is divided in two parts: additives for motor fuels and additives for lubricants. Concerning additives for gasoline, one will find here in Chapter 9 some useful complements to Chapter 5, especially regarding the synthesis of additives and their modes of action.

9.1 **Additives for Gasolines**

If one talks henceforth about the necessity of matching an engine and its fuel, the demand for quality in motor fuels has, however, never ceased to be a preoccupation for refiners ever since gasoline became a commodity item. Two main classes of products are added to gasoline coming from refining: octane number improvers and detergents.

Since the 1960's, two ideas have gained our attention: the struggle against pollution before the first oil crisis of 1973 and the diminution of consumption since. One can consider, in fact, that the two objectives are linked. Indeed, any maladjustment of a fuel admission system will modify the equivalence ratio of the mix. The consequences are modifications, on one hand, of the consumption and on the other, of the nature and the quantity of pollutants emitted: CO, NO_x, and unburned hydrocarbons.

The continuous cleaning of the admission system by an additive contained in the gasoline will help maintain the setting at its optimum value and will prevent the engine operation from drifting from its original settings.

Motor fuels are submitted to strict regulations concerning physical properties and properties of combustion, for which the octane number is the most representative characteristic.

Refining alone can not, economically, provide a fuel to meet specifications; the role of organic lead derivatives has long been to make up the difference between the octane number of the clear fuel and the octane required by legislation.

The development of catalytic converters for combustion of unburned hydrocarbons prohibits a return to lead compounds and henceforth refiners are turning to oxygenated compounds that must be used as a gasoline component; therefore, in amounts much greater than those of lead compounds.

The picture we see now is that of new lead-free fuels having lower aromatics content but containing a variety of oxygenates. It is thus likely that the additives entering into the composition of gasoline will be also modified.

9.1.1 **Detergent Additives**

A primary class of additives is that of the detergents. We will examine in turn the role they play in motor fuels and the chemical structures that are necessary.

9.1.1.1 **Role of Detergent Additives** (Ranney, 1974)

The first detergents for gasoline were used in order to avoid engine stalling during start-ups in cold and humid weather. Evaporation of gasoline in fuel admission systems caused condensation and crystallization of water on the still cold engine walls during the start-up.

Plugging due to ice, although causing power loss or even motor seizing, was reversible; however the development of recycled crankcase gas resulted in irreversible solid organic deposits on the walls of the system, especially in the carburetor.

The introduction of low quantities of surfactants (50 to 125 ppm) helps solve these two problems. The surfactant molecule has a lipophilic organic tail and a polar head that is adsorbed selectively on the metal walls of the admission system. These products have a double action:

- on clean and new systems, they coat the metal wall with an organic layer that helps prevent crystals of ice and organic deposits from adhering
- on systems that have already been contaminated, they dissolve the organic deposits.

Today the surfactants are required to have an ability to clean hot parts of the admission system as far as the valve seats.

Concerning detergent additives, the *anti-ORI* additives are particularly noteworthy.

The Octane Requirement Increase, ORI, is a phenomenon manifested by the appearance of knocking and is due to the increase in engine octane demand with time. This phenomenon is correlated with the increase of solid deposits in the combustion chamber. Although the causes have not been determined with certainty, some companies have patented additives which modify the deposits. The effect is to limit the increase in octane demand (Bert et al., 1983; Chevron, 1988; Nelson et al., 1989).

9.1.1.2 **Chemical Structure of Detergent Additives for Gasoline**

There are between 1000 and 2000 different detergent additive formulations for gasoline.

The skeleton of the molecule can be drawn showing an oleophilic hydrocarbon part, R, to which is attached a polar hydrophilic part, X:

$$R-X$$

The organophilic part R can come from a natural fatty acid whose carbon number is around 18 and whose chain contains a number of unsaturated bonds. Dimers of fatty acids (C_{36} diacids) have also been used.

A second family is based on isobutene polymers (PIB) having molecular weights from 600 to 2000 that are equally important raw materials for detergent additives. So as to render them reactive with the hydrophilic part, they can be chlorinated or condensed with the maleic anhydride. A third way is based on the utilization of polypropylphenols of molecular weights between 600 and 3000.

The hydrophilic parts can contain oxygenated groups (glycol ether types) or amines. The first detergents used amine and phosphoric acid salts or phosphoric esters. It is evident that modern engines equipped with catalytic converters exclude such compounds.

Figure 9.1 shows three chemical families of representative surfactants effective for admission systems.

The fatty acid amides (**a.** in Figure 9.1) do not allow variations in the lipophilic part.

Derivatives of polyisobutylene (**b.** in Figure 9.1) offer the advantage of control over the molecular weight of the polyisobutylene obtained by cationic polymerization of isobutylene. Condensation on maleic anhydride can be done directly either by thermal activation ("ene-synthesis" reaction) (2.1), or by chlorinated polyisobutylene intermediates (2.2). The condensation of the PIBSA on polyethylene polyamines leads to succinimides. Note that one can obtain mono- or disuccinimides. The mono-succinimides are used as

---

a. Propylenediamine amides

$$C_{17}H_{33} — CO — NH —(CH_2)_3 —NH_2$$

$$C_{17}H_{33} — CO — NH —(CH_2)_3 —NH —CO — C_{17}H_{33}$$

b. Derivatives of polyisobutene succinic anhydride

1. Cationic polymerization

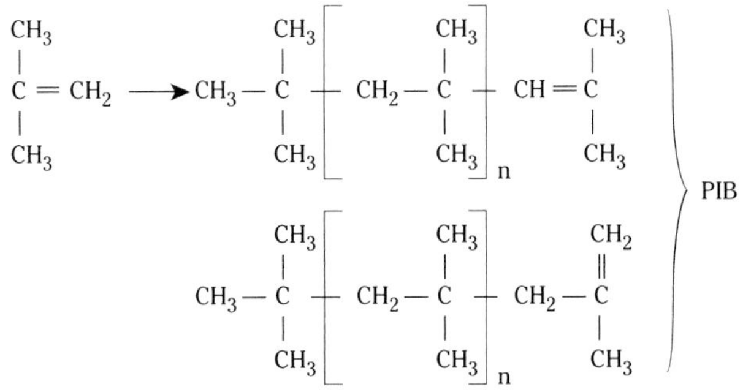

2. Action of maleic anhydride

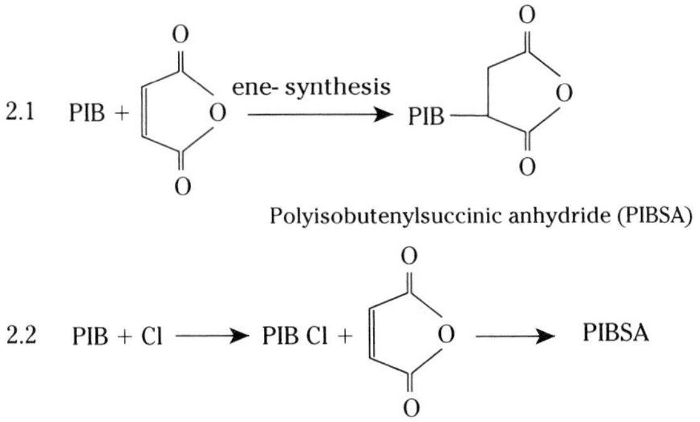

2.1   PIB + [maleic anhydride] $\xrightarrow{\text{ene- synthesis}}$ PIB—[succinic anhydride]

Polyisobutenylsuccinic anhydride (PIBSA)

2.2   PIB + Cl $\longrightarrow$ PIB Cl + [maleic anhydride] $\longrightarrow$ PIBSA

additives for motor fuels and the disuccinimides are dispersant additives used in lubricating oils.

The third family (**c.** in Figure 9.1) less widespread, derived from the alkylphenols, offers as with the succinimides several possibilities of modification to the ratio of hydrophilic and lipophilic groups. Mannich's reaction of the alkyl-phenols also provides additives for lubricating oils.

9.1.2 **Additives for Improving the Octane Number**

A second series of additives comprises those that improve the octane number (see Chapter 5). We will examine in succession the role played by these additives, substances currently used, and the future prospects concerning additives in this area.

Figure 9.1

Chemical families typical of surfactants that are effective in engine admission systems.

3. Action of the polyamines

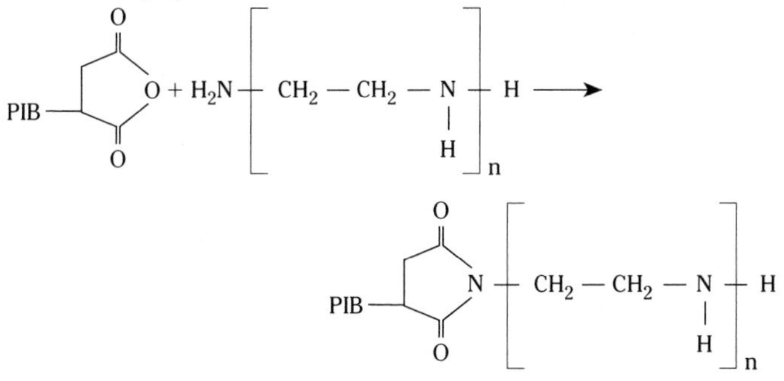

Polyisobutenylsuccinimide

c. Derivatives of polypropylphenols

Action of amines and formaldehyde (Mannich reaction)

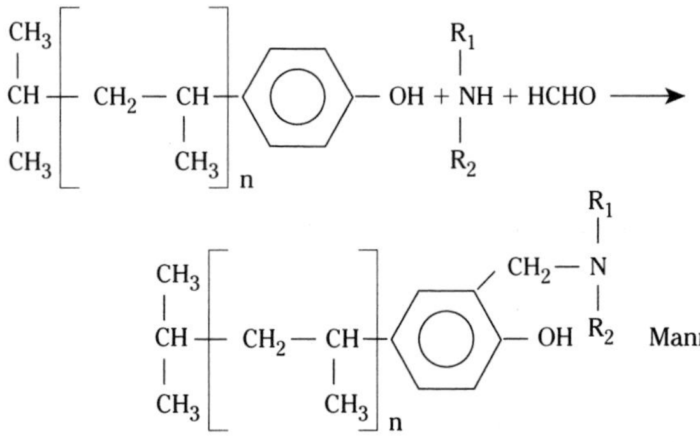

Mannich Base

9.1.2.1 **Role of Additives for Improving the Octane Number**

The octane number is a measure of a fuel's ability to resist auto-ignition during the compression phase prior to ignition.

Additives function by reacting with hydrocarbon partial oxidation products by stopping the oxidation chain reaction that would otherwise drive the combustion.

In relatively small doses (see Chapter 5), additives made it possible for the refiner to gain several points in octane number and thereby to allow the premium gasoline to meet specifications.

9.1.2.2 **Chemical Structures Effective in Improving the Gasoline Octane Number**

Products that have been the most widely used are organo-metallic compounds and especially the lead derivatives that have been commercialized since 1920.

Two derivatives are used to ensure constant lead content throughout the gasoline boiling range: tetraethyl- and tetramethyl lead and their mixtures in variable proportions.

These compounds are obtained by action of halogenated organic derivatives on lead alloys (magnesium or alkaline metal alloys).

Other organo-metallic structures (based on manganese in particular), based on the chemistry of π complexes with aromatic structures, can also be used to improve the octane number (Guibet, 1987, p. 276).

Concerning non-metallic compounds, the antiknocking properties of nitrogen compounds such that derivatives of aniline, indole and quinoline, and certain phenol derivatives have been mentioned.

9.1.2.3 **Future Prospects for Additives to Improve the Octane Number**

The protection of the environment implies the elimination of lead compounds, first of all because of their individual toxicities and second because these derivatives or their products of decomposition poison catalytic converter catalysts.

Refiners will turn to reformulated motor fuels where the octane number will be increased by alkylate or oxygenated compounds. It has indeed been shown for a long time that oxygenated compounds, alcohols, ethers and ketones improved the octane number of hydrocarbon-based blends (Whitcomb, 1975).

9.1.3 **Biocide Additives**

In addition to their antiknock properties, organic lead compounds possess bactericidal properties and motor fuels with lead are known to inhibit bacterial growth during storage in contact with water. With the disappearance of lead-based compounds, it is necessary to incorporate biocides from the cyclic imine family, (piperidine, pyrrolidine, hexamethyleneimine), alkylpropylene diamines or imidazolines (Figure 9.2).

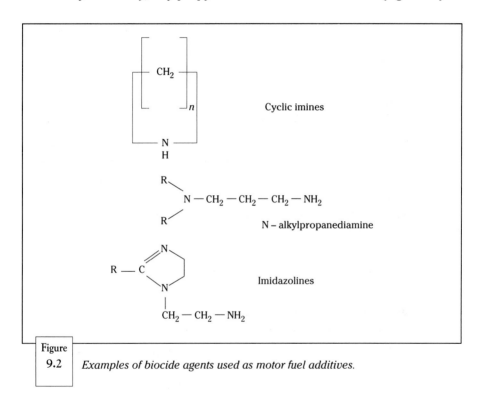

Figure	
9.2	*Examples of biocide agents used as motor fuel additives.*

9.1.4 **Antistatic Additives**

Hydrocarbons generally have very low electrical conductivities and manipulation of these fluids creates electrostatic charges that can result in fire or explosions. This problem is encountered with gasoline and kerosene.

There are many compounds that act to increase the conductivity of hydrocarbons. Among them are organic salt compounds such as organic amine salts and metallic salts of carboxylic acid, and amine derivatives that also have surfactant detergent properties, such as fatty acid and polyamine amides. Furthermore, some filming polymers containing polar groups have been proposed for this application (see Figure 9.3).

R — CO — NH — CH$_2$ — CH$_2$ — NH$_2$ R = unsaturated C$_{18}$ chain

$$\left[CH_2 - CH - CH_2 - CH_2 \right]_n$$

with side groups O—C$_{16}$H$_{33}$ and N-pyrrolidone (=O)

Cetylvinylether-vinylpyrrolidone copolymer

Figure 9.3	*Examples of antistatic fuel additives.*

9.2 **Additives for Diesel Fuels**

Diesel fuels, like gasoline, are formulated with additives that affect the process of combustion, in this case to improve the cetane number. Diesel fuels also contain detergents for injection systems as well as compounds for improving the fuel's low temperature rheology. Finally, decreasing particulate emissions is a problem of increasing concern, but the mechanism of action to promote this effect is not clearly understood.

9.2.1 **Additives for Improving the Cetane Number of Diesel Fuels**

Unlike spark-induced combustion engines requiring fuel that resists auto-ignition, diesel engines require motor fuels, for which the reference compound is cetane, that are capable of auto-igniting easily. Additives improving the cetane number will promote the oxidation of paraffins. The only compound used is ethyl-2-hexyl nitrate.

9.2.2 **Detergent Additives for Diesel Fuels**

The role of detergent additives is to maintain clean injectors so as to insure good distribution of diesel fuel in the cylinder. The structure of these compounds is similar to that of detergents for gasoline engine admission systems. Commercialized compounds are from the succinimide family (see Figure 9.1).

Moreover, the same surfactant structures that favor dispersion of fuel droplets in the combustion chamber most likely play a role in reducing particulate emissions.

9.2.3 **Additives for Improving Combustion and for Reducing Smoke and Soot Emissions**

The reaction mechanism for these products is not clearly understood, but the introduction of organo-metallic compounds (barium or iron salts in colloidal suspension) has been shown to have a beneficial action on the combustion of diesel fuel in engines and reduce smoke. However, these products cause deposits to form because they are used in relatively large proportions (on the order 0.6 to 0.8 weight %) to be effective.

9.2.4 **Additives for Improving the Cold Behavior of Diesel Fuel** (Coley, 1989)

Straight run diesel fuels have a high paraffin content, which is desirable, incidentally, for obtaining high cetane numbers. The higher the distillation end point, the higher is the heavy paraffin content (with a carbon number greater than C_{24}).

The nature of these paraffins and their concentration in diesel fuel affect the three temperatures that characterize the cold behavior. The *cloud point* is the temperature at which crystals of paraffins appear when the temperature is lowered. The *cold filter plugging point* is defined as the temperature under which a suspension no longer flows through a standard filter. Finally, the *pour point* is the temperature below which the diesel fuel no longer flows by simple gravity in a standard tube. These three temperatures are defined by regulations and the refiner has three types of additives to improve the quality of the diesel fuel of winter.

Additives that affect the cloud point are no longer in frequent use; however, it has been shown that certain polymers having branched paraffins can "recognize" paraffins of equivalent size and keep them in solution. It is therefore possible to "complex" the longest paraffins selectively and to decrease the cloud point by 3 to 4°C (Damin et al., 1986).

Additives affecting the cold filter plugging point are also copolymers. The most common are olefin copolymers of vinyl esters (vinyl ethylene-acetate for example) or olefin and alkyl fumarate or copolymers of alpha-olefins. The action mechanism for these copolymers is based on the fact that they crystallize slightly before the surrounding paraffins; they have therefore an effect on the germination of crystals. These formulations help control the growth and crystalline morphology by their adsorption on one of the crystal faces. Modified paraffin crystals are smaller and can pass through the filter media.

Additives acting on the pour point also modify the crystal size and, in addition, decrease the cohesive forces between crystals, allowing flow at lower temperatures. These additives are also copolymers containing vinyl esters, alkyl acrylates, or alkyl fumarates. In addition, formulations containing surfactants, such as the amides or fatty acid salts and long-chain dialkyl-amines, have an effect both on the cold filter plugging point and the pour point.

9.2.5 **Conclusion**

The composition of motor fuels is undergoing constant change in view of improving performance while protecting the environment.

The elimination of lead, the reduction of aromatics in gasoline, and the desulfurization of diesel fuels are going to require significant reformulations of these products that will imply development of specific additives that allow the refiner to optimize costs while meeting the required specifications.

9.3 **Additives for Lubricants**

The modern lubricant is a highly technicological product containing 15 to 20 weight per cent of various additives that enable the engine to function under increasingly severe conditions. For example, the reduction in the automobile Cx coefficient (which lowers the engine cooling capacity) and the longer oil-change interval both of which are sought by the owner, force the lubricant to operate at higher temperatures and for longer periods of time.

The development of a line of lubricant additives is an expensive and slow undertaking; the market for these products — on the order of 10 billion dollars in 1992 — is very large and is dominated by a few companies.

One can distinguish three important additive classes according to their modes of action:

- those whose role is purely physical; they affect the rheology of the lubricant at low and high temperatures
- those whose action occurs on the interfaces
- those that act by chemical mechanism; that is the case for antioxidants and extreme pressure additives, for example.

9.3.1 **Additives Modifying the Rheological Properties of Lubricating Oils**

A lubricant should be characterized by its viscosity at low temperature $(-18°C)$ to account for its behavior during cold weather starting and by its viscosity at high temperature $(100°C)$. Values of these two viscosities are based on the SAE classification. See Chapter 6.

The viscosity of a hydrocarbon mixture, as with all liquids, decreases when the temperature increases. The way in which lubricant viscosities vary with temperature is quite complex and, in fact, charts proposed by ASTM D 341 or by Groff (1961) (Figure 6.1) are used that provide a method to find the viscosity index for any lubricant system. Remember that a high viscosity index corresponds to small variation of viscosity between the low and high temperatures. That is the desired situation because it corresponds to a smaller variation in behavior of the lubricant film between moving parts with respect to operating temperature.

A refinery lubricant base stock is obtained having an viscosity index around 100, certain hydrotreatments result in VI's of 130, and paraffin hydroisomerization provides oils with a VI close to 150.

Nevertheless, a great majority of lubricating oils is obtained having bases whose VI is around 100.

The function of viscosity additives is to improve the viscosity index so as to obtain "multigrade" oils. The problem is to use materials that, by only slightly increasing the low temperature viscosity, are capable of counterbalancing the decrease in viscosity when the temperature increases.

Additives for improving the viscosity index are added in concentrations of five to ten weight per cent of the oil.

9.3.1.1 **Mechanism of Action for Viscosity Additives**

Viscosity additives are aliphatic polymers of high molecular weight whose main chain is flexible. It is known that in a "poor" solvent, interactions between the elements making up the polymer chain are stronger than interactions between the solvent and the chain (Quivoron, 1978), to the point that the polymer chain adopts a "ball of yarn" configuration. The macromolecules in this configuration occupy a small volume. The viscosity of a solution being related to the volume occupied by the solute, the effect of polymers on the viscosity in a "poor" solvent will be small.

Consider that at low temperatures, a lubricant is a "poor solvent" for polymer chains. When the temperature increases, interactions between polymer chains decrease; the space occupied by the polymer ball takes on greater volume and consequently, the viscosity decrease due to the lubricant temperature increase is compensated by the unfolding of the polymer chain and the result is a reduction of the difference between the viscosities at low and high temperature, and therefore an increase in viscosity index.

One of the problems generally associated with the utilization of additives is the continuous action under the engine's operating conditions. That is particularly important for polymers that are sensitive to mechanical deterioration due to shear effects.

Shearing causes polymer chains to break, therefore a decrease in molecular weight and, consequently, in thickening power. It has been shown that the higher its molecular weight, the more the polymer is sensitive to mechanical shearing (Briant et al., 1985).

9.3.1.2 **Main Chemical Families Used in Viscosity Additives**
(Briant et al., 1985)

Two main viscosity additive families are used: hydrocarbon polymers and polymers containing ester functional groups.

*a. **Hydrocarbon Polymers*** (Figure 9.4)

Polyisobutylene is obtained through cationic polymerization of isobutylene at low temperature.

The copolymers of ethylene and propylene (OCVP) are obtained by coordination catalysis using a derivative of vanadium and a derivative of an aluminum alkyl. Molar compositions of ethylene and propylene are usually on the order 45 and 55%.

Other products such as butadiene and styrene copolymers have been commercialized.

Generally speaking, hydrocarbon polymers are compatible with mineral oils and are undergoing considerable development.

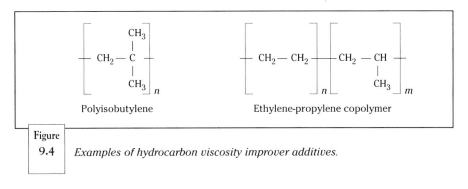

Polyisobutylene   Ethylene-propylene copolymer

Figure
9.4   *Examples of hydrocarbon viscosity improver additives.*

*b. **Polymers Containing Ester Groups*** (Figure 9.5)

These materials are obtained through free-radical polymerization of acrylic or methacrylic monomers, or of fumarates.

The monomers used are second generation petrochemical products. The polymethacrylates are in fact copolymers based on methyl methacrylate and up to C_{20} molecular weight alcohol methacrylate. The properties of the additive are controlled based on the molecular ratio of these different monomers and their molecular weight.

Polymethacrylates are less soluble than hydrocarbon polymers in mineral oils, they thicken less at low temperatures and are more effective at high temperatures. In this respect, it is important to note that the modification of rheological properties is based on interactions between polymer and oil; it will therefore be always dependent of the nature of oil.

R = C_1 to C_{20} alkyl chain
Copolymethacrylate

Polymaleate

Figure 9.5 *Examples of viscosity additives containing esters.*

c. ***Modified Polymers*** (Briant et al., 1985)

The introduction of monomers containing polar groups such as: tertiary amines, imidazoles, pyrrolidones, pyridines, etc., gives the polymer dispersant properties that will be discussed in the article on dispersant additives for lubricants.

9.3.2 **Pour Point Depressants** (Satriana, 1982) (Figure 9.6)

Although lubricant base stocks have been subjected to dewaxing processes, they still contain large amounts of paraffins that result in a high pour point for the oil. In the paragraph on the cold behavior of diesel fuels, additives were mentioned that modify the paraffin crystalline system and oppose the precipitation of solids.

The problem is similar to the case of lubricating oils; polyalkylnaphthalenes or alkyl polymethacrylates called pour point depressants have been commercialized to lower the pour point.

Polyalkylnaphthalene
(Reaction of a chlorinated paraffin on naphthalene)

Polyalkylmethacrylates

Figure 9.6 *Examples of pour point depressants for lubricating oils.*

These products have molecular weights between 2000 and 10,000, well below those of additives improving the viscosity index (100,000). They are added in very small concentrations (0.01 to 0.3 weight percent) and at these concentrations they can lower the pour point 30°C.

9.3.3 **Antioxidant Additives for Lubricants**

The crankcase of a gasoline or diesel engine is in reality a hydrocarbon oxidation reactor; oil is submitted to strong agitation in the presence of air at high temperature (120°C); furthermore, metals such as copper and iron, excellent catalysts for oxidation, are present in the surroundings.

Oxidation first produces soluble oxygenated compounds of molecular weights between 500 and 3000 that increase the viscosity of oil; then they polymerize, precipitate, and form deposits. Oxidation also causes formation of low molecular weight organic acids which are very corrosive to metals.

The lubricant oxidation mechanism is free-radical in nature and the additives act on the kinetic oxidation chain by capturing the reactive species either by decomposition of the peroxides, or by deactivation of the metal.

Organic sulfur compounds such as sulfurized spermaceti oil, terpene sulfides, and aromatic disulfides have been used. Encumbered phenols such as di-tertiary-butylphenols and amines of the phenyl-alphanaphthylamine type are effective stopping the kinetic oxidation chain by creating stable radicals.

However, the most widely used materials are the zinc dialkyl-dithiophosphates that have an anti-wear effect in addition to their antioxidant power and, besides, offer an attractive cost/effectiveness ratio.

The antioxidants (Figure 9.7) are applied in concentrations between 1 and 10 weight %.

9.3.4 **Dispersant and Detergent Additives for Lubricants**

9.3.4.1 **Role and Action Mode**

Antioxidant additives can not totally prevent the oxidation phenomenon, especially with the modern trend in oil-change intervals; at the end of the interval, oil contains a significant quantity of insoluble oxidized material.

It is necessary to keep these materials suspended in the oil to avoid the formation of varnishes on the engine walls and deposits in the crankcase.

Two types of compounds having different functions are used: detergents and dispersants.

Dispersants and detergents are surfactants in organic media and contain an oleophilic hydrocarbon part and a hydrophilic polar part.

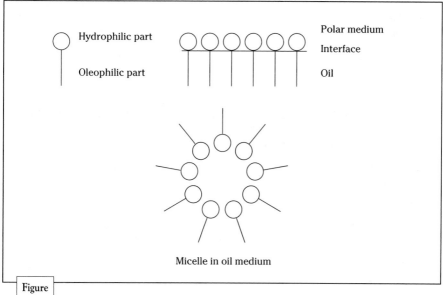

OH

CH₃

Ditertiarybutylphenol

HN

Alphanaphthylamine

R — O
|
R — O
    P
    ‖ O
    \ S
Zn
S /
    P
    ‖ O
    / O — R
    \ O — R

Zinc dialkyldithiophosphates

Figure 9.7   *Examples of antioxidant additives for lubricating oils.*

This character, called amphiphilic, produces two characteristic sets of behavior, adsorption on the interfaces and auto-association in the form of micelles that extend into the oily surroundings as illustrated in Figure 9.8.

Hydrophilic part

Oleophilic part

Polar medium

Interface

Oil

Micelle in oil medium

Figure 9.8   *Principle of surfactant structures in a lubricating oil.*

In the case of lubricant detergents, the hydrophilic or polar part is a metallic salt (calcium, magnesium) and at the center of the micelle it is possible to store a reserve of a metal base (lime or magnesia); the detergent will be able therefore to neutralize the acids produced by oxidation of the oil as soon as they are created.

The basicity of a detergent is an essential additive characteristic. It is expressed as Total Basic Number (T.B.N.) (ASTM D 664).

The second detergent function is to prevent formation of varnishes that come from polymerization of deposits on hot surfaces of the cylinder and the piston. Finally, by adsorption on metallic surfaces, these compounds have anti-corrosion effects.

In the case of lubricant dispersants, the polar part is organic (amine, polyamine, heterocyclic nitrogen compounds, polyglycol).

Dispersants are particularly important when engines operate below their normal operating temperature (as is the case of a short urban trip). Dispersants act by keeping oxidation products in suspension.

Detergent and dispersant additives are used in concentrations between 5 and 15 weight per cent.

9.3.4.2 **Typical Structures Used for Detergent Additives And Dispersants**

a. Lubricant Detergents

The two principal raw material classes used to make detergents are sulfonic acids and phenols.

Sulfonic acids can come from the sulfonation of oil cuts from white oil production by sulfuric acid treatment. Sodium salts of alkylaromatic sulfonic acids are compounds whose aliphatic chains contain around 20 carbon atoms. The aromatic ring compounds are mixtures of benzene and naphthalene rings.

These sodium sulfonates or so-called natural sulfonates are transformed into simple or "overbase" calcium or magnesium sulfonates (Satriana, 1982).

The second source of sulfonic acid uses the following reaction scheme: alkylation of benzene by a propylene oligomer then sulfonation of the alkylbenzene.

The technique of neutralization and the production of superbase systems from synthetic sulfonates are similar to those employed for the natural sulfonates.

In using the alkylphenols, it is possible to obtain three types of detergents: the alkylphenols themselves transformed as salts of calcium, the alkylphenol-sulfides conferring antioxidant properties and finally the alkylphenol-sulfides transformed by action of CO_2 into alkylsalicylate-sulfides (Figure 9.9).

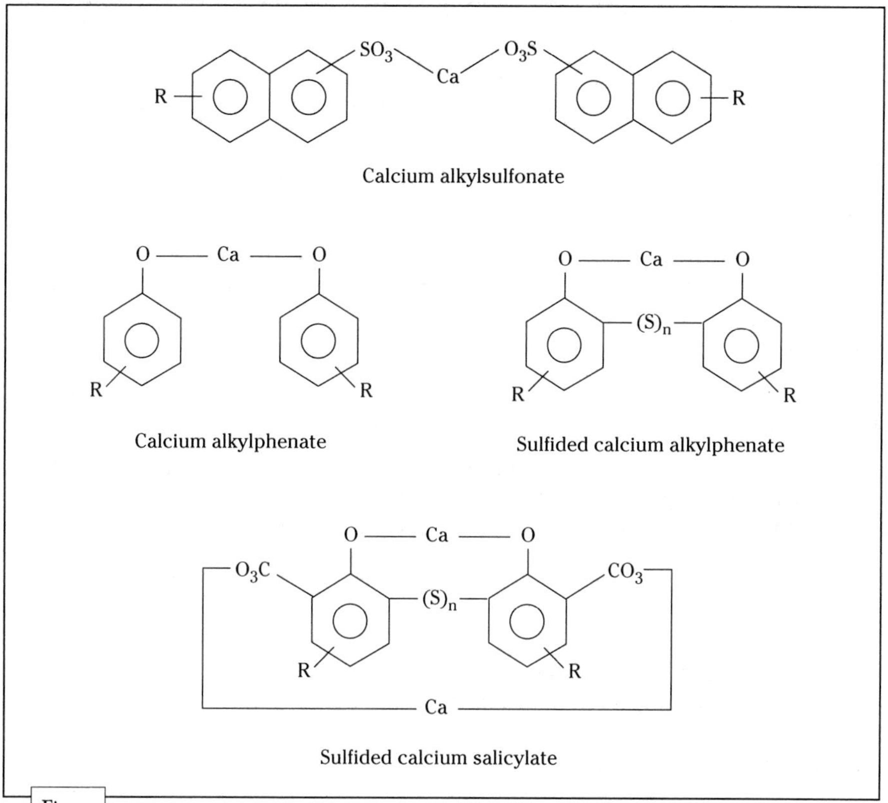
SO3
Ca
O3S
R
R
Calcium alkylsulfonate
O
Ca
O
R
R
Calcium alkylphenate
O
Ca
O
(S)n
R
R
Sulfided calcium alkylphenate
O
Ca
O
O3C
CO3
(S)n
R
R
Ca
Sulfided calcium salicylate

Figure 9.9 *Simplified structures of detergent additives for lubricating oils.*

b. Lubricant Dispersants

The organophilic part of commercial dispersants is obtained by cationic oligomerization of isobutylene.

The polyisobutylenes (PIB) having molecular weights ranging from 1000 to 2000 are substituted by maleic anhydride, and the polyisobutylene succinic anhydride (PIBSA) formed is neutralized by a polyethylene-polyamine as indicated in Figure 9.10.

c. Dispersant Polymers for Lubricants (Figure 9.11)

Polymers for improving the viscosity index of the copolymethacrylate type can be made into dispersants by copolymerization with a nitrogen monomer. The utilization of these copolymers allows the quantity of dispersant additives in the formulation to be reduced.

Figure
9.10 | *Structures of commercial detergent additives for lubricating oils.*

Figure
9.11 | *Examples of polymethacrylate dispersants.*

9.3.5 **Extreme-Pressure and Anti-Wear Additives**

When two metallic surfaces are lubricated in a hydrodynamic regime, the oil film is stable and problems of wear are not very important. In severe service, the film can be destroyed; from then on the metallic parts rubbing on each other can cause first metal loss and then even the seizing of the parts by welding.

The role of anti-wear and extreme-pressure additives is to create a solid lubricant at the interface of the metal by chemical reaction.

These additives must thus be capable of decomposing under heat action by liberating the species that react with the moving metal or metals by creating an interphase more fusible than the metal itself.

Commercial compounds are oil-soluble organic molecules containing chlorine, sulfur or phosphorus atoms (Figure 9.12).

| Figure 9.12 | *Examples of dithiophosphoric derivatives used as anti-wear and extreme pressure additives.* |

For example, olefin sulfurization products (Lubrizol, 1980), dithiophosphomolybdates (Mobil, 1980), or more simply the dithiophosphates of alcohol (Shell, 1980) whose anti-oxidant properties have been announced, are used in oil formulations for their anti-wear properties.

9.3.6 **Conclusion**

The increase in the oil-change interval has already been a strong incentive for improving lubricant formulations. The increase in engine operating temperatures and the development of catalytic converters are without doubt two orientations that will have consequences on lubricant additives.

Reducing ash content and increasing anti-oxidant and dispersant efficiencies are among the goals needed to improve the quality of future lubricants.

10

Introduction to Refining

Gérard Heinrich

10.1 Historical Survey of Refining

Since the discovery of petroleum, the rational utilization of the fractions that compose it has strongly influenced the development of refining processes as well as their arrangement in refining flowsheets.

At the end of the 1960's, oil refining underwent significant transformation linked to the continuous increase in the need for light products (gasoline-diesel oil) at the expense of heavy products (fuel-oils) as shown in Table 10.1.

	1973	**1990**	**2000**
Light products	29.5	35	37–39
Middle distillates	30.0	36	39–41
Heavy products	40.5	29	20–24

Table 10.1 *World demand for oil products (weight %).*

The trend in demand has also been accompanied by improvements in product quality illustrated by the increases in gasoline octane numbers and diesel oil cetane numbers.

With the introduction of new antipollution standards as well as limitations envisaged for the chemical composition of finished products, current refining flowsheets and especially those beyond the year 2000 will have to adapt to the new specifications using new processes.

Table 10.2 shows expected trends in specifications for some major products.

	Situation in 1991	Foreseeable trend 2000–2020
Gasoline		
Clear RON	89–94	95–98
Clear MON	80–84	85–88
Benzene, vol. %	3–5	1–2
Aromatics, vol. %	30–50	20–30
Olefins, vol. %	10–20	5–10
Sulfur, ppm	300–500	50–100
Diesel oil		
Sulfur, wt. %	0.2–0.5	0.05
Cetane number	45–50	50–53
Aromatics, vol. %	25–35	10–20
Fuel oil		
Sulfur, wt. %	3–4	0.5–1
Nitrogen, wt. %	0.5–0.7	0.3–0.5

Table 10.2   *Finished product specifications and future constraints for Europe.*

Owing to its flexibility, the refining industry is able to meet the changes in demand and quality:

Early 1970's: simple refinery (motor fuels, heavy fuels)

End of the 1980's: first stage in the introduction of heavy ends' conversion

For the long term, 2010–2020: refinery complex meeting environmental regulations and ensuring total conversion of the heavy ends

To adapt to this trend, refining calls on a great variety of processes. The most important are as follows:

- Separation Processes that split a feed into simpler or narrower fractions.
- Conversion Processes that generate new molecules having properties adapted to the product's end use.
- Finishing Processes that eliminate (most often by hydrogenation) undesirable compounds.
- Environmental Protection Processes that treat the refinery gases (fuel and tail gas), stack gas, and water effluents.

10.2 **Separation Processes**

The main separation processes are:

• distillation
• absorption
• extraction
• crystallization
• adsorption.

The foremost separation process is crude distillation and in second place, if deeper conversion is envisaged, solvent extraction (deasphalting).

10.2.1 **Primary Distillation (Atmospheric Pressure) of Crude Oil**

The first process step a crude oil undergoes after its production is distillation.

A first operation on the crude, desalting (washing by water and caustic), extracts salts (NaCl, KCl and the $MgCl_2$ that is converted to NaCl by the caustic), reduces acid corrosion as well as it minimizes fouling and deposits.

Next the crude is distilled into well defined fractions according to their end uses.

The main distillation products are:

• refinery gases
• liquefied petroleum gases (propane/butane)
• gasolines (light/heavy)
• kerosenes, lamp oils, jet fuels
• diesel oils and domestic heating oils
• heavy industrial fuel-oils.

10.2.2 **Secondary Distillation or Vacuum Distillation**

Vacuum distillation of the atmospheric residue complements primary distillation, enabling recovery of heavy distillate cuts from atmospheric residue that will undergo further conversion or will serve as lube oil bases. The vacuum residue containing most of the crude contaminants (metals, salts, sediments, sulfur, nitrogen, asphaltenes, Conradson carbon, etc.) is used in asphalt manufacture, for heavy fuel-oil, or for feed for others conversion processes.

Figure 10.1 presents the part of the refining diagram that includes the atmospheric and reduced pressure distillations.

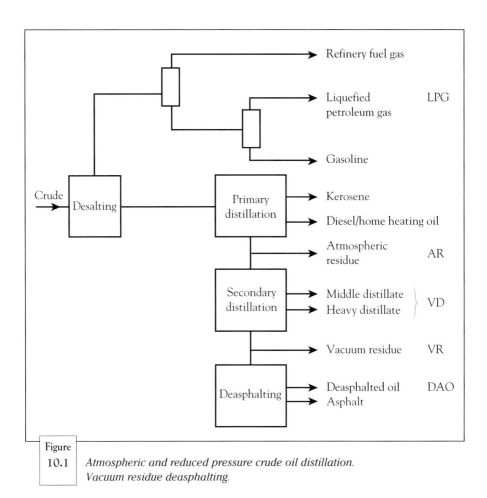

Figure 10.1 *Atmospheric and reduced pressure crude oil distillation. Vacuum residue deasphalting.*

Table 10.3 provides, for a typical crude, the yields and average properties of the various cuts obtained from a conventional distillation operation.

10.2.3 **Processing Vacuum Residue by Solvent Extraction (Deasphalting)** (Biedermann et al., 1987)

Deasphalting is a liquid-liquid separation operation that extracts the last of the easily convertible hydrocarbons from the vacuum residue. Solvents employed are light paraffins: propane, butane, and pentane. The yield in deasphalted oil increases with the molecular weight of the solvent, but its quality decreases.

Asphalt makes up the residue of the deasphalting operation and concentrates the major portion of the impurities such as metals, sediment, salts and asphaltenes. Asphalt fluidity decreases as the molecular weight of

Cuts	Weight % of crude	Volume % of crude
Refinery gas	0.28	–
LPG	1.09	1.70
Light gasoline	3.87	5.18
Heavy gasoline	13.85	16.33
Kerosene	6.74	7.44
Diesel oil – home heating fuel	24.37	25.06
Vacuum distillate	23.50	22.00
Vacuum residue	26.30	22.20
Total	100.00	99.91

Properties of cuts	Light gasoline	Heavy gasoline	Kerosene	Diesel oil	AR	VD	VR
TBP cut, °C	C_5–80	80–180	180–225	225–375	375+	375–550	550+
Sp. gr. d_4^{15}	0.654	0.742	0.793	0.851	0.986	0.935	1.037
Sulfur, wt. %	0.003	0.035	0.15	1.4	3.95	2.8	5.0
Nitrogen, ppm						$\cong 1000$	$\cong 3500$
Paraffins, vol %		73					
Naphthenes, vol %		15					
Aromatics, vol %		12	20.5				
Smoke point, mm			19				
Freezing point, °C			−50				
Cloud point, °C				−5			
Cetane number				53			
RON clear							
Visc. at 50°C, mm²/s				2.4			
Visc. at 100°C, mm²/s				1.1	85	9	3300
Conr. Carb., wt. %					12.5	1.2	22.6
C_7 insol., wt. %					5.6		10.6
Ni, ppm					25		47
V, ppm					73		138

Table 10.3 *Material balance and properties of the main fractions resulting from primary and secondary fractionation of a 50/50 volume % mixture of Arabian Light and heavy crude oil (specific gravity $d_4^{15} = 0.875$).*

the solvent used increases. The use of heavy solvent results in hard asphalt being produced whose ultimate utilization is combustion in power plants, or partial oxidation (production of city gas, hydrogen, methanol, etc.). Table 10.4 gives the yields and product properties resulting from the deasphalting of a typical crude, depending on the solvent used. Figure 10.1 shows the position of the deasphalting unit in the refining flowsheet.

Solvent	Propane	Butane	Pentane	Heavy Arabian VR feed 538°C+
Deasphalted oil				
Yield, wt. %	32	50	66	–
Density	0.945	0.963	0.985	1.048
Sulfur, wt. %	3.8	4.45	4.9	5.78
Nitrogen, ppm	1000	2000	2100	3500
Ni + V, ppm	6	24	67	290
Conr. Carb., wt. %	2.3	5.2	11	23
C_7 Insol., wt. %	< 0.05	< 0.05	< 0.05	12.6
Visc. at 100°C, mm²/s	60	100	250	4500
Asphalt				
Yield, wt. %	68	50	34	
Density	1.105	1.15	1.197	
Sulfur, wt. %	6.71	7.11	7.5	
Nitrogen, wt. %	0.47	0.50	0.62	
Ni + V, ppm	422	556	720	
Softening Pt, °C	60	110	150	
Visc. at 300°C mm²/s	–	40	500	
Visc. at 200°C mm²/s	50	–	6.10^4	

Table 10.4 *Effect of the solvent on the yields and product properties resulting from a deasphalting operation.*

10.2.4 Other Separation Processes

Among the other separation processes, one can cite the following:

- solvent extraction of lubricating oil base stocks
- absorption of refinery gases (H_2S by amines)
- adsorption, as in the purification of hydrogen (PSA) or the demercurization of natural gas.

10.3 Conversion Processes

These are the major processes in refining and petrochemicals.

In this large group of processes, the following are distinguished:

Processes for the Improvement of Properties (see 10.3.1)

By molecular rearrangement:

- catalytic reforming
- isomerization.

Using co-reactants:

- alkylation

- ether synthesis
- oligomerization.

Conversion Processes (see 10.3.2)

Thermal processes:
- visbreaking
- coking.

Catalytic processes:
- catalytic cracking
- steam reforming
- hydroconversion.

Finishing Processes (see 10.3.3)

- hydrotreatment/hydrogenation
- sweetening.

Environmental Protection Processes (see 10.3.4)

- acid gas processing (sulfur recovery)
- stack gas processing
- waste water treatment.

We will examine hereafter these four large process categories.

10.3.1 **Processes for the Improvement of Properties**

10.3.1.1 **Property Improvement Processes Using Molecular Rearrangement**

a. Catalytic Reforming

A key process in the production of gasoline, catalytic reforming is used to increase the octane number of light crude fractions having high paraffin and naphthene contents (C_7–C_8–C_9) by converting them to aromatics.

The modern reforming process operates with continuous regeneration of the catalyst, at low pressure (2 to 5 bar) and high temperature (510–530°C).

In addition to reformate, reforming provides hydrogen, an important by-product, and a small quantity of gas and LPG.

The main feedstock for catalytic reforming is heavy gasoline (80 to 180°C) available from primary distillation. If necessary, reforming also converts by-product gasoline from processes such as visbreaking, coking, hydroconversion and heart cuts from catalytic cracking.

Upstream of the reforming unit, the feedstock undergoes hydrotreatment so as to eliminate impurities such as S, N, olefins, and metals which are all catalyst poisons.

Table 10.5 gives yields and properties of reformer products coming from a typical feedstock.

Feedstock (after pretreatment)		Catalytic reforming	Isomerization
TBP cut	°C	80–180	C_5–80
Specific gravity d_4^{15}		0.742	0.65
n-Paraffins + iso-Paraffins	vol %	73	50 + 39
Naphthenes	vol %	15	9
Aromatics	vol %	12	2
Sulfur	ppm	< 0.5	< 0.5
Nitrogen	ppm	< 0.5	< 0.5
Water	ppm	< 4	< 1
RON clear			67–72
MON clear			65–67

Product		C_5+ Reformate	C_5+ Isomerate
Specific gravity d_4^{15}		0.810	0.647
RON clear		98–104	88
MON clear		88–92	86
RVP	bar	0.3	

Yield, weight % for RON =	102	88
H_2	3.00	−0.2
$C_1 + C_2$	3.75	
C_3	3.50	2.2
iC_4	1.75	
nC_4	2.50	
C_5+	85.50	98.0
Total	100.00	100.00

Table 10.5   *Typical feedstock composition, yields and characteristics of effluents from reforming and isomerization processes.*

b. Isomerization

As a complementary process to reforming, isomerization converts normal paraffins to iso-paraffins, either to prepare streams for other conversions: $nC_4 \longrightarrow iC_4$ destined for alkylation; or to increase the motor and research octane numbers of light components in the gasoline pool, i.e., the C_5 or C_5–C_6 fractions from primary distillation of the crude, or light gasoline from conversion processes, having low octane numbers.

Ultimate products are the iso-paraffins: isopentane and the C_6 isomers, mainly 22 and 23 dimethylbutane.

Isomerization can be done in a single pass or with recycle of the unconverted fraction.

Figure 10.2 shows the locations of reforming and isomerization units in refinery configurations.

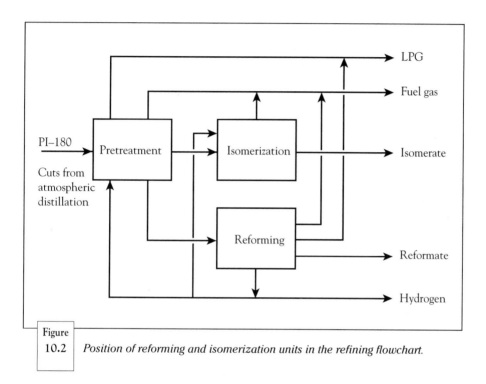

Figure	
10.2	*Position of reforming and isomerization units in the refining flowchart.*

10.3.1.2 **Property Improvement Processes using Co-reactants**

a. Alkylation

Alkylation is a process that produces high octane number (RON and MON) components from light olefins ($C_3^=$, $C_4^=$, $C_5^=$) by addition of isobutane.

The highly exothermic reaction is catalyzed by strong acids: sulfuric, hydrofluoric.

The feedstock usually comes from catalytic cracking, sometimes from steam cracking. The reaction products are C_7–C_8 isoparaffins. The by-products are the C_3–C_4 n-paraffins which do not react.

Table 10.6 gives yields and properties of alkylate for a typical feedstock.

Compound	Feedstock, weight %	Alkylate, weight %	LPG, weight %	Losses, weight %
C3=	0.80	–		
iC_4	39.98	0.04	0.40	
nC_4	11.14	2.23	8.92	
iC_4=	0.76	–	0.03	
1 C_4=	15.15	0.08	0.71	
2 C_4=	31.67	1.85	7.39	
C_5=	0.50	0.16		
iC_5	–	1.66	0.45	
Alkylate	–	75.40	0.38	
Total	100.00	81.42	18.28	0.30

Alkylate properties	
Sp. gr. d_4^{15}	0.710
RON	97.6
MON	94.4
RVP bar	0.4

ASTM D 86, °C	
IP	32
10 vol. %	72
30 vol. %	100
50 vol. %	106
70 vol. %	110
90 vol. %	126
EP	198

Table 10.6 *Typical feedstock, yields and product properties from an alkylation unit.*

b. Synthesis of Ethers from Isobutene

MTBE–ETBE (methyl- or ethyl-tertiary butyl ether)

Ethers result from the selective addition of methanol or ethanol to the isobutene contained in C_4 olefin fractions. Ethers are used as components in gasoline because of their high octane blending value (RON and MON).

These synthesis processes are generally associated with an alkylation process.

Table 10.7 shows the typical composition of a feedstock to an etherification unit and the average product properties obtained.

C_4's that are inert with respect to the reaction constitute the raffinate. These can serve as feedstock to an alkylation unit.

Feedstock: FCC C$_4$ cut	Composition, weight %
Propane	0.1
Isobutane	34.6
n-Butane	11.0
Isobutene	15.0
1 Butene	12.6
2 Butene	25.2
1-3 Butadiene	0.5
C$_5$	1.0
Total	100.0

Performance:		
Isobutene conversion	to MTBE: 96 to 98%	
	to ETBE: 92 to 95%	

Properties	MTBE	ETBE
% Purity	98–99	97–98
Sp. gr. d_4^{15}	0.74	0.75
Boiling point, °C	55	72
Net heating value, kJ/kg	35,100	35,100

Blending octane number		
RON	118	118
MON	102	102
Blending RVP, bar	0.55	0.4

Table 10.7   *Typical feedstock composition and product properties for the synthesis of MTBE–ETBE.*

c. Synthesis of Ethers from C$_5$ Olefins

TAME (Tertiary amylmethylether)

One can react methanol with the tertiary olefins having five carbon atoms (isoamylenes). This process increases the octane number of FCC olefinic C$_5$ fractions, in order to reduce the concentration of olefins and to increase gasoline production.

Table 10.8 shows a typical feedstock composition and the average properties of crude or pure TAME.

This process can also be associated with an alkylation process.

Typical feedstock: FCC C_5 fraction	Composition, wt. %	
1-Pentene	5	
2-Pentene	18	
2 Methyl 1 butene	9	isoamylenes
2 Methyl 2 butene	16	
3 Methyl 1 butene	1	
Cyclopentene	2	
Isopentane	40	
n-Pentane	8	
Dienes	1	
Total	100	

RON clear	92
MON clear	82.5

PERFORMANCE	
Conversion of isoamylenes to TAME: 70%	

Properties of crude mixture of TAME + residual C_5 cut at 4% methanol	
Sp. gr. d_4^{15}	0.69
RVP, bar	0.9
RON clear	94–94.5
MON clear	84–84.5

Properties of pure TAME	
Sp. gr. d_4^{15}	0.75
Boiling point, °C	86.3
Net heating value, kJ/kg	36,575
RON	115
MON	102
Blending RVP, bar	0.1

Table 10.8 | *Typical feedstock composition used for TAME synthesis. Product performance and properties.*

Figure 10.3 shows the position of MTBE, ETBE, TAME and alkylation units in a refining flowsheet with the objective of upgrading light olefins originating from catalytic cracking. Sweetening and selective hydrogenation will be seen in 10.3.3.2.

d. Oligomerization (Dimerization)

In refining, the oligomerization process produces gasoline from C_3 fractions containing approximately 75% propylene or fuel-gas containing ethylene and propylene.

This process thus enables gasoline production to be increased if the propylene can not be used for petrochemical manufacture. It recovers ethylene economically from fuel-gas.

Feedstocks come mainly from catalytic cracking. The catalyst system is sensitive to contaminants such as dienes and acetylenes or polar compounds such as water, oxygenates, basic nitrogen, organic sulfur, and chlorinated compounds, which usually require upstream treatment.

Table 10.9 provides some general data concerning this process.

Table 10.10 is an evaluation of the improvement resulting from the addition of alkylate and ethers to an FCC gasoline.

Feedstocks	Propylene fraction from FCC	Fuel gas from FCC
Cryogenic recovery before oligomerization		
Ethylene		95%
Propylene		100%
Conversion, wt. %		
Ethylene	–	99
Propylene	90–95	95
Products	Dimer 71%	C_5+ 83.7%
	LPG (C_3) 29%	LPG 16.3%

Properties of dimer or $C_5{}^+$ fraction		
Sp. gr. d_4^{15}	0.7	0.72
RON clear	96	93–94
Blending value	99–113	97–110
MON clear	80	79–80
RVP, bar	0.48	0.4

Table 10.9 *Typical oligomerization feedstock compositions. Performance and product properties.*

Properties	FCC Gasoline alone, after sweetening (1)	FCC Gasoline (1) + C_4 alkylate (2)	FCC Gasoline (5) + C_4/C_5 alkylates (2) + MTBE (3) + TAME (4)
TBP cut, °C	C_5–220	C_5–220	C_5–220
RON clear	93	93.5	94.8
MON clear	80.5	82.9	85.0
Aromatics, weight %	33	26.8	25.7
Olefins, weight %	31	25.2	17.4
Oxygen, weight %	–	–	1.37
RVP, bar	0.35	0.44	0.41
Gasoline production	100	123	128

Proportions correspond to the material balance for catalytic cracking in Figure 10.3 showing streams (1)(2)(3)(4) and (5).

Table 10.10 *Properties of FCC gasoline alone compared with the same "improved" gasoline.*

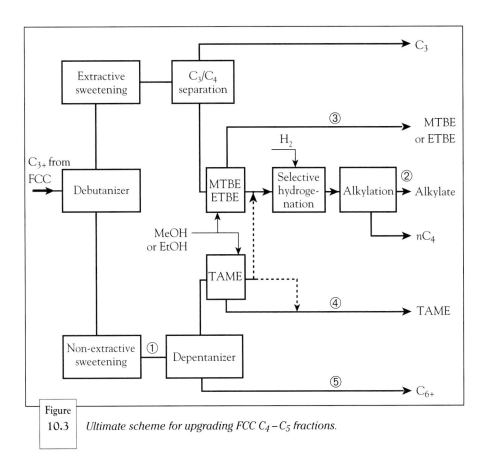

Figure 10.3 *Ultimate scheme for upgrading FCC C₄ – C₅ fractions.*

10.3.2 **Conversion Processes**

10.3.2.1 **Thermal Conversion Processes**

a. Visbreaking (Périès et al., 1988)

The visbreaking process thermally cracks atmospheric or vacuum residues. Conversion is limited by specifications for marine or industrial fuel-oil stability and by the formation of coke deposits in equipment such as heaters and exchangers.

Applied to atmospheric residue, its purpose is to produce maximum diesel oil and gasoline cuts while meeting viscosity and thermal stability specifications for industrial fuels.

Applied to vacuum residue, its purpose is to reduce the viscosity of the feedstock to a maximum so as to minimize the addition of light diluents for production of fuel-oil for industrial uses.

Visbreaking conversion products are unstable, olefinic, and very high in sulfur and nitrogen. They must be upgraded by processing before they can be incorporated into finished products.

Vacuum flashing of an effluent from thermal conversion allows recovery of a distillate that is sent to the FCC and replaced as diluent by a product of lesser quality coming from the FCC, (HCO or LCO).

Table 10.11 provides some general performance data on a visbreaking process applied on a typical VR.

Figure 10.4 shows the position of visbreaking units in a refining flowsheet.

Feedstock	Heavy Arabian VR
TBP Cut, °C	538+
Sp. gr. d_4^{15}	1.048
Sulfur, wt. %	5.78
Nitrogen, ppm	3500
Ni + V, ppm	290
Conr. Carb., wt. %	23
C_7 insol., wt. %	12.6
Visc. at 100°C, mm²/s	4500

Yields	Wt. %
H_2S	0.35
C_1–C_4	1.60
C_5–80	1.20
80–150	2.40
150–350	12.45
350+	82.00
Total	**100.00**

Product properties	Light gasoline	Heavy gasoline	Diesel oil	Residue
TBP cut, °C	C_5–80	80–150	150–350	350+
Sp. gr. d_4^{15}	0.68	0.745	0.865	1.065
Sulfur, wt. %	0.5	1.4	3.35	6.1
Visc. at 50°C, mm²/s			2.2	$\cong$ 400,000
Visc. at 100°C, mm²/s			1.1	2600
Bromine number, g/100 g	85	70	20	
MAV, mg/g	20	12		
Nitrogen, ppm	5	25	300	4100
Cetane number			42	
Ash, wt. %				0.25
Carbone Conr., wt. %				28
Ni + V, ppm				350
Stability (ASTM D 1661)				≤ 2

Table 10.11 *Typical composition of a visbreaking feedstock. Performance and product properties.*

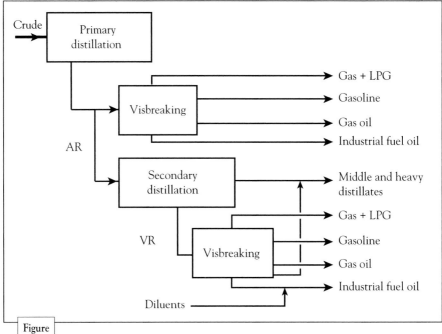

Figure 10.4 | *Refining flowchart with two visbreaking units, one for atmospheric residue, the other for vacuum residue.*

b. Coking

The coking process produces electrode quality coke from vacuum residues of good quality (low metal and sulfur contents) or coke for fuel in the case of heavy crude or vacuum residue conversion having high impurity levels.

Available processes are:

- delayed coking to produce either electrode quality coke, or coke for fuel
- fluid coking producing only the coke destined for combustion or gasification (Cavanaugh et al., 1978).

Liquid products from coking are very unstable (high diene contents), very olefinic, and highly contaminated by sulfur and nitrogen. The production of gas is considerable.

Liquid products must undergo hydrogen processing before joining equivalent crude oil fractions and continuing the normal process property improvement steps.

Table 10.12 contains some general data on coking processes and the resulting products.

Figure 10.5 indicates the position of a coking unit in a refining flowsheet.

Coking	Delayed	Fluid
Feedstock	**VR**	**VR**
Sp. gr. d_4^{15}	1.025	1.025
Sulfur, wt. %	2.9	2.9
Nitrogen, wt. %	0.3	0.3
Conr. carb., wt. %	22	22
Ni + V, ppm	250	250

Yields	Weight %	Weight %
H_2S	1.1	0.7
C_1-C_4	11.1	11.6
Light gasoline	4.3	3.7
Heavy gasoline	6.1	5.0
Diesel oil	32.4	21.5
VD	12.0	29.7
Coke	33.0	27.8
Total	100.0	100.0

Net coke	33.0	20.0
Sulfur wt. %	3.7	4.1

Product properties	Delayed coking	Fluid coking
Light gasoline	**$C_5-80°C$**	**$C_5-80°C$**
Sp. gr. d_4^{15}	0.710	0.715
Sulfur, wt. %	0.25	0.3
Nitrogen, ppm	10	15
Bromine number, g/100 g	77	110
RON clear	83	85
MAV, mg/g	30	40
Heavy gasoline	**80–150°C**	**80–150°C**
Sp. gr. d_4^{15}	0.765	0.780
Sulfur, wt. %	0.7	1.0
Nitrogen, ppm	100	120
Bromine number, g/100 g	61	90
RON clear	72	77
MAV, mg/g	10	15
Diesel oil	**150–350°C**	**150–350°C**
Sp. gr. d_4^{15}	0.850	0.845
Sulfur, wt. %	0.95	1.3
Nitrogen, ppm	600	900
Bromine number, g/100 g	30	45
Cetane number	45	46
VD	**350°C$^+$**	**350°C$^+$**
Sp. gr. d_4^{15}	0.960	0.980
Sulfur, wt. %	1.9	2.5
Nitrogen, ppm	2200	2800
Conr. carb., wt. %	2.5	4.5
Bromine number, g/100 g	10	20

Table 10.12   *Typical composition of a coking feedstock. Yields and product properties.*

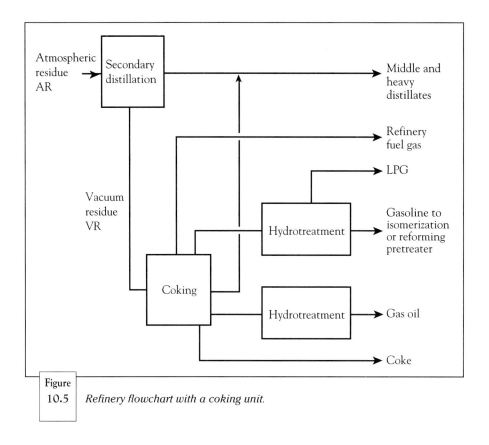

Figure 10.5 *Refinery flowchart with a coking unit.*

c. Steam Cracking

Properly speaking, steam cracking is not a refining process. A key petrochemical process, it has the purpose of producing ethylene, propylene, butadiene, butenes and aromatics (BTX) mainly from light fractions of crude oil (LPG, naphthas), but also from heavy fractions hydrotreated or not (paraffinic vacuum distillates, residue from hydrocracking HDC).

The conversion takes place at high temperature (820–850°C) and very short residence time (hundredth of seconds) in the presence of steam. The by-products are hydrogen, methane and a highly aromatic residual fuel-oil.

Table 10.13 provides some characteristic data relative to the steam cracking process.

d. Partial Oxidation (Gasification) (Strelzoff, 1974)

Here again, this is not a refining process, properly speaking. Partial oxidation is one of the processes for the ultimate conversion of heavy residues, asphalts, coke and even coal.

Feedstock	Naphtha	Gas oil	HDC residue
Sp. gr. d_4^{15}	0.70	0.840	0.832
ASTM D 86 distillation, °C			
IP	37	195	
50 vol %	83	270	
EP	196	355	
Sulfur, wt. %	0.03	0.15	0.005
Carbon, wt. %	84.77	86.55	
Hydrogen, wt. %	15.20	13.3	14.1
Bureau of Mines Correlation index[1]	8.4	30	8.2
P, wt. %	72.8		48
N, wt. %	20.8		47.3
A, wt. %	6.4		4.7

Yields	Weight %	Weight %	Weight %
Hydrogen (H_2)	0.9	0.6	0.9
Methane (CH_4)	15.8	10.0	12.0
Acetylene (C_2H_2)	0.4	0.3	1.1
Ethylene (C_2H_4)	28.6	24.4	30.3
Ethane (C_2H_6)	3.9	3.1	2.8
Propadiene (C_3H_4)	0.7	0.5	1.2
Propylene (C_3H_6)	15.0	14.1	11.9
Propane (C_3H_8)	0.4	0.5	0.3
Butadiene (C_4H_6)	4.4	5.4	6.2
Butenes (C_4H_8)	4.2	5.2	2.4
Butanes (C_4H_{10})	0.5	0.1	0.2
C_5–200	21.7	15.5	24.0
Fuel oil (residue)	3.5	20.3	6.7

Table 10.13   *Typical steam cracking feedstock composition and yields.*

---

*
$$\text{BMCI} = \frac{48,640}{\text{VABP}} + 473.7\,S - 456.8 \qquad \text{(Smith, 1940)}$$

where VABP = Volume average boiling point [K]   (see Chapter 3)
        S = Standard specific gravity   (see Chapter 4)

The basic conversion produces essentially carbon monoxide (CO) and hydrogen.

Variations of the process enable production of the following:

- high purity hydrogen
- methanol, ammonia
- methane or low heating value gas.

Table 10.14 provides some essential information concerning the production of hydrogen by partial oxidation of a VR. By-products are carbon dioxide and hydrogen sulfide.

Feedstock				Product	
VR		Oxygen			
Composition	Weight %	Composition	Vol. %	Composition	Weight %
C	84.60	O_2	95	H_2	97.5
H	11.30	N_2	2	CO	(10 ppm)
S	3.50	Argon	3	CH_4	1.0
Ash	0.07			N_2	0.7
N	0.40			Argon	0.8
O	0.13				

Table 10.14 *Hydrogen production by partial oxidation of a VR.*

10.3.2.2 **Catalytic Conversion Processes**

a. *Fluid Catalytic Cracking* (FCC)

Catalytic cracking is a key refining process along with catalytic reforming and alkylation for the production of gasoline. Operating at low pressure and in the gas phase, it uses the catalyst as a solid heat transfer medium. The reaction temperature is 500–540°C and residence time is on the order of one second.

Feedstocks for this very flexible process are usually vacuum distillates, deasphalted oils, residues (hydrotreated or not), as well as by-products from other processes such as extracts, paraffinic slack waxes, distillates from visbreaking and coking, residues from hydrocracking, converted in mixtures with the main feedstock.

Products of conversion from catalytic cracking are largely olefinic for light fractions and strongly aromatic for the heavy fractions.

Cracking reactions are endothermic; the energy balance is obtained by the production of coke that deposits on the catalyst and that is burned in the regenerator.

Contaminants such as the sulfur and nitrogen contained in the effluent gas are directly dependent on the feedstock properties.

The main products are the following:

- Liquefied gas fractions (propane, propylene, butanes, butenes) that will be able to provide feedstocks to units of MTBE, ETBE, alkylation, dimerization, polymerization after sweetening and/or selective hydrogenation.

- A gasoline fraction of good octane number (RON 91–93, MON 79–81) which is sent to the gasoline pool after sweetening. The light C_5 fraction can be etherified (TAME); the lower quality heart cut (75–125°C) is sent to catalytic reforming while the heavy fraction (125–210°C), strongly aromatic with a high octane number, is sent to the gasoline pool.

- A light distillate cut (light cycle oil – LCO) similar to gas oil but having high aromaticity and low cetane number.

The FCC by-products are:

- Refinery gases.

- Residue (slurry) or clarified oil (CLO) used as refinery fuel or as a base in the manufacture of carbon black.

- Coke (deposited on the catalyst) which is burned in the regenerator producing energy (electricity, steam) and the necessary heat for the reaction. Produced gases are cleansed when necessary of SO_x and NO_x as well as particles of entrained catalyst.

Table 10.15 provides some general data concerning FCC feedstocks and products.

Feedstock 50/50 Arabian light and heavy crude	VD	VD hydrotreated	Atm. residue hydrotreated
TBP Cut, °C	375–550	375–550	375+
Sp. gr. d_4^{15}	0.935	0.923	0.933
Sulfur, wt. %	2.80	0.15	0.35
Nitrogen, ppm	1000	300	1000
Visc. at 100°C, mm²/s	9	8.5	23
Conr. Carb., wt. %	1.2	0.4	6
C_7 insol., wt. %	< 0.02	–	1.5
Ni, ppm	< 1	< 0.5	2
V, ppm	< 1	< 0.5	4

Table 10.15   *Typical FCC feedstock composition. Yields and product properties (to be continued).*

Yields, weight %	VD	VD hydrotreated	Atm. residue hydrotreated
H_2S	1.35	0.09	0.18
Dry gas (C_1–C_2)	3.50	2.80	3.90
C_3	1.16	1.32	1.25
$C_3^=$	4.62	5.28	5.00
iC_4	1.05	2.51	2.36
nC_4	0.72	0.89	0.72
$iC_4^=$	1.93	1.85	1.75
$1\text{-}C_4^=$	1.26	1.38	1.32
$2\text{-}C_4^=$	2.76	3.40	3.20
LPG	13.50	16.63	15.60
Gasoline C_5–220°C	42.70	49.00	45.80
LCO 220–360°C	20.95	17.55	17.55
Slurry 360°C$^+$	12.80	9.33	9.37
Coke	5.20	4.60	7.60
Total	100.00	100.00	100.00

Products properties	VD	VD hydrotreated	Atm. residue hydrotreated
LPG			
Sp. gr. d_4^{15}	0.563	0.563	0.563
RSH, ppm	275	35	70
COS, ppm	5	5	5
Total sulfur, ppm	400	50	100
Gasoline			
Sp. gr. d_4^{15}	0.763	0.752	0.750
Sulfur, wt. %	0.13	0.004	0.010
of which RSH, ppm	350	10	25
RON clear	93.2	92.9	93.9
MON clear	80.4	80.1	81.0
RVP, bar	0.33	0.35	0.35
P, wt. %	26	28	27
O, wt. %	40	41	43
N, wt. %	5	5	5
A, wt. %	29	26	25
LCO			
Sp. gr. d_4^{15}	0.961	0.942	0.941
Sulfur, wt. %	3.08	0.16	0.33
Nitrogen, ppm	350	100	350

Table 10.15 *Typical FCC feedstock composition.*
Yields and product properties (to be continued).

Products properties	VD	VD hydrotreated	Atm. residue hydrotreated
LCO			
Cetane number	18.3	23.3	23.3
Visc. at 100°C, mm²/s	1.22	1.14	1.14
Visc. at 50°C, mm²/s	2.79	2.53	2.53
Cloud point, °C	−11	−9	−9
Slurry			
Sp. gr. d_4^{15}	1.096	1.067	1.067
Sulfur, wt. %	5.16	0.26	0.6
Visc. at 100°C, mm²/s	9.2	8.3	17.2
Conr. carb., wt. %	11.2	10,4	16
Catalyst content, ppm	≈ 1000	≈ 1000	≈ 1000

Table 10.15	*Typical FCC feedstock composition. Yields and product properties (continued and end).*

Figure 10.6 shows the position of an FCC and its related units in a refining flowsheet.

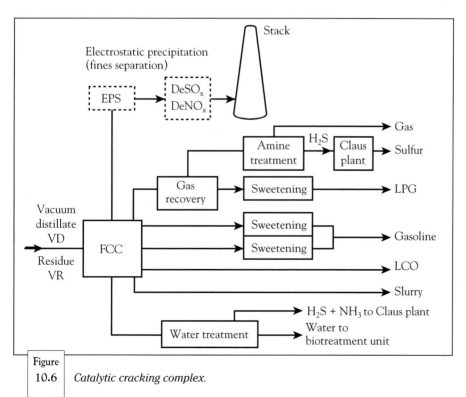

Figure 10.6	*Catalytic cracking complex.*

Figure 10.7 presents the case of an FCC feedstock comprising a mixture of vacuum distillate and light atmospheric residue, and the case of an FCC feedstock composed of vacuum distillate and DAO, as well as the constraints of such configurations.

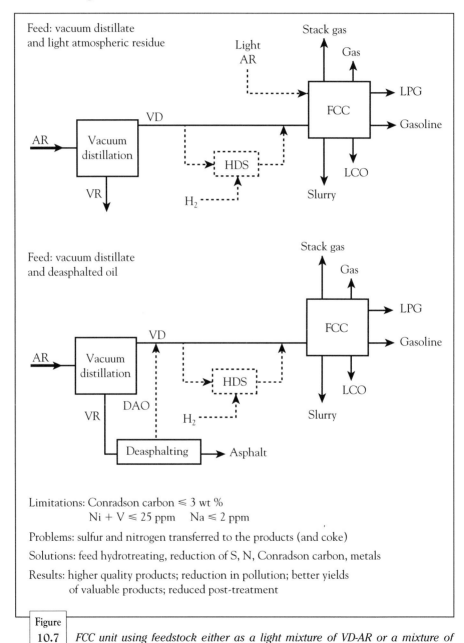

Limitations: Conradson carbon ≤ 3 wt %

              Ni + V ≤ 25 ppm   Na ≤ 2 ppm

Problems: sulfur and nitrogen transferred to the products (and coke)

Solutions: feed hydrotreating, reduction of S, N, Conradson carbon, metals

Results: higher quality products; reduction in pollution; better yields of valuable products; reduced post-treatment

Figure	
10.7	*FCC unit using feedstock either as a light mixture of VD-AR or a mixture of VD-DAO.*

Figure 10.8 presents a variant of the FCC process, the RCC (Residue Catalytic Cracking) capable of processing heavier feedstocks (atmospheric residue or a mixture of atmospheric residue and vacuum distillate) provided that certain restrictions be taken into account (Heinrich et al., 1993).

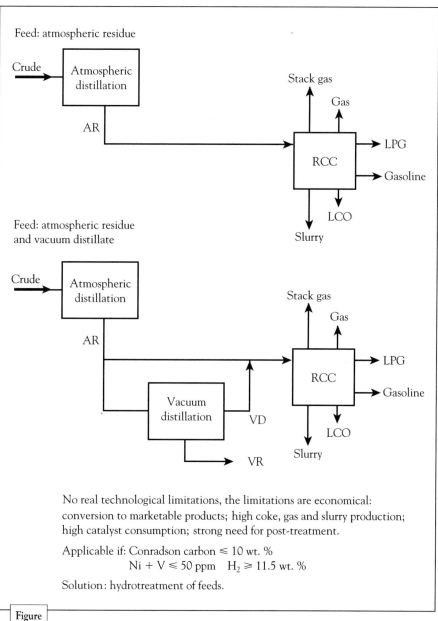

No real technological limitations, the limitations are economical: conversion to marketable products; high coke, gas and slurry production; high catalyst consumption; strong need for post-treatment.

Applicable if: Conradson carbon ≤ 10 wt. %
$\quad\quad\quad\quad\quad$ Ni + V ≤ 50 ppm $\quad$ H$_2$ ≥ 11.5 wt. %

Solution: hydrotreatment of feeds.

Figure 10.8 *FCC (RCC) unit treating either an AR or a mixture of VD-AR.*

Figure 10.9 shows advantages of a hydrodesulfurization unit upstream of an FCC or RCC unit.

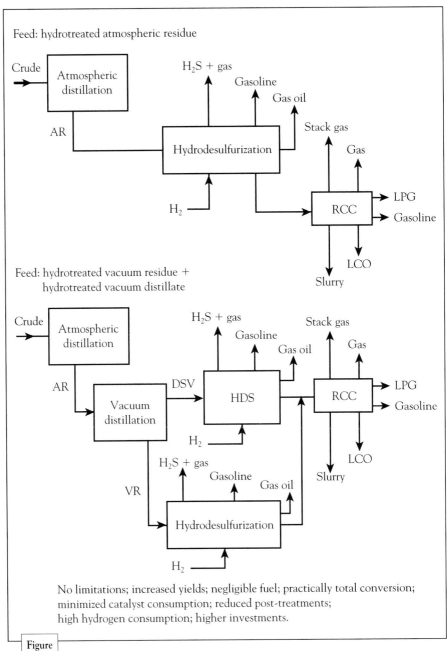

Figure
10.9 *FCC (RCC) unit preceded by a desulfurization unit.*

b. Steam Reforming

Steam reforming is, along with catalytic reforming, a process that can produce the additional hydrogen needed for upgrading and converting the heavy fractions of crude oil.

Feedstocks are natural gas, refinery fuel gas, LPG and paraffinic naphthas. After elimination of CO_2, the last traces of contaminants are converted to methane (methanation) or eliminated by adsorption on molecular sieves (PSA process).

The reactions take place at high temperature in the presence of catalyst and steam.

The hydrogen product obtained has a purity between 97 and 99.9 volume %. The balance is methane, and the by-product of the process is CO_2.

c. Hydroconversion

Among hydroconversion processes we will distinguish:

- processes that transform vacuum distillates partially or totally into lighter products:
  - hydrocracking, total or partial
  - hydrorefining
- limited conversion processes for atmospheric and vacuum residues that prepare feedstocks for more elaborate conversions, e.g., catalytic cracking, coking (Mariette et al., 1988).

1. Hydrocracking (Maier et al., 1988)

Hydrocracking is the preeminent process for making high quality kerosene and diesel oil (Figure 10.10).

Feedstocks are light vacuum distillates and/or heavy ends from crude distillation or heavy vacuum distillates from other conversion processes: visbreaking, coking, hydroconversion of atmospheric and vacuum residues, as well as deasphalted oils.

This highly flexible process allows the best optimization of yields in desirable products and it features a high degree of selectivity.

In a single stage, without liquid recycle, the conversion can be optimized between 60 and 90%. The very paraffinic residue is used to make lubricant oil bases of high viscosity index in the range of 150 N to 350 N; the residue can also be used as feedstock to steam cracking plants providing ethylene and propylene yields equal to those from paraffinic naphthas, or as additional feedstock to catalytic cracking units.

In a single stage with liquid recycle, total conversion to products lighter than the feedstock is possible. The yield of kerosene plus diesel is between 70 and 73 weight %.

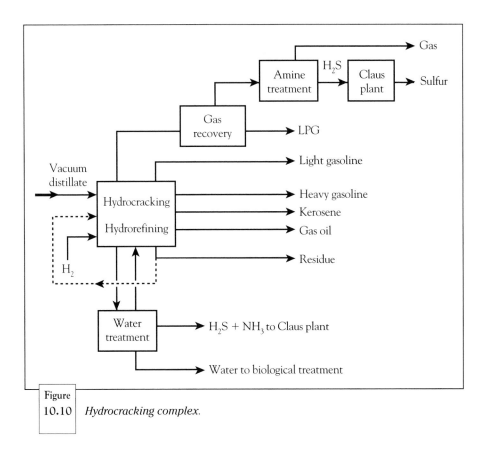

Figure 10.10 *Hydrocracking complex.*

In two stages with recycle to the second stage, the conversion per pass is approximately 50 wt. % and the selectivity to middle distillates is maximal: 75 to 80 wt. %. However, the investment is clearly higher and is justified only when feedstocks are difficult to convert and that their content in nitrogen is high. Figure 10.11 represents two variants of the hydrocracking process.

The hydrocracking process is characterized by a very low gas production and a low LPG yield especially when operated for maximum distillates. By-products in this operating mode are:

- light gasoline of excellent quality: RON 78 to 81
- heavy gasoline, a very good feedstock for catalytic reforming.

With regard to the unconverted residue, the VI after dewaxing is 120 to 135 and the BMCI is between 10 and 15, which makes it an excellent feed for steam cracking units.

Table 10.16 gives a typical feedstock composition, as well as yields and product properties.

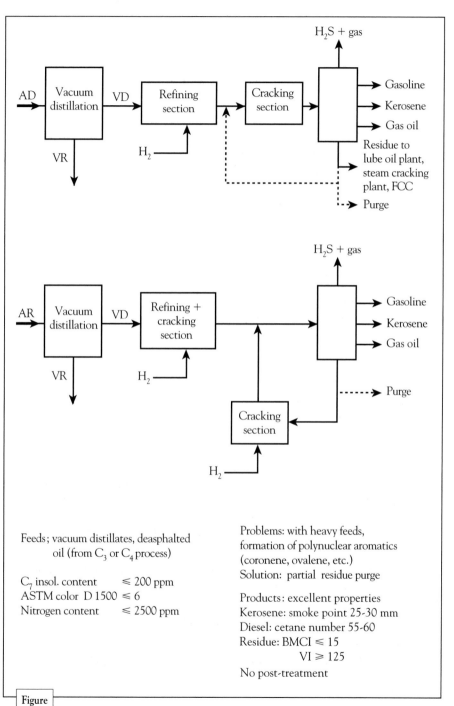

Feeds; vacuum distillates, deasphalted oil (from C_3 or C_4 process)

C_7 insol. content   ≤ 200 ppm
ASTM color D 1500 ≤ 6
Nitrogen content   ≤ 2500 ppm

Problems: with heavy feeds, formation of polynuclear aromatics (coronene, ovalene, etc.)
Solution: partial residue purge

Products: excellent properties
Kerosene: smoke point 25-30 mm
Diesel: cetane number 55-60
Residue: BMCI ≤ 15
                VI ≥ 125

No post-treatment

Figure 10.11 *Refinery flowchart with a hydrocracking unit.*

Feedstock			
Sp. gr. d_4^{15}	0.935	Ni, ppm	< 1
Sulfur, wt. %	2.80	V, ppm	< 1
Nitrogen, ppm	1000	TBP, °C, 5 vol. %	375
Conr. carb., wt. %	1.2	TBP, °C, 50 vol. %	472
Visc. at 100°C, mm²/s	9	TBP, °C, 95 vol. %	550
C_7 insol., wt. %	< 0.02		

Yields at middle-of-run, wt. %			
	1 stage without recycle	**1 stage with recycle**	**2 stages**
H_2S	2.97	2.97	2.97
NH_3	0.12	0.12	0.12
C_1	0.15	0.17	0.15
C_2	0.20	0.25	0.23
C_3	0.60	0.90	0.75
iC_4	1.25	1.60	0.95
nC_4	1.15	1.30	0.80
C_5–85°C	7.90	6.20	6.25
85–150°C	18.50	22.70	17.80
150–260°C	31.60	36.00	34.80
260–380°C	25.55	27.69	37.93
380°C+	12.86	3.00	–
Total	102.85	102.90	102.75
or 85–132°C	13.50	14.10	11.50
132–260°C	36.60	44.60	41.10
260–380°C	25.55	27.69	37.93
Chemical H_2 consumption, as weight % of feed	2.85	2.90	2.75

Products properties	**1 stage without recycle**	**1 stage with recycle**	**2 stages**
Light gasoline TBP cut, °C	C_5–85	C_5–85	C_5–85
Sp. gr. d_4^{15}	0.667	0.665	0.668
Sulfur, ppm	< 1	< 1	< 1
Nitrogen, ppm	< 1	< 1	< 1
RON	80	80	80
MON	78	78	78
P/N/A, vol. %	78/21/1	79/20/1	77/21/2

Table 10.16 *Typical feedstock composition, yields and product properties for a hydrocracking unit (to be continued).*

Products properties	1 stage without recycle	1 stage with recycle	2 stages
Heavy gasoline			
TBP cut, °C	85–132 85–150	85–132 85–150	85–132 85–150
Sp. gr. d_4^{15}	0.742 0.751	0.740 0.749	0.744 0.755
Sulfur, ppm	< 1	< 1	< 1
Nitrogen, ppm	< 1	< 1	< 1
P/N/A, vol. %.	45/50/5 40/55/5	46/48/6 42/52/6	44/50/6 39/55/6
Kerosene			
TBP cut, °C	132–260 150–260	132–260 150–260	132–260 150–260
Sp. gr. d_4^{15}	0.798 0.803	0.795 0.801	0.800 0.807
Sulfur, ppm	< 20	< 20	< 20
Smoke point, mm	30 29	31 30	29 27
Pour point, °C	< −60	< −60	< −60
Flash point, °C	≥ 40 > 55	≥ 40 > 55	≥ 40 > 55
Aromatics, wt. %	10 11	6 7	12 13
Diesel oil			
TBP cut, °C	260–380	260–380	260–380
Sp. gr. d_4^{15}	0.820	0.820	0.825
Sulfur, ppm	< 20	< 20	< 20
Nitrogen, ppm	< 2	< 2	< 2
Cloud point, °C	−6	−3	−6
Pour point, °C	−12	−9	−12
Cetane number	65	65	62

Residue	380⁺	380⁺ purge	
Sp. gr. d_4^{15}	0.830	0.842	
Sulfur, ppm	< 50	< 50	
Nitrogen, ppm	< 10	< 5	
Visc. at 100°C, mm²/s	3.7	3.5	
Visc. at 50°C, mm²/s	10.9	10.3	
VI after dewaxing	≥ 125	120	
BMCI	8	10.5	
Single-ring aro (UV)	3.0		
Two-ring aro (UV)	0.5		
Polynucl. aro (UV)	0.1		
Coronene, ppm	≤ 20		
Pyrene, ppm	300		
Ovalene, ppm	< 1		
Aromatic carbons by NMR, wt. %	< 1		

Table 10.16 *Typical feedstock composition, yields and product properties for a hydrocracking unit (continued and end).*

2. Hydrorefining (Hennico et al., 1993)

This form of limited-conversion hydrocracking is a process that selectively prepares high quality residues for the special manufacture of base oils of high viscosity index or treating residues having low BMCI for the conversion of heavy fractions to ethylene, propylene, butadiene and aromatics.

In the particular framework for lubricating oil bases, the operation takes place batchwise, generally using distillates selected according to the desired base, so as to minimize by-products and to maximize lubricating oils and their qualities.

Hydrorefining can substitute for extraction processes such as furfural where it integrates perfectly into the conventional process scheme.

It is used to prepare oil bases with good yields, even from the crudes having a "non oil" reputation.

Table 10.17 provides some typical data.

Feedstocks	Vacuum distillate (VD)	Deasphalted oil (DAO)
Sp. gr. d_4^{15}	0.926	0.925
Visc. at 50°C, mm^2/s	46.1	884
Visc. at 100°C, mm^2/s	8.4	41.6
Sulfur, wt. %	2.6	2.0
Nitrogen, ppm	1500	650
Conr. carb., wt. %	0.3	1.5
C_7 insol., ppm	100	200

ASTM D 1160, °C		
IP	310	463
50 vol. %	451	586
EP	580	–

Yields	VD 1	VD 2	DAO
$H_2S + NH_3$	2.9	2.9	2.2
$C_1–C_4$	1.3	2.5	1.7
Light gasoline	4.4	8.3	5.0
Heavy gasoline	7.5	12.4	8.2
Kerosene	21.5	34.0	24.3
Diesel oil	25.6	24.3	19.4
Lubricant base	39.0	18.0	41.0
Total	102.2	102.4	101.8

Table 10.17 *Typical feedstock composition, yields and product properties for a hydrorefining unit (to be continued).*

Properties	VD 1	VD 2	DAO
Light gasoline			
Sp. gr. d_4^{15}	0.670	0.669	0.665
Sulfur, ppm	< 10	< 10	< 10
RON	79	80	80
P/N/A, wt. %	65/31/4	70/27/3	79/19/2
Heavy gasoline			
Sp. gr. d_4^{15}	0.742	0.740	0.745
Sulfur, ppm	< 10	< 5	< 10
P/N/A, wt. %	24/66/10	29/64/7	35/61/4
Kerosene			
Sp. gr. d_4^{15}	0.811	0.798	0.796
Sulfur, ppm	< 10	< 5	< 10
Smoke point, mm	25	27	29
Aromatics, wt. %	12	10	6
Diesel oil			
Sp. gr. d_4^{15}	0.842	0.823	0.820
Sulfur, ppm	< 10	< 5	< 10
Aromatics (M.S.), wt. %	5	3	2
Cetane number	61	63	64

380°C$^+$ oily residue			
Yield after dewaxing, wt. %	30	13.5	30.8
Sp. gr. d_4^{15}	0.854	0.833	0.866
Visc. at 100°C, mm^2/s	5.5	4.5	13.8
Viscosity index (V.I.)	119	127	112
Pour point, °C	−18	−18	−18
Noack volatility, %	< 20	< 20	

After vacuum distillation of 380°C$^+$ residues								
	From VD1			From VD2			From DAO	
Oil	100 N	200 N	350 N	100 N	200 N	350 N	300 N	150 BS
Yield, wt. %	11.1	5.1	10.8	7.6	2.7	2.6	18.5	9.2
Sp. gr. d_4^{15}	0.848	0.854	0.858	0.830	0.833	0.844	0.860	0.875
Visc. at 100°C, mm^2/s	4.5	6.0	9.0	4.0	6.0	9.0	8.8	32
Viscosity index (V.I.)	115	120	118	125	127	127	120	107
Pour point, °C	−18	−18	−15	−18	−18	−15	−18	−15
Noack volatility, %	< 20	< 10	< 10	< 20	< 10	< 10	< 9	< 3

Table 10.17 *Typical feedstock composition, yields and product properties for a hydrorefining unit (continued and end).*

Figures 10.12 and 10.13 show, respectively, a flow diagram for lubricant oil production by hydrorefining and an integrated lubricating oil production unit using both extraction and hydrorefining.

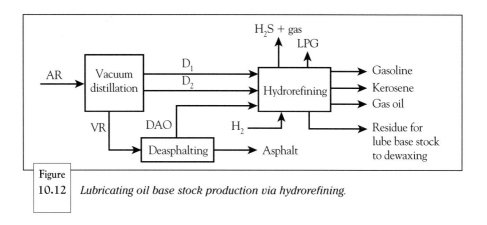

Figure 10.12 *Lubricating oil base stock production via hydrorefining.*

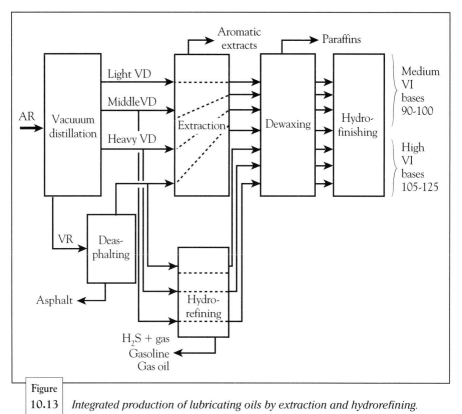

Figure 10.13 *Integrated production of lubricating oils by extraction and hydrorefining.*

3. Mild Hydrocracking

Mild hydrocracking prepares the feedstock for catalytic cracking or for the conventional lubricant production scheme.

The feedstocks can be vacuum distillate and deasphalted oils (DAO).

Conversion is approximately 15–20% in the HDS mode and 25–35% in the hydrocracking mode.

Table 10.18 presents some useful data on the mild hydrocracking process and resulting products.

Feedstock	Vacuum distillate (VD)	Deasphalted oil (DAO)
TBP cut, °C	375-550	538+
Sp. gr. d_4^{15}	0.935	0.985
Sulfur, wt. %	2.80	4.9
Nitrogen, ppm	1000	2100
Viscosity at 100°C, mm²/s	10	250
Conr. Carb., wt. %	1.2	11
Ni + V, ppm	< 1	67
C_7 insol., wt. %	< 0.02	< 0.05

Yields, weight %	HDS	HDC	HDS
$H_2S + NH_3$	2.90	3.00	5.23
C_1–C_4	1.10	1.50	1.20
Light gasoline	0.35	0.80	0.40
Heavy gasoline	0.75	1.60	0.90
Diesel oil	15.00	25.60	12.00
375+ residue	80.70	68.45	81.92
Total	100.80	100.95	101.65

Properties	HDS	HDC	HDS
Diesel oil			
Sp. gr. d_4^{15}	0.865	0.860	0.860
Sulfur, ppm	300	200	400
Cetane number	42	43	43

Residue			
Sp. gr. d_4^{15}	0.910	0.905	0.930
Sulfur, wt. %	0.15	0.1	0.15
Nitrogen, ppm	600	400	700
Conr. Carb., wt. %	< 0.4	< 0.4	2.7
Ni + V, ppm	< 0.5	< 0.5	< 2
Visc. at 100°C, mm²/s	9	8.5	42

Table 10.18	*Typical Feedstocks composition, performance and product properties from mild hydrocracking.*

4. Residue Hydroconversion (Billon et al., 1988) (Mariette et al., 1988)

The residue hydroconversion process applies to both atmospheric and vacuum residues.

Its purpose is to partially convert heavy fractions highly contaminated by natural compounds such as sulfur, nitrogen, metals: Ni, V, and asphaltenes and to prepare feedstocks for deeper conversion or to produce low-sulfur fuel-oil.

Operating at high pressure (150 to 200 bar) in the presence of hydrogen, the process is a large consumer of catalyst because of the high amount of metals in the feedstock which deposit on the catalyst.

The conversion to lighter products is limited by the asphaltenes content (C_7 insolubles). At high conversions, the residual asphaltenes —no longer being soluble in their environment— tend to precipitate, resulting in the production of unstable residues that are unmarketable.

The conversion products, other than gas and hydrogen sulfide (H_2S), are essentially a gasoline fraction that, after pretreatment, will be converted by catalytic reforming; an average quality distillate fraction to be sent to the gas oil pool; and an atmospheric residue or vacuum distillate and vacuum residue whose properties and impurity levels (S, N, Conr. Carb.) depend on the crude origin and the processing severity. Metals, having been 90 to 95% retained on the catalysts, are found in low quantities in residues.

Conversions of atmospheric residues are 20 to 35 weight % and 50 to 65% on vacuum residue.

Table 10.19 summarizes these conclusions.

Feedstock: 50/50 vol. % light and heavy Arabian crudes	Atm. residue	Vacuum residue
TBP cut, °C	375+	550+
Sp. gr. d_4^{15}	0.986	1.037
Sulfur, wt. %	3.95	5.0
Nitrogen, ppm	2300	3500
Conr. carb., wt. %	12.5	22.6
C_7 insolubles, wt. %	5.6	10.6
Visc. at 100°C, mm²/s	85	3300
Ni, ppm	25	47
V, ppm	73	138
Yields, wt. %		
H_2S + NH_3	3.88	4.84
C_1 + C_2	0.86	1.00
C_3	0.53	0.66
i C_4	0.33	0.40
n C_4	0.43	0.50
Gasoline	3.47	4.13
Gas oil	21.55	20.47
Vacuum distillate	42.95	34.25
Vacuum residue	27.55	35.50
Total	101.55	101.75
Chemical consumption of H_2, wt. %	1.55	1.75

Table 10.19
Typical Feedstocks. Performance and product properties from residue hydroconversion (to be continued).

Products properties					
Atmospheric residue case	**Gasoline**	**Gas oil**	**Atm. residue**	**Vacuum distillate**	**Vacuum residue**
TBP cut, °C	C$_5$-150	150-375	375+	375-550	550+
Sp. gr. d_4^{15}	0.730	0.866	0.934	0.908	0.978
Sulfur, wt. %	45 ppm	0.06	0.5	0.2	0.95
Nitrogen, ppm	15	450	2000	$\cong 1000$	$\cong 3500$
P/N/A, vol. %	67/23/10				
Cetane number		43			
Pour point, °C		−12			
Visc. at 50°C, mm^2/s		2.6			
Visc. at 100°C, mm^2/s		1.2	25	9	320
Conr. Carb., wt. %			6 − 6.5	0.3	15 − 16
C$_7$ insol., wt. %			1.5		
Ni, ppm		2	< 0.5	5	
V, ppm		4	< 0.5	10	

Vacuum residue case	**Gasoline**	**Gas oil**	**Atm. residue**	**Vacuum distillate**	**Vacuum residue**
TBP cut, °C	C$_5$-150	150-375	375+	375-550	550+
Sp. gr. d_4^{15}	0.735	0.860	0.965	0.938	0.992
Sulfur, wt. %	50 ppm	0.08	0.8	0.5	1.1
Nitrogen, ppm	20	600	3200	$\cong 2000$	$\cong 4400$
P/N/A, vol. %	65/25/10				
Cetane number		43			
Pour point, °C		− 12			
Visc. at 50°C, mm^2/s		2.6			
Visc. at 100°C, mm^2/s		1.2	60	11	800
Conr. Carb., wt. %			10.3	0.3	20
C$_7$ insol., wt. %			4.5		
Ni, ppm			5	< 0.5	10
V, ppm			9	< 0.5	18

Table 10.19 *Typical Feedstocks.*
Performance and product properties from residue hydroconversion (continued and end).

Figure 10.14 gives the position of this process in the refinery flowscheme.

- ARDS signifies hydroconversion of atmospheric residue
- VRDS signifies hydroconversion of vacuum residue.

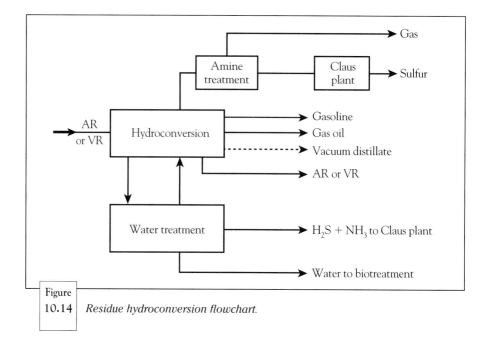

Figure 10.14   *Residue hydroconversion flowchart.*

10.3.3 **Finishing Processes**

10.3.3.1 **Hydrotreating**

Hydrotreating processes are applied to finished products to improve their characteristics: sulfur content, cetane number, smoke point and the aromatics and olefins contents.

The feedstocks in question are primary distillation streams and some conversion products from catalytic cracking, coking, visbreaking, and residue conversion units.

In regard to kerosene, the hydrotreating processes are used to reduce aromatics in order to improve the smoke point.

For gas oil: sulfur and aromatics reduction serves to increase the cetane number and to improve color and thermal stability.

By-products of these processes of hydrotreating are gases, H_2S, and some naphtha. The hydrogen consumption is relatively high as a function of the required performance.

Table 10.20 summarizes the main characteristics of hydrotreating processes.

Improving the cetane number as well as lowering the aromatics content requires higher partial pressures as well as higher hydrogen consumption.

Feedstocks	SR gas oil	SR gas oil + LCO 80/20	LCO
Sp. gr. d_4^{15}	0.846	0.863	0.941
Sulfur, wt. %	1.31	1.50	2.24
Nitrogen, ppm	70	245	940
Aniline point, °C	68	59	15
Visc. at 50°C, mm2/s	3	2.8	2.4
Aromatics, wt. %	26.7	37	78
Cetane number	53	47	20
ASTM D 86, °C			
10%	257	253	236
50%	294	288	276
90%	341	332	346
EP	358	353	374

Yields wt. %			
$H_2S + NH_3$	1.34	1.55	2.45
$C_1 - C_4$	0.30	0.35	0.50
$C_5 - 150°C$	3.50	3.60	4.50
150°C+	95.26	95.00	93.90
Total	100.40	100.50	101.35

Chemical H_2 consumption, wt. %	0.40	0.50	1.35

Product properties	Gas oil SR	Gas oil SR + LCO	LCO
Sp. gr. d_4^{15}	0.833	0.848	0.898
Sulfur, wt. %	0.05	0.05	0.05
Nitrogen, ppm	45	150	120
Visc. at 50°C, mm2/s	2.8	2.7	2.1
Aromatics, wt. %	25.7	38	71
Cetane number	56	51	32
ASTM color	< 0.5	< 1.0	2
Color stability	good	good	good

Table 10.20   *Typical hydrotreating feedstock composition.*
Performance and product properties.

Table 10.21 gives a typical example of hydrotreating a straight run (SR) gas oil.

Aromatics wt. %	Partial pressure H$_2$ bar	Chemical H$_2$ consumption, wt. %
25.7	30	0.40
20.0	40	0.60
10	65	0.90
5	85	1.10

Table 10.21 | *Relationship between the residual aromatics content, the hydrogen partial pressure, and the chemical hydrogen consumption (for a SR gas oil).*

For gas oil from catalytic cracking (LCO), reducing the aromatics content to 20 wt. % results in a chemical hydrogen consumption of 3.4 wt % and a cetane number of 40.

10.3.3.2 Sweetening Processes

Mercaptans are naturally present in crude oil (Chapters 1 and 8), or they result from the decomposition of other sulfur compounds during thermal or catalytic cracking operations.

The sweetening operation consists of converting the mercaptans to disulfides by air oxidation in the presence of a catalyst in a caustic environment.

The purpose of the operation is to obtain the following:

• a negative doctor test (sweetening)

• a reduction of the sulfur content in light fractions (sweetening plus extraction of converted sulfur compounds).

Fractions treated by this process are light products from the primary distillation: LPG to Kerosene, or light products from thermal and catalytic cracking (visbreaking, coking, FCC).

The by-products are:

• spent caustic

• disulfides.

10.3.4 Environmental Protection Processes

10.3.4.1 Acid Gas Treatment

Acid gases are mainly hydrogen sulfide (H$_2$S) originating essentially from hydrotreating units' off-gas. Smaller quantities are also produced in thermal and catalytic cracking units.

Amine Washing

Hydrogen sulfide concentrates in refinery off gases. Before being used as fuel gas, the gas undergoes an amine (MEA, DEA, etc.) washing step in order to extract the H_2S.

The rich amine loaded with H_2S is regenerated and recycled to the contacting column.

The concentrated hydrogen sulfide gas is then sent to the sulfur production unit (Claus process).

Claus Process

The Claus process converts the H_2S to sulfur by controlled combustion of the acid gas and Claus reaction on a catalyst.

The sulfur vapor is condensed and recovered in the liquid or solid form.

Yields are from 90 to 97%.

Tail gas containing traces of SO_2, H_2S, COS and CS_2 are usually sent to a finishing processing before being incinerated.

The overall yield of the operation is 99.5 to 99.8 wt % depending on the type of finishing process employed.

10.3.4.2 **Wastewater Treatment**

Contaminated water comes from primary distillation (desalting), hydrotreating, thermal cracking and catalytic cracking units.

These water streams contain mainly dissolved salts: ammonium chloride and sulfide, sodium chloride, traces of cyanide, phenols for water coming from catalytic and thermal cracking operations.

All the process water streams are collected, the entrained hydrocarbons decanted, and the water is sent to the waste water stripper.

Practically all of the H_2S and NH_3 is stripped from the water, and, with a small amount of phenols remaining, the gas effluent is sent to the Claus unit.

The treated water containing sodium chloride, cyanides, phenols and traces of H_2S and NH_3 is recycled to the crude desalting unit and used as wash water for the hydrotreaters and FCC units.

The purge stream is sent to the biological treatment plant.

Figure 10.15 shows a simplified diagram for effluent gas and wastewater treatment.

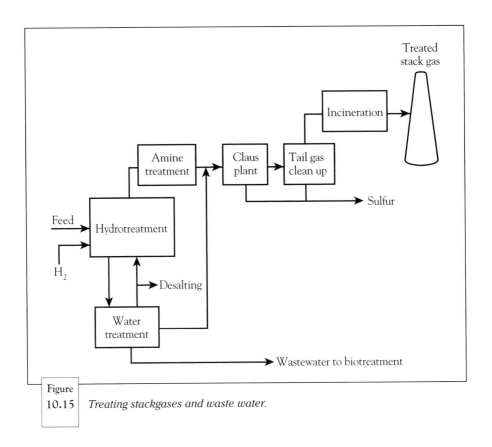

Figure 10.15 *Treating stackgases and waste water.*

10.4 **The Evolving Refinery Flowscheme**

10.4.1 **The Refinery from 1950 to 1970**

The flowscheme of the typical refinery during the period 1950–1970 was essentially focused on the production of gasoline, diesel oil, domestic heating oil and industrial fuel-oil. Except for heavy naphtha, the product streams underwent no deep conversion.

The main elements of this flowscheme are as follows:

- primary distillation
- catalytic reforming accompanied by a pretreatment step
- partial hydrodesulfurization of the gas oil fraction
- LPG and kerosene sweetening.

Purification of refinery gases by elimination of hydrogen sulfide as well as Claus units for sulfur recovery began to make their appearance.

The treatment of aqueous wastes was also taken in account.

Residual fuel-oil represented from 40 to 50% of the crude.

This highly simplified diagram highlights the emerging importance of catalytic reforming in the process of converting oil fractions. Catalytic reforming improves the gasoline octane number and produces hydrogen, an essential product for improving other crude oil fractions.

Figure 10.16 presents the refining flowsheet of the two decades between 1950 and 1970. An optional unit, visbreaking, appears there.

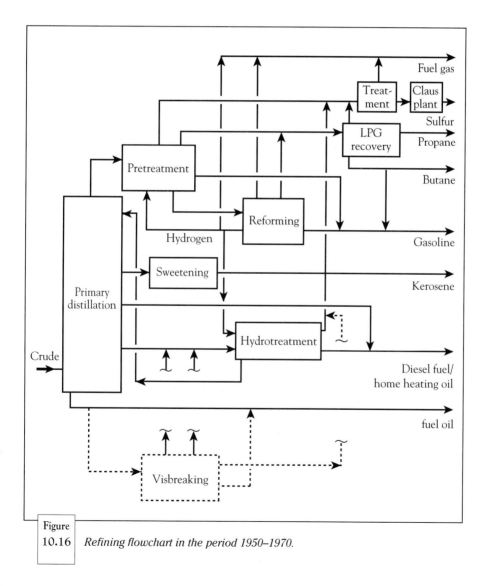

Figure 10.16   *Refining flowchart in the period 1950–1970.*

10.4.2 **The Refining Flow Diagram of the 1980s**

Following the energy crisis of the 1970s, the introduction of catalytic cracking and visbreaking allowed refiners to meet the growing demand for gasoline and distillates while minimizing crude oil imports and reducing industrial fuel-oil production. The demand for the latter is decreasing because it is being superseded by other energy sources such as nuclear and coal.

During the 1970s, the following units were added:

- secondary, or vacuum, distillation
- catalytic cracking
- visbreaking.

Residual fuel-oil represented more than 20 to 25% of the crude and the content in pollutants (sulfur, nitrogen, metals) increased.

This flowsheet can not meet newly required environmental regulations:

- elimination of lead in gasoline
- lowering the sulfur content in diesel and domestic heating oil
- reduction of SO_x emissions.

Figure 10.17 presents a refinery flow diagram of the 1980s.

10.4.3 **The Refining Flowsheet of the 1990s**

In a first phase, the diagram for processing oil fractions features the addition of complementary units that enable the production of unleaded gasoline such as:

- isomerization
- etherification (MTBE, ETBE, TAME)
- alkylation.

Furthermore, deeper hydrotreating is increasingly necessary to reduce SO_x emissions and to improve product quality:

- hydrodesulfurization of FCC feedstock
- hydrodesulfurization of diesel oils and domestic heating oil.

Heavy residues are not always converted. The use of low sulfur light crude and crudes having a reduced ultimate residue (higher ratio of gasoline + distillates / vacuum residue) as well as natural gas utilization has been intensified.

The increase in demand for good quality white products and the reduced consumption of fuel-oil related to pollution controls are going to be important factors in residue processing and heavy oil conversion in the years to come.

The addition of hydrogen and the diminution of carbon are priorities; demand for hydrogen is becoming a determinant factor.

Figure 10.18 presents the refining diagram for 1990s.

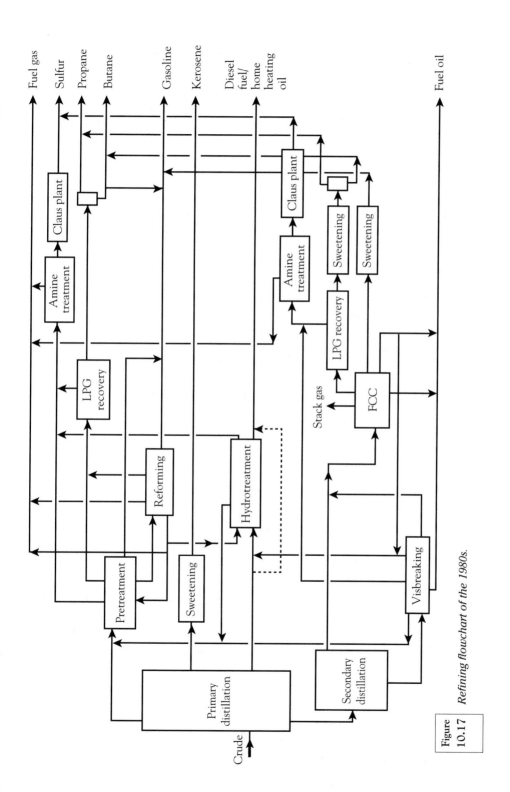

Figure 10.17 *Refining flowchart of the 1980s.*

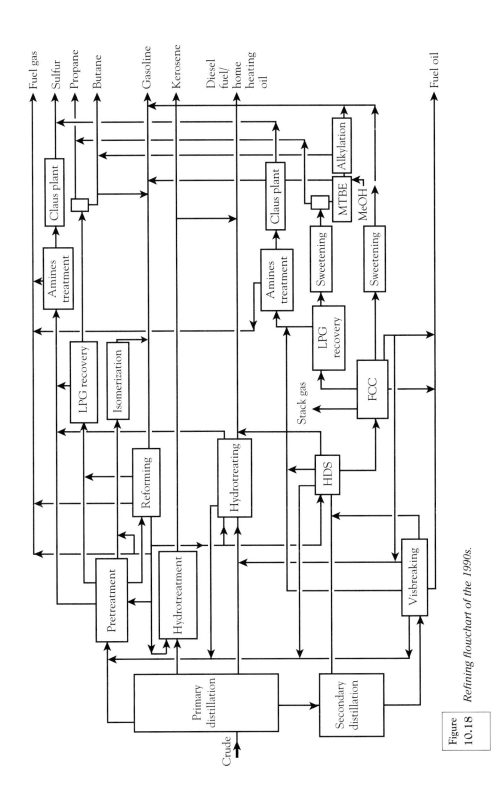

Figure 10.18   *Refining flowchart of the 1990s.*

10.4.4 **The Refining Configuration Beyond the Year 2000**

(Heinrich et al., 1991) (Convers et al., 1992)

Refining after year 2000 will be characterized by heavy residue conversion and the reduction in aromatics content.

Heavy residue conversion is linked to the demand for high quality diesel motor fuel (aromatics content $\leq 10\%$, cetane number ≥ 55) as well as to the demand for production of light fuel-oil having very low sulfur, nitrogen and metal contents.

Hydrocracking is a major process for the production of diesel motor fuel; catalytic cracking is its counterpart for the gasoline production.

The question then lies in the selection of more appropriate feedstocks for these two processes. The cost of hydrocracking leads to selecting feedstocks that are the easiest to convert; as for catalytic cracking, its flexibility and extensive capabilities lead to selection of heavier feedstocks.

These respective choices are dictated by our current knowledge, the state of the art in research and the projection of specifications in the future.

The needs for hydrogen being considerably accentuated, the introduction of partial oxidation of at least a part of the ultimate residues is foreseen, in spite of its high cost.

Intermediate feedstock preparation processes such as direct hydroconversion of vacuum residues, solvent deasphalting, improved coking will also make their appearance.

Furthermore, the major problem of reducing aromatics is focused around gasoline production. Catalytic reforming could decrease in capacity and severity. Catalytic cracking will have to be oriented towards light olefins production. Etherification, alkylation and oligomerization units will undergo capacity increases.

New processes such as isomerization and the dehydrogenation of *n*-butane will make their appearance.

Figure 10.19 shows one of the possible configurations for a refinery of the year 2000.

The refining configuration for the future and its evolution are largely dependent on the cost of crude oil; the increase in demand for white products; the shrinking market for fuel-oil; gasoline: diesel motor fuel ratio; final product specifications; and the margin that the refiner can expect, taking into account the enormous investments he must agree to make.

Table 10.22 indicates the historical trend in refining costs over the years.

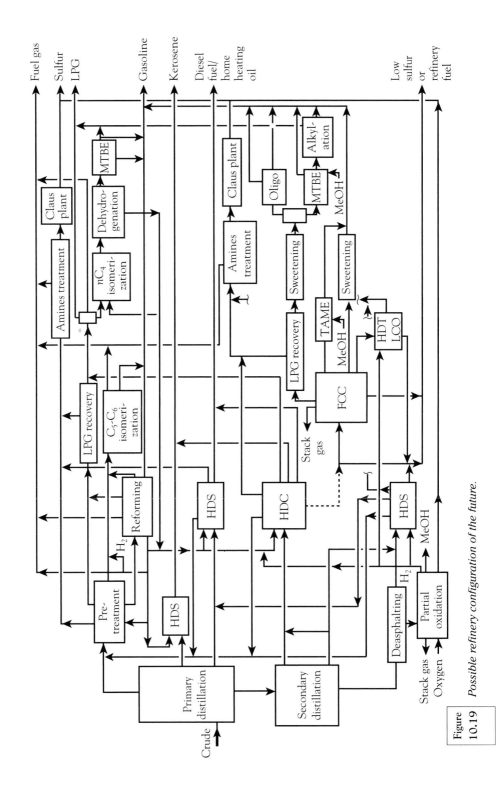

Figure 10.19   *Possible refinery configuration of the future.*

	Investment, 10^9 \$	Total cost, \$/bbl*	Internal consumption and losses, weight %
Refinery of the years 1950–1970	0.5	1.5	4–4,5
Refinery of the years 1980–1990	0.9	3.0	7–8
Refinery after the year 2000	2	7.5	11–13

Table 10.22 *Trends in refining costs for a capacity of $8 \cdot 10^6$ t/an.*
* Excluding fuel consumption.

Appendix 1

Principal Characteristics of Pure Components

Appendix 1 comprises a series of tables giving the principal characteristics of pure components most commonly found in the petroleum industry and supplying data for calculation of some useful properties.

	N_c	Normal boiling point K	Standard specific gravity	Molecular weight kg/kmol	Liquid viscosity at 100°F mm²/s	Liquid viscosity at 210°F mm²/s	Critical temperature K	Critical pressure bar
H_2	0	20.3		2.016			33.0	12.9
N_2	0	77.4		28.013			126.2	33.9
CO	1	81.7		28.010			132.9	35.0
O_2	0	90.2		31.999			154.6	50.4
CO_2	1	185.0	0.8179	44.010	0.0828	0.0881	304.1	73.8
COS	1	223.0	1.0198	60.076	0.1447	0.1578	378.8	63.5
H_2S	0	213.5	0.8021	34.082	0.1990	0.1372	373.2	89.4
NH_3	0	239.7	0.6177	17.031	0.2650	0.2136	405.7	112.8
CH_3SH	1	279.1	0.8736	48.109	0.3141	0.2307	470.0	72.3
C_2H_5SH	2	308.2	0.8458	62.136	0.6811	0.2867	499.2	54.9
H_2O	0	373.2	1.0000	18.015			647.3	221.2

Table A1.1   *Principal characteristics of common light components.*

	N_c	Triple point temperature K	Heat of fusion kJ / kg	Heat of vaporization kJ / kg	Liquid conductivity at T_1 W / (m·K)	Liquid conductivity at T_2 W / (m·K)	Temperature T_1 K	Temperature T_2 K
H_2	0	13.95	58.09	444.74	0.0754	0.0994	13.95	20.9
N_2	0	63.15	25.70	198.77	0.1595	0.1357	63.15	77.37
CO	1	68.15	30.02	214.14	0.1639	0.1397	68.15	81.7
O_2	0	54.36	13.88	212.05	0.1991	0.1496	54.36	90.19
CO_2	1	216.58	204.93	381.86	0.1769	0.1443	216.58	243.37
COS	1	134.3	150.13	311.67	–	–	–	–
H_2S	0	187.68	69.73	549.78	0.262	0.2322	187.68	212.8
NH_3	0	195.41	332.16	1369.8	0.7168	0.6143	195.41	239.72
CH_3SH	1	150.18	122.72	510.81	–	–	–	–
C_2H_5SH	2	125.26	80.07	432.44	–	–	–	–
H_2O	0	273.16	333.15	2264.7	0.5991	0.6764	293.15	373.15

Table A1.2   *Additional characteristics of common light components.*

	N_c	A	B	C $\cdot 10^4$	D $\cdot 10^7$	E $\cdot 10^{11}$	F $\cdot 10^{15}$
H_2	0	12.32674	3.199617	3.927862	−2.934520	10.900690	−13.878670
N_2	0	−0.93401	0.255204	−0.177935	0.158913	−0.322032	0.158927
CO	1	−0.97557	0.256524	−0.229112	0.222803	−0.563256	0.455878
O_2	0	−0.98176	0.227486	−0.373050	0.483017	−1.852433	2.474881
CO_2	1	4.77805	0.114433	1.011325	−0.264936	0.347063	−0.131400
COS	1	0	0.051642	1.714520	−1.067637	3.864938	−5.834600
H_2S	0	−0.61782	0.238575	−0.244571	0.410673	−1.301258	1.448520
NH_3	0	−0.94695	0.480156	−0.862580	1.749520	−6.542850	8.558870
CH_3SH	1	0	0.113823	1.267246	0.081457	−1.340404	2.507788
C_2H_5SH	2	0	0.065084	2.217950	−0.220541	−0.918771	2.471704
H_2O	0	−2.46342	0.457392	−0.525117	0.645939	−2.027592	2.363096

Table A1.3   *Coefficients for calculation of the enthalpy of an ideal gas (equation 4.77) for common light components.*

	N_c	Soave m coefficient	Solubility parameter at 25°C $(kJ/m^3)^{1/2}$	Temperature T_3 °C	Interfacial tension at T_3 mN/m	Lee Kesler acentric factor
H_2	0	0.104831	210.2	−259	2.99	−0.2185
N_2	0	0.534491	287.2	−210	12.20	0.0366
CO	1	0.557689	202.4	−205	12.38	0.0511
O_2	0	0.506394	258.7	−183	13.15	0.0197
CO_2	1	0.821396	460.6	−57	16.69	0.2236
COS	1	0.626156	574.9	20	8.59	0.0941
H_2S	0	0.639960	569.2	20	10.08	0.1028
NH_3	0	0.850923	924.2	20	21.36	0.2442
CH_3SH	1	0.709239	640.7	20	24.69	0.1472
C_2H_5SH	2	0.767881	577.7	20	23.42	0.1870
H_2O	0	0.972217	1512.0	20	73.82	0.3217

Table A1.4 *Calculational Coefficients related to a change of state of common light components.*

	N_c	Boiling point K	Standard specific gravity	Molecular weight kg/kmol	Liquid viscosity at 100°F mm²/s	Liquid viscosity at 210°F mm²/s	Critical temperature K	Critical pressure bar
Methane	1	111.63		16.043			190.56	45.96
Ethylene	2	169.44		28.054			282.36	50.33
Ethane	2	184.54	0.3560	30.070			305.33	48.72
Acetylene	2	189.15	0.4179	26.038			308.32	61.41
Propylene	3	225.45	0.5210	42.080	0.1801		364.76	46.14
Propane	3	231.07	0.5070	44.097	0.1858		369.85	42.48
Propyne	3	249.93	0.6212	40.065	0.2214	0.1913	402.39	56.29
Propadiene	3	238.65	0.5997	40.065	0.2305	0.1953	393.15	54.71
iso-Butane	4	261.36	0.5629	58.123	0.2586	0.1675	407.85	36.41
n-Butane	4	272.64	0.5840	58.123	0.2773	0.1873	425.16	37.97
1 Butene	4	266.89	0.6005	56.107	0.2224	0.1789	419.59	40.21
iso-Butene	4	266.25	0.6013	56.107	0.2853	0.2488	417.90	40.00
cis 2 Butene	4	276.87	0.6286	56.107	0.2675	0.2374	435.58	42.07
trans 2 Butene	4	274.03	0.6112	56.107	0.2759	0.2478	428.63	41.03
1 3 Butadiene	4	268.74	0.6273	54.092	0.2033	0.1274	425.37	43.31
1 Butyne	4	281.22	0.6565	54.092	0.2933	0.2304	443.20	49.50
Vinylacetylene	4	278.25	0.6918	52.076			454.00	48.61

Table A1.5   *Principal characteristics of common light hydrocarbons.*

	N_c	Triple point temperature K	Heat of fusion kJ/kg	Heat of vaporization kJ/kg	Liquid conductivity at T_1 W/(m·K)	Liquid conductivity at T_2 W/(m·K)	Temperature T_1 K	Temperature T_2 K
Methane	1	90.68	58.68	510.22	0.2240	0.1876	90.67	111.66
Ethylene	2	104.00	119.45	479.74	0.2681	0.1874	104.00	169.41
Ethane	2	90.35	95.08	490.90	0.2570	0.1617	90.35	184.55
Acetylene	2	192.35	144.62	640.44	–	–	–	–
Propylene	3	87.90	71.36	439.24	0.1872	0.1304	87.89	225.46
Propane	3	85.52	79.91	425.50	0.2165	0.1314	85.44	231.11
Propyne	3	170.45	–	555.05	0.1729	0.1384	170.45	249.94
Propadiene	3	136.85	–	513.92	0.1767	0.1347	136.87	238.65
iso-Butane	4	113.55	78.11	365.56	0.1635	0.1086	113.54	261.43
n-Butane	4	134.79	80.19	385.79	0.1845	0.1180	134.86	272.65
1 Butene	4	87.86	68.58	399.85	0.1828	0.1201	87.80	266.90
iso-Butene	4	132.80	105.71	395.74	0.1873	0.1171	132.81	266.25
cis 2 Butene	4	134.24	130.28	417.41	0.1697	0.1206	134.26	276.87
trans 2 Butene	4	167.60	173.91	408.06	0.1561	0.1184	167.62	274.03
1 3 Butadiene	4	164.23	147.60	415.41	0.1621	0.1238	164.25	268.74
1 Butyne	4	147.43	111.46	451.77	0.1715	0.1245	147.43	281.22
Vinylacetylene	4	–	–	456.68	0.1278	–	278.25	–

Table A1.6   *Additional characteristics of common light hydrocarbons.*

	N_c	A	B	C $\cdot 10^4$	D $\cdot 10^7$	E $\cdot 10^{11}$	F $\cdot 10^{15}$
Methane	1	58.40160	0.571700	−2.943122	4.231568	−15.267400	19.452610
Ethylene	2	173.77800	0.144963	1.710121	0.761974	−4.503085	6.664928
Ethane	2	163.05960	0.264878	−0.250140	2.923341	−12.860530	18.220570
Acetylene	2	274.09960	0.022636	5.459217	−2.976556	8.738401	−10.045390
Propylene	3	193.22930	0.030810	3.512242	−0.494661	−0.226171	1.125539
Propane	3	165.72380	0.172601	0.940410	2.155433	−10.709860	15.927940
Propyne	3	172.51340	0.080387	3.240066	−0.879745	1.599146	−1.400978
Propadiene	3	229.26380	0.033745	3.715168	−1.062607	1.864230	−1.435039
iso-Butane	4	162.08110	0.046682	3.348013	0.144230	−3.164196	5.428928
n-Butane	4	164.44400	0.098571	2.691795	0.518202	−4.201390	6.560421
1 Butene	4	187.76640	−0.018519	4.263451	−0.940582	1.072240	−0.349830
iso-Butene	4	179.26680	0.033009	3.782637	−0.733312	0.697566	−0.174830
cis 2 Butene	4	210.45030	−0.042795	4.034318	−0.684280	0.134493	0.878860
trans 2 Butene	4	187.24240	0.037032	3.551222	−0.560440	0.158471	0.444673
1 3 Butadiene	4	208.81850	−0.100603	5.651872	−2.123463	4.830541	−4.738449
1 Butyne	4	185.60900	0.053336	3.461000	−0.803301	1.052757	−0.533994
Vinylacetylene	4	0	0.152619	1.293917	1.015455	−7.396445	14.425954

Table A1.7   Coefficients for calculation of the enthalpy of an ideal gas (equation 4.77) for common light hydrocarbons.

	N_c	Soave m coefficient	Solubility parameter at 25°C $(kJ/m^3)^{1/2}$	Temperature T_3 °C	Interfacial tension at T_3 mN/m	Lee Kesler acentric factor
Methane	1	0.485801	367.4	−161	13.34	0.0103
Ethylene	2	0.611674	393.3	−104	16.41	0.0883
Ethane	2	0.627417	391.4	−89	15.98	0.0984
Acetylene	2	0.785634	594.9	−84	19.38	0.2008
Propylene	3	0.699302	408.4	20	6.88	0.1448
Propane	3	0.711519	414.0	20	7.02	0.1529
Propyne	3	0.806292	581.6	20	11.51	0.2148
Propadiene	3	0.692237	443.1	20	9.50	0.1399
iso-Butane	4	0.760275	397.4	20	9.81	0.1853
n-Butane	4	0.783557	426.6	20	11.87	0.2007
1 Butene	4	0.770672	432.1	20	12.12	0.1919
iso-Butene	4	0.776179	431.9	20	11.69	0.1956
cis 2 Butene	4	0.788141	465.5	20	13.99	0.2035
trans 2 Butene	4	0.804933	449.3	20	12.72	0.2148
1 3 Butadiene	4	0.772083	493.3	20	13.41	0.1927
1 Butyne	4	0.853584	523.5	20	17.06	0.2469
Vinylacetylene	4	0.684767	538.7	20	17.79	0.1352

Table A1.8   *Calculational coefficients related to a change of state of common light hydrocarbons.*

	N_c	Boiling point K	Standard specific gravity	Molecular weight kg/kmol	Liquid viscosity at 100°F mm²/s	Liquid viscosity at 210°F mm²/s	Critical temperature K	Critical pressure bar
n-Pentane	5	309.2	0.6311	72.151	0.3397	0.2647	469.7	33.70
n-Hexane	6	341.9	0.6638	86.178	0.4095	0.3023	507.5	30.13
n-Heptane	7	371.6	0.6882	100.205	0.5050	0.3521	540.2	27.37
n-Octane	8	398.8	0.7070	114.232	0.6372	0.3995	568.8	24.88
n-Nonane	9	424.0	0.7219	128.259	0.8070	0.4697	594.7	22.88
n-Decane	10	447.3	0.7342	142.286	1.0130	0.5525	617.6	21.05
n-Undecane	11	469.0	0.7445	156.313	1.2580	0.6394	638.8	19.66
n-Dodecane	12	489.5	0.7527	170.340	1.5446	0.7467	658.2	18.24
n-Tridecane	13	508.6	0.7617	184.367	1.7546	0.8299	675.8	17.23
n-Tetradecane	14	526.7	0.7633	198.394	2.2511	0.9930	692.4	16.21
n-Pentadecane	15	543.8	0.7722	212.421	2.4916	1.0950	706.8	15.20
n-Hexadecane	16	560.0	0.7772	226.448	2.9193	1.2460	720.6	14.19
n-Heptadecane	17	575.2	0.7797	240.475	3.5813	1.4462	733.4	13.17
n-Octadecane	18	589.5	0.7820	254.502	4.1314	1.5974	745.3	12.14
n-Nonadecane	19	603.1	0.7869	268.529	4.6990	1.7939	755.9	11.17
n-Eicosane	20	617.0	0.7924	282.556	5.3926	1.9846	767.0	10.40

Table A1.9   *Principal characteristics of n-paraffins.*

N_c		Triple point temperature K	Heat of fusion kJ/kg	Heat of vaporization kJ/kg	Liquid conductivity at T_1 W/(m·K)	Liquid conductivity at T_2 W/(m·K)	Temperature T_1 K	Temperature T_2 K
n-Pentane	5	143.4	116.44	357.1	0.1782	0.1085	143.4	309.2
n-Hexane	6	177.8	151.78	334.7	0.1621	0.1041	177.8	341.9
n-Heptane	7	182.6	140.22	316.2	0.1597	0.1024	182.6	371.6
n-Octane	8	216.4	181.56	301.1	0.1518	0.0980	216.4	398.8
n-Nonane	9	219.7	120.56	289.1	0.1510	0.0971	219.7	424.0
n-Decane	10	243.5	201.77	278.2	0.1454	0.0945	243.5	447.3
n-Undecane	11	247.6	141.90	271.9	0.1454	0.0940	247.6	469.1
n-Dodecane	12	263.6	216.27	261.0	0.1434	0.0908	263.6	489.5
n-Tridecane	13	267.8	154.59	251.0	0.1433	0.0940	267.8	508.6
n-Tetradecane	14	279.0	227.18	239.1	0.1401	0.0907	279.0	526.7
n-Pentadecane	15	283.1	162.85	233.4	0.1435	0.0878	283.1	543.8
n-Hexadecane	16	291.3	235.63	226.6	0.1423	0.0809	291.3	560.0
n-Heptadecane	17	295.1	168.25	217.5	0.1439	0.0818	295.1	575.3
n-Octadecane	18	301.3	243.48	217.0	0.1458	0.0809	301.3	589.9
n-Nonadecane	19	305.0	170.61	208.7	0.1451	0.0796	305.3	603.1
n-Eicosane	20	309.6	247.29	208.8	0.1487	0.0800	309.6	616.9

Table A1.10   *Additional characteristics of n-paraffins.*

	N_c	A	B	C $\cdot 10^4$	D $\cdot 10^7$	E $\cdot 10^{11}$	F $\cdot 10^{15}$
n-Pentane	5	173.46090	-0.002795	4.400733	-0.862875	0.817644	-0.197154
n-Hexane	6	133.19390	0.229107	-0.815691	4.527826	-25.231790	47.480200
n-Heptane	7	134.12590	0.180209	0.347292	3.218786	-18.366030	33.769380
n-Octane	8	130.57280	0.173084	0.488101	3.054008	-17.365470	31.248310
n-Nonane	9	126.71600	0.169056	0.581255	2.926114	-16.558500	29.296090
n-Decane	10	118.42310	0.203347	-0.349035	4.070565	-23.064410	42.968970
n-Undecane	11	156.57930	-0.023843	4.607729	-0.998387	1.084149	-0.331217
n-Dodecane	12	152.44400	-0.018522	4.538933	-0.964642	1.013931	-0.296646
n-Tridecane	13	151.99910	-0.022933	4.595173	-0.997582	1.083507	-0.330908
n-Tetradecane	14	150.25060	-0.022048	4.580788	-0.991639	1.071259	-0.325375
n-Pentadecane	15	148.84100	-0.024114	4.607172	-1.007675	1.104474	-0.341474
n-Hexadecane	16	146.85880	-0.022825	4.590237	-1.000209	1.089122	-0.333900
n-Heptadecane	17	144.59410	-0.023563	4.599069	-1.006645	1.103066	-0.340757
n-Octadecane	18	141.79860	-0.023616	4.599502	-1.008040	1.105867	-0.342047
n-Nonadecane	19	139.17470	-0.022153	4.579824	-0.998648	1.086444	-0.332753
n-Eicosane	20	141.05750	-0.022726	4.586627	-1.003133	1.095161	-0.336584

Table
A1.11   *Coefficients for calculation of the enthalpy of an ideal gas (equation 4.77) for n-paraffins.*

	N_c	Soave m coefficient	Solubility parameter at 25°C $(kJ/m^3)^{1/2}$	Temperature T_3 °C	Interfacial tension at T_3 mN/m	Lee Kesler acentric factor
n-Pentane	5	0.860773	454.4	20	15.47	0.2530
n-Hexane	6	0.932953	470.2	20	17.98	0.3024
n-Heptane	7	1.007679	481.6	20	19.78	0.3539
n-Octane	8	1.078127	488.3	20	21.08	0.4038
n-Nonane	9	1.145923	492.0	20	22.38	0.4517
n-Decane	10	1.211964	496.3	20	23.37	0.4986
n-Undecane	11	1.277030	500.9	20	24.24	0.5458
n-Dodecane	12	1.335354	503.1	20	24.94	0.5861
n-Tridecane	13	1.403470	502.8	20	25.55	0.6354
n-Tetradecane	14	1.461162	505.5	20	26.32	0.6771
n-Pentadecane	15	1.527496	503.4	20	26.68	0.7246
n-Hexadecane	16	1.579018	504.0	20	27.09	0.7609
n-Heptadecane	17	1.617797	503.6	20	27.52	0.7884
n-Octadecane	18	1.642516	502.4	20	27.97	0.8049
n-Nonadecane	19	1.669170	500.0	20	28.22	0.8234
n-Eicosane	20	1.703665	498.7	20	28.54	0.8461

Table A1.12 *Calculational coefficients related to a change of state for n-paraffins.*

	N_c	Boiling point K	Standard specific gravity	Molecular weight kg/kmol	Liquid viscosity at 100°F mm²/s	Liquid viscosity at 210°F mm²/s	Critical temperature K	Critical pressure bar
iso-Pentane	5	301.0	0.6247	72.150	0.3175	0.2346	460.4	33.82
2 Methylpentane	6	333.4	0.6548	86.177	0.3914	0.2583	497.5	30.11
3 Methylpentane	6	336.4	0.6689	86.177	0.3925	0.2670	504.4	31.25
2 2 Dimethylbutane	6	322.9	0.6535	86.177	0.4719	0.2628	488.8	30.81
2 3 Dimethylbutane	6	331.1	0.6670	86.177	0.4443	0.1314	500.0	31.28
2 Methylhexane	7	363.2	0.6823	100.200	0.4764	0.3065	530.4	27.34
3 Methylhexane	7	365.0	0.6928	100.200	0.4479	0.2884	535.3	28.15
2 2 Dimethylpentane	7	352.3	0.6821	100.200	0.5361	0.3017	520.5	27.74
2 3 Dimethylpentane	7	362.9	0.7024	100.200	0.5627	0.3445	537.4	29.09
2 4 Dimethylpentane	7	353.7	0.6764	100.200	0.5265	0.3303	519.8	27.38
3 3 Dimethylpentane	7	359.2	0.6961	100.200	0.5625	0.3460	536.4	29.46
2 Methylheptane	8	390.8	0.7037	114.230	0.5585	0.3400	559.6	24.85
2 Methyloctane	9	416.4	0.7177	128.260	0.6582	0.3865	586.8	22.91

Table
A1.13 *Principal characteristics of isoparaffins.*

	N_c	Triple point temperature K	Heat of fusion kJ/kg	Heat of vaporization kJ/kg	Liquid conductivity at T_1 W / (m·K)	Liquid conductivity at T_2 W / (m·K)	Temperature T_1 K	Temperature T_2 K
iso-Pentane	5	113.3	71.39	342.1	0.1685	0.1087	113.3	301.0
2 Methylpentane	6	119.5	72.73	324.1	0.1598	0.0999	119.6	333.4
3 Methylpentane	6	110.3	61.54	334.3	0.1644	0.1009	110.3	336.4
2 2 Dimethylbutane	6	173.3	6.72	306.4	0.1344	0.0949	174.3	322.9
2 3 Dimethylbutane	6	144.6	9.27	318.8	0.1420	0.0967	145.2	331.1
2 Methylhexane	7	154.9	91.66	310.9	0.1552	0.0991	154.9	363.2
3 Methylhexane	7	153.8	94.41	311.1	0.1566	0.1000	153.8	365.0
2 2 Dimethylpentane	7	149.3	58.12	291.9	0.1371	0.0903	149.3	352.3
2 3 Dimethylpentane	7	255.4		304.2	0.1403	0.0904	160.0	362.9
2 4 Dimethylpentane	7	153.9	68.31	297.2	0.1350	0.0897	153.9	353.6
3 3 Dimethylpentane	7	138.7	70.53	296.6	0.1402	0.0913	138.7	359.2
2 Methylheptane	8	164.1	103.98	292.6	0.1571	0.0985	164.2	390.8
2 Methyloctane	9	192.8	140.34	283.9	0.1533	0.0979	192.8	416.4

Table
A1.14   *Additional characteristics of isoparaffins.*

	N_c	A	B	C $\cdot 10^4$	D $\cdot 10^7$	E $\cdot 10^{11}$	F $\cdot 10^{15}$
iso-Pentane	5	169.01630	-0.031504	4.698836	-0.982825	1.029852	-0.294847
2 Methylpentane	6	152.96710	0.041484	3.116936	0.622252	-6.385257	12.592750
3 Methylpentane	6	148.17180	0.095013	1.605403	2.399189	-16.146920	32.727430
2 2 Dimethylbutane	6	166.76920	-0.119500	6.005360	-1.780500	2.920860	-1.344890
2 3 Dimethylbutane	6	0	-0.054622	4.854010	-0.983122	0.829638	0.266461
2 Methylhexane	7	140.72270	0.096697	1.770980	2.121378	-14.382430	28.531670
3 Methylhexane	7	149.50080	0.039990	3.001828	0.786641	-7.296178	14.037090
2 2 Dimethylpentane	7	0	0.024189	3.412306	0.474432	-6.010310	12.246112
2 3 Dimethylpentane	7	0	-0.081589	5.221015	-1.185969	1.154055	0.355282
2 4 Dimethylpentane	7	166.21450	0.022285	3.447680	0.464800	-5.849540	10.396080
3 3 Dimethylpentane	7	0	-0.010050	4.067768	-0.179990	-2.894518	6.689286
2 Methylheptane	8	141.97100	0.072874	2.424130	1.285768	-9.465965	17.639110
2 Methyloctane	9	0	0.129746	1.074314	2.784486	-17.228020	32.478120

Table A1.15 *Coefficients for calculation of the enthalpy of an ideal gas (equation 4.77) for isoparaffins.*

	N_c	Soave m coefficient	Solubility parameter at 25°C $(kJ/m^3)^{1/2}$	Temperature T_3 °C	Interfacial tension at T_3 mN/m	Lee Kesler acentric factor
iso-Pentane	5	0.825736	438.3	20	14.45	0.2290
2 Methylpentane	6	0.901698	456.0	20	16.87	0.2812
3 Methylpentane	6	0.894051	470.3	20	17.58	0.2758
2 2 Dimethylbutane	6	0.833460	435.4	20	15.80	0.2348
2 3 Dimethylbutane	6	0.854622	434.1	20	16.87	0.2495
2 Methylhexane	7	0.977879	472.3	20	18.79	0.3340
3 Methylhexane	7	0.968046	475.3	20	19.30	0.3272
2 2 Dimethylpentane	7	0.914862	449.3	20	17.53	0.2908
2 3 Dimethylpentane	7	0.927598	469.3	20	19.47	0.2996
2 4 Dimethylpentane	7	0.939537	454.1	20	17.64	0.3067
3 3 Dimethylpentane	7	0.885525	459.6	20	19.07	0.2703
2 Methylheptane	8	1.049892	477.0	20	20.15	0.3836
2 Methyloctane	9	1.111634	484.5	20	21.44	0.4281

Table
A1.16   *Calculational coefficients related to a change of state of isoparaffins.*

	N_c	Boiling point	Standard specific gravity	Molecular weight	Liquid viscosity at 100°F	Liquid viscosity at 210°F	Critical temperature	Critical pressure
		K		kg/kmol	mm²/s	mm²/s	K	bar
Cyclopentane	5	322.4	0.7603	70.130	0.4927	0.3193	511.8	45.03
Methylcyclopentane	6	345.0	0.7540	84.160	0.5646	0.3487	532.8	37.86
Cyclohexane	6	353.9	0.7835	84.160	0.9407	0.5447	553.5	40.76
Methylcyclohexane	7	374.1	0.7748	98.190	0.7660	0.4343	572.2	34.72
Ethylcyclopentane	7	376.6	0.7712	98.190	0.6199	0.3901	569.5	33.99
Ethylcyclohexane	8	404.9	0.7921	112.210	0.8634	0.5123	609.2	30.41
n-Propylcyclopentane	8	404.1	0.7811	112.210	0.7256	0.4613	603.0	30.01
n-Propylcyclohexane	9	429.9	0.7981	126.240	1.0010	0.5759	639.2	28.07
n-Butylcyclopentane	9	429.8	0.8103	126.240	–	–	621.3	27.24

Table A1.17   *Principal characteristics of naphthenes.*

	N_c	Triple point temperature K	Heat of fusion kJ/kg	Heat of vaporization kJ/kg	Liquid conductivity at T_1 W / (m·K)	Liquid conductivity at T_2 W / (m·K)	Temperature T_1 K	Temperature T_2 K
Cyclopentane	5	179.3	8.68	388.5	0.1583	0.1197	179.3	322.4
Methylcyclopentane	6	130.7	82.33	348.4	0.1603	0.1069	130.7	344.9
Cyclohexane	6	279.7	32.56	355.0	0.1281	0.1095	279.7	353.9
Methylcyclohexane	7	146.6	68.75	324.0	0.1455	0.0934	146.6	374.1
Ethylcyclopentane	7	134.7	69.96	328.3	0.1533	0.0994	134.7	376.6
Ethylcyclohexane	8	161.8	74.27	304.3	0.1440	0.0951	161.8	405.0
n-Propylcyclopentane	8	155.8	89.41	308.2	0.1435	0.0936	155.8	404.1
n-Propylcyclohexane	9	178.3	82.16	289.0	0.1367	0.0900	178.3	429.9
n-Butylcyclopentane	9	165.2		409.5				

Table
A1.18   *Additional characteristics of naphthenes.*

	N_c	A	B	C $\cdot 10^4$	D $\cdot 10^7$	E $\cdot 10^{11}$	F $\cdot 10^{15}$
Cyclopentane	5	229.11130	−0.174553	4.878999	−0.790213	−0.259001	1.873384
Methylcyclopentane	6	203.57300	−0.163500	5.315238	−1.239759	1.465505	−0.497681
Cyclohexane	6	209.80430	−0.149848	4.572747	−0.387392	−1.791242	3.793529
Methylcyclohexane	7	191.58840	−0.168390	5.444843	−1.126886	0.751131	0.606023
Ethylcyclopentane	7	196.95860	−0.152454	5.279018	−1.232110	1.467753	−0.499779
Ethylcyclohexane	8	173.54180	−0.084958	4.278138	−0.283095	−2.216911	4.555816
n-Propylcyclopentane	8	181.63590	−0.134564	5.169364	−1.192975	1.398243	−0.467750
n-Propylcyclohexane	9	177.45520	−0.169580	5.867914	−1.509927	1.978016	−0.741553
n-Butylcyclopentane	9	172.24100	−0.121404	5.094686	−1.169329	1.360705	−0.451765

Table
A1.19   *Coefficients for calculation of the enthalpy of an ideal gas (equation 4.77) for naphthenes.*

	N_c	Soave m coefficient	Solubility parameter at 25°C $(kJ/m^3)^{1/2}$	Temperature T_3 °C	Superficial tension at T_3 mN/m	Lee Kesler acentric factor
Cyclopentane	5	0.775629	527.6	20	21.71	0.1954
Methylcyclopentane	6	0.831328	511.5	20	21.65	0.2323
Cyclohexane	6	0.802374	528.1	20	24.65	0.2127
Methylcyclohexane	7	0.839095	514.5	20	23.30	0.2381
Ethylcyclopentane	7	0.890157	516.7	20	23.33	0.2728
Ethylcyclohexane	8	0.860121	517.4	20	25.05	0.2533
n-Propylcyclopentane	8	0.899869	518.5	20	24.42	0.2798
n-Propylcyclohexane	9	0.884080	518.0	20	26.07	0.2694
n-Butylcyclopentane	9	1.041826	–	20	25.39	0.3779

Table A1.20 *Calculational coefficients related to a change of state for naphthenes.*

	N_c	Boiling point K	Standard specific gravity	Molecular weight kg/kmol	Liquid viscosity at 100°F mm²/s	Liquid viscosity at 210°F mm²/s	Critical temperature K	Critical pressure bar
1 Pentene	5	303.1	0.6458	70.13	0.2822	0.2179	464.8	35.27
cis 2 Pentene	5	310.1	0.6598	70.13	0.3060	0.2460	475.9	36.55
trans 2 Pentene	5	309.5	0.6524	70.13	0.3090	0.2465	475.4	36.55
2 Methyl 1 butene	5	304.3	0.6563	70.13	0.3068	0.2407	465.4	34.00
2 Methyl 2 butene	5	311.7	0.6683	70.13	0.2881	0.2181	470.9	34.00
Cyclopentene	5	317.4	0.7773	68.12	0.2656	0.2001	507.0	47.91
2 Methyl 1 3 butadiene	5	307.2	0.6864	68.12	0.2788	0.1954	484.0	38.51
Cyclopentadiene	5	314.7	0.8041	66.10	0.3094	0.2411	507.0	51.51
1 Hexene	6	336.6	0.6769	84.16	0.3423	0.2398	503.8	31.42
Cyclohexene	6	356.1	0.8157	82.15	0.6605	0.3654	560.4	43.51
1 Heptene	7	366.8	0.7015	98.19	0.4318	0.3044	537.5	28.31
1 Octene	8	394.4	0.7193	112.21	0.5618	0.3599	567.1	25.61
1 Nonene	9	420.0	0.7336	126.24	0.7009	0.4314	593.3	23.31

Table A1.21 *Principal characteristics of olefins.*

	N_c	Triple point temperature K	Heat of fusion kJ/kg	Heat of vaporization kJ/kg	Liquid conductivity at T_1 W / (m·K)	Liquid conductivity at T_2 W / (m·K)	Temperature T_1 K	Temperature T_2 K
1 Pentene	5	107.9	82.8	365.7	0.1744	0.1143	107.9	303.1
cis 2 Pentene	5	121.8	101.4	378.2	0.1723	0.1141	121.8	310.1
trans 2 Pentene	5	132.9	119.1	374.1	0.1665	0.1135	132.9	309.5
2 Methyl 1 butene	5	135.6	112.8	364.6	0.1551	0.1062	135.6	304.3
2 Methyl 2 butene	5	139.4	108.3	377.4	0.1558	0.1057	139.4	311.7
Cyclopentene	5	138.1	49.4	405.1	0.1754	0.1230	138.1	317.4
2 Methyl 1 3 butadiene	5	127.2	72.3	376.0	0.1835	0.1179	127.3	307.2
Cyclopentadiene	5	188.2	–	389.5	0.1683	0.1284	188.2	314.7
1 Hexene	6	133.3	111.1	340.2	0.1856	0.1070	133.4	336.6
Cyclohexene	6	169.7	40.1	378.5	0.1615	0.1162	169.7	356.1
1 Heptene	7	154.3	126.3	323.1	0.1730	0.1030	154.3	366.8
1 Octene	8	171.4	136.4	306.3	0.1671	0.0999	171.5	394.4
1 Nonene	9	191.8	142.5	293.2	0.1446	0.0898	191.8	420.0

Table
A1.22   *Additional characteristics of olefins.*

	N_c	A	B	C $\cdot 10^4$	D $\cdot 10^7$	E $\cdot 10^{11}$	F $\cdot 10^{15}$
1 Pentene	5	175.27340	−0.006874	4.210531	−0.908305	1.003804	−0.315910
cis 2 Pentene	5	0	−0.002027	3.341622	0.121528	−3.904030	8.190906
trans 2 Pentene	5	0	0.049991	3.043060	0.123096	−3.358723	6.801698
2 Methyl 1 butene	5	0	0.003130	4.098669	−0.725377	−0.149655	2.074780
2 Methyl 2 butene	5	0	−0.006115	4.897993	−1.577338	3.626100	−4.131536
Cyclopentene	5	210.53920	−0.059928	2.957608	0.332898	−3.885328	6.485819
2 Methyl 1 3 butadiene	5	0	−0.088139	5.763155	−2.244854	5.502070	−5.935700
Cyclopentadiene	5	0	0.038615	1.110083	1.829800	−11.898873	22.388840
1 Hexene	6	167.42860	−0.004262	4.196656	−0.882105	0.925317	−0.270520
Cyclohexene	6	0	−0.084231	4.096340	−0.646804	0.073416	0.511888
1 Heptene	7	162.87650	−0.007807	4.259363	−0.904956	0.959607	−0.284711
1 Octene	8	157.50340	−0.012888	4.341314	−0.942250	1.030563	−0.318544
1 Nonene	9	154.12580	−0.015388	4.387824	−0.962266	1.066200	−0.333519

Table
A1.23   *Coefficients for calculation of the enthalpy of an ideal gas (equation 4.77) for olefins.*

	N_c	Soave m coefficient	Solubility parameter at 25°C $(kJ/m^3)^{1/2}$	Temperature T_3 °C	Interfacial tension at T_3 mN/m	Lee Kesler acentric factor
1 Pentene	5	0.833312	461.8	20	15.46	0.2346
cis 2 Pentene	5	0.846912	479.4	20	16.80	0.2430
trans 2 Pentene	5	0.842268	473.3	20	16.41	0.2404
2 Methyl 1 butene	5	0.828945	464.7	20	15.40	0.2317
2 Methyl 2 butene	5	0.895769	482.2	20	17.06	0.2766
Cyclopentene	5	0.773821	542.6	20	21.51	0.1936
2 Methyl 1 3 butadiene	5	0.740378	484.9	20	16.38	0.1722
Cyclopentadiene	5	0.765512	535.2	20	32.76	0.1875
1 Hexene	6	0.907106	475.9	20	17.89	0.2854
Cyclohexene	6	0.802079	557.0	20	26.09	0.2129
1 Heptene	7	0.976433	488.1	20	19.81	0.3329
1 Octene	8	1.040256	494.6	20	21.29	0.3779
1 Nonene	9	1.105314	500.7	20	22.56	0.4235

Table
A1.24   *Calculational coefficients related to a change of state of olefins.*

	N_c	Boiling point K	Standard specific gravity	Molecular weight kg/kmol	Liquid viscosity at 100°F mm²/s	Liquid viscosity at 210°F mm²/s	Critical temperature K	Critical pressure bar
Benzene	6	353.3	0.8829	78.11	0.5959	0.3335	562.2	48.99
Toluene	7	383.8	0.8743	92.14	0.5605	0.3429	591.8	41.10
Ethylbenzene	8	409.3	0.8744	106.17	0.6537	0.3971	617.2	36.10
o-Xylene	8	417.6	0.8849	106.17	0.7415	0.4238	630.4	37.35
m-Xylene	8	412.3	0.8694	106.17	0.5934	0.3642	617.1	35.42
p-Xylene	8	411.5	0.8666	106.17	0.6152	0.3700	616.3	35.12
Styrene	8	418.3	0.9087	104.15	0.6643	0.3766	636.0	38.40
Isopropylbenzene	9	425.6	0.8685	120.19	0.7376	0.4367	631.2	32.10
n-Propylbenzene	9	432.4	0.8683	120.19	0.7966	0.4524	638.4	32.00
1 2 4 Trimethylbenzene	9	442.5	0.8806	120.19	0.8547	0.4383	649.1	32.33
1 3 5 Trimethylbenzene	9	437.9	0.8699	120.19	0.8449	0.4139	637.4	31.28
n-Butylbenzene	10	456.4	0.8660	134.22	0.9433	0.5165	660.6	28.88
Naphtalene	10	491.1	1.0305	128.17	1.5521	0.7737	748.4	40.52

Table
A1.25   *Principal characteristics of aromatics.*

	N_c	Triple point temperature K	Heat of fusion kJ/kg	Heat of vaporization kJ/kg	Liquid conductivity at T_1 W / (m·K)	Liquid conductivity at T_2 W / (m·K)	Temperature T_1 K	Temperature T_2 K
Benzene	6	278.7	126.3	393.4	0.1503	0.1266	278.68	353.24
Toluene	7	178.2	72.0	364.4	0.1617	0.1126	178.18	383.78
Ethylbenzene	8	178.2	86.5	338.1	0.1588	0.1033	178.15	409.35
o-Xylene	8	248.0	128.1	348.4	0.1429	0.1039	247.98	417.58
m-Xylene	8	225.3	109.0	342.1	0.1474	0.1034	225.30	412.27
p-Xylene	8	286.4	161.2	337.3	0.1325	0.1030	286.41	411.51
Styrene	8	242.5	105.1	351.5	0.1488	0.1101	242.54	418.31
Isopropylbenzene	9	177.1	61.0	316.6	0.1494	0.0964	177.14	425.56
n-Propylbenzene	9	173.6	77.1	319.0	0.1525	0.0993	173.55	432.39
1 2 4 Trimethylbenzene	9	229.3	109.8	325.1	0.1439	0.0990	229.38	442.53
1 3 5 Trimethylbenzene	9	228.5	79.2	325.7	0.1513	0.1037	228.46	437.89
n-Butylbenzene	10	185.2	83.6	300.7	0.1510	0.0922	185.30	456.45
Naphtalene	10	353.4	148.1	338.7	0.1332	0.1278	353.43	491.14

Table
A1.26   *Additional characteristics of aromatics.*

	N_c	A	B	C $\cdot 10^4$	D $\cdot 10^7$	E $\cdot 10^{11}$	F $\cdot 10^{15}$
Benzene	6	225.05180	−0.122662	4.310824	−1.138140	1.494985	−0.564766
Toluene	7	206.73640	−0.101151	4.225723	−1.061438	1.337653	−0.484075
Ethylbenzene	8	193.42630	−0.093633	4.390639	−1.126299	1.458215	−0.543200
o-Xylene	8	186.29360	−0.014950	3.342431	−0.484082	−0.460172	1.705569
m-Xylene	8	184.78120	−0.030090	3.300522	−0.394355	−0.821163	2.173243
p-Xylene	8	194.82020	−0.068902	3.995003	−0.924756	1.059785	−0.345267
Styrene	8	0	−0.076968	4.070007	−0.974180	0.591315	1.389206
Isopropylbenzene	9	180.41750	−0.084771	4.425572	−1.112942	1.399989	−0.509541
n-Propylbenzene	9	187.36470	−0.099907	4.668002	−1.269276	1.767887	−0.694428
1 2 4 Trimethylbenzene	9	163.28510	0.078874	1.845273	0.650586	−4.327797	6.661838
1 3 5 Trimethylbenzene	9	186.49760	−0.055925	3.915613	−0.869152	0.956028	−0.296279
n-Butylbenzene	10	170.38210	−0.074034	4.378250	−1.077189	1.334904	−0.477754
Naphtalene	10	0	0.038615	1.110083	1.829800	−11.898873	22.388840

Table
A1.27   *Coefficients for calculation of the enthalpy of an ideal gas (equation 4.77) for aromatics.*

	N_c	Soave m coefficient	Solubility parameter at 25°C $(kJ/m^3)^{1/2}$	Temperature T_3 °C	Interfacial tension at T_3 mN/m	Lee Kesler acentric factor
Benzene	6	0.801679	591.5	20	28.21	0.2118
Toluene	7	0.879203	580.2	20	27.93	0.2646
Ethylbenzene	8	0.937342	570.6	20	28.59	0.3055
o-Xylene	8	0.948748	583.4	20	29.60	0.3126
m-Xylene	8	0.970967	572.0	20	28.26	0.3282
p-Xylene	8	0.962623	564.2	20	27.92	0.3227
Styrene	8	0.922175	604.9	20	30.85	0.2942
Isopropylbenzene	9	0.972845	553.8	20	27.69	0.3296
n-Propylbenzene	9	0.998748	563.1	20	28.43	0.3478
1 2 4 Trimethylbenzene	9	1.045328	567.5	20	29.19	0.3802
1 3 5 Trimethylbenzene	9	1.077131	567.0	20	28.05	0.4026
n-Butylbenzene	10	1.067972	551.8	20	28.63	0.3968
Naphtalene	10	0.936404	606.8	80	34.22	0.3041

Table A1.28 *Calculational coefficients related to a change of state of aromatics.*

Appendix
2

Principal Standard Test Methods
for Petroleum Products

Alphabetical List

A specification or standard for product characteristics is valid only if it is matched with references to well defined and recognized test methods, such that quality control tests conducted by the parties involved —client and supplier, for example— are comparable even if they are performed at different locations.

In light of the above criterion, many test methods are standardized. The following pages contain a non-exhaustive list of the main test methods.

Alphabetical list of the principal standardized test methods

Characteristic	Method	Principle
Aniline point	NF M 07-021 ISO 2977 ASTM D 611	Phase separation temperature of a hydrocarbon/ aniline mixture
Antirust properties (greases)	NF T 60-135	Rust spots on bearing after test
Antirust properties (inhibited mineral oils)	NF T 60-151 ISO 7120 ASTM D 665	Spots on a test tube after agitation with oil + water
Apparent viscosity (greases)	NF T 60-139 ASTM D 1092	Forced passage of the grease in a capillary tube
Aromatics	ASTM D 2269	UV absorption
Ash content	EN 7 (M07-045) ISO 6245 ASTM D 482	Weight of residue from total combustion
Asphaltenes	NF T 60-115 IP 143	Precipitation
Benzene content	EN 238 (M 07-062) ASTM D 2267	Gas chromatography
Carbonizable substances (paraffin)	NF T 60-134 ASTM D 612	Coloration after treatment in concentrated sulfuric acid
Cetane index	ASTM D 4737 ISO 4264	Correlation between density and distillation
Cetane number	NF M 07-035 ISO 5165 ASTM D 613	Combustion in a variable compression ratio motor
Chlorides	NF M 07-023	Volumetric analysis of an aqueous extract
Cloud point	NF T 60-105 ISO 3016 ASTM D 2500	Observation during gradual cooling
Color	NF T 60-104 ISO 2049 ASTM D 1500	Comparison with colored glass standard references
Cone penetration (waxes)	NF T 60-119 ISO/DIS 3986 ASTM D 937	Penetration of weighted cone
Cone penetration (greases)	NF T 60-132 ISO 2137 ASTM D 217	Penetration of weighted cone
Congealing temperature (waxes)	NF T 60-128 ISO 2207 ASTM D 938	Observation during cooling

Characteristic	Method	Principle
Cold filter plugging point	EN 116 (NF M 07-042)	Vacuum filtration through a calibrated filter
Conradson carbon	NF T 60-116 ISO 6615 ASTM D 189 ASTM D 4530 ISO 10370	Weighing after combustion then pyrolysis Micro method
Copper strip corrosion	NF M 07-015 ISO 2160 ASTM D 130	Appearance of a copper blade after immersion
Copper strip corrosion (LPG)	NF M 41-007 ISO 6251 ASTM D 1838	As above
Corrosive sulfur	NF T 60-131 ISO 5662 ASTM D 1275	Observation of copper strip after immersion for 19 h at 140°C
Demulsibility of petroleum oils and synthetic fluids	NF T 60-125 ISO 6614 ASTM D 1401	Time necessary for separation of phases
Density Density and specific gravity	NF T 60-101 ISO 3675 ASTM D 1298	Hydrometer
Density Density and specific gravity	NF T 60-172 ISO 12185 ASTM D 4052	Oscillating frequency
Density of LPG	NF M 41-008 ISO 3993 ASTM D 1657	Hydrometer
Distillation	NF M 07-002 ISO 3405 ASTM D 86	Rapid distillation. Temperature at every 10% distilled point
Distillation of cutbacks and fluxed bituminous products	NF T 66-003 ASTM D 402	Distillation
Dropping point for greases	NF T 60-102 ISO 2176 ASTM D 566	Heating up to the fall of the first droplet
Ductility of bituminous materials	NF T 66-006 ASTM D 113	Test-sample elongation at the point of rupture
Existent gums	NF EN 26246 ISO 6246 ASTM D 381	Weight of evaporated residue

Characteristic	Method	Principle
Flash point – Abel	NF M 07-011	Cyclic approaches of a flame over a heated closed cup
Flash point – Abel-Pensky	EN 57 (NF M 07-036)	Cyclic approaches of a flame over a heated closed cup
Flash point (bitumen)	NF T 66-009	As above, closed cup (modified Abel)
Flash point – Cleveland	NF EN 22592 ISO 2592 ASTM D 92	As above, open cup
Flash point – Luchaire	T 60-103	Cyclic approaches of a flame over a heated closed cup
Flash point – Pensky-Martens	NF EN 22719 ISO 2719 ASTM D 93	Cyclic approaches of a flame above a heated closed cup
Freezing point	NF M 07-048 ISO 3013 ASTM D 2386	Temperature of disappearance of cloud on reheating
Foaming characteristics of lubricating oils	NF T 60-129 ISO/DIS 6247 ASTM D 892	Air sparging and measurement of foam after standing
Gas bubble separation time of petroleum oils	NF T 60-149 ASTM D 3427	Time for air liberation after supersaturation (measurement of density)
Hydrocarbon families	NF T 60-155 ASTM D 2007	Chromatography on clay and silica gel
Hydrocarbon groups (FIA method)	NF M 07-024 ISO/DIS 3837 ASTM D 1319	Chromatography on silica gel with fluorescent indicators
Insolubles content	NF M 07-063	Filtration and weighing
Kinematic viscosity	NF T 60-100 ISO 3104 and 3105 ASTM D 445 and D 446	Measurement of time required to flow between 2 marks in a tube
Loss on heating of oil and asphaltic compounds	NF T 66-011 ASTM D 6	Weight after heating
Lead (content) 0.03 to 1 g/l	NF EN 23830 (M 07-043) ISO 3830 ASTM D 3341	Reaction with iodine mono-chloride
2.5 to 25 mg/l	EN 237 (M 07-061) ASTM D 3237	Atomic absorption
Melting point for waxes	NF T 60-114 ISO 3841 ASTM D 87	Observations under standardized cooling conditions

Characteristic	Method	Principle
Mercaptan sulfur	NF M 07-022 ISO 3012 ASTM D 3227	Silver nitrate analysis
Needle penetration (bitumen)	NF EN 1426 ISO/DIS 3997 ASTM D5	Penetration of weighted needle
Needle penetration (waxes)	NF T 60-123 ASTM D 1321	Penetration of weighted needle
Neutralization index	NF T 60-112 ISO 6618 ASTM D 974	Titration in presence of colored indicators
Noack volatility (lubricants)	T 60-161	Weight of a crucible before and after evaporation
Octane number	NF EN 25163 and EN 25164 ISO 5163 and 5164 ASTM D 2700 and 2699	Combustion in a variable compression ratio motor
Odor (paraffins)	NF T 60-158 ASTM D 1833	Odor panel
Oil content (waxes)	NF T 60-120 ISO 2908 ASTM D 721	Soluble portion in methylethylketone at $-32°C$
Oxidation stability (distillate fuel oil)	NF M 07-047 ISO/DIS 12205 ASTM D 2274	Measurement of precipitate after 16 h of oxygen sparging at $95°C$
Oxidation stability (gasoline) (induction period)	NF M 07-012 ISO/DIS 7536 ASTM D 525	Time necessary for a sample bomb under oxygen pressure to reach the critical induction point
Oxygenates content	NF M 07-054	Gas chromatography
Paraffin content (bitumen)	NF T 66-015	Insolubility at $-20°C$ in an alcohol-ether mixture
Pour point	NF EN 23015 ISO 3015 ASTM D 97	Observation during gradual cooling
Pseudo-viscosity (bitumen)	NF T 66-005 IP 72	measurement of flow duration
Reid vapor pressure	NF EN 12 (M 07-007) ISO 3007 ASTM D 323	Pressure in a sample bomb held at $37.8°C$
Relative density (specific gravity) of bituminous materials	NF T 66-007 ASTM D 70	Pycnometer

Characteristic	Method	Principle
Saponification (number)	NF T 60-110 ISO 6293 ASTM D 94	Reaction with potash and analysis of excess
Saybolt color	NF M 07-003 ASTM D 156 standard	Height of liquid column for equality with colored glass
Sediments	NF M 07-010 ISO 3735 ASTM D 473	Extraction with toluene and weighing of residue
Smoke point	NF M 07-028 ISO 3014 ASTM D 1322	Maximum flame height with no smoking
Softening point for bitumen (ring and ball method)	NF T 66-008 (future NF EN 1427) ASTM D 36	Temperature at which a ball passes through an asphalt sample disk attached to a ring
Solubility (bituminous products)	NF T 66-012 ASTM D 4	Dissolving, filtration, weighing
Sulfated ash in lubricating oils	NF T 60-143 ISO 3987 ASTM D 874	Weight of residue after treatment of the ash by sulfuric acid and calcination
in greases	NF T 60-144 ASTM D 128	As above
Sulfur/butane (qualitative test)	NF M 41-006	Sodium plumbite test (interface coloration)
Sulfur (detection of H_2S and mercaptans) ("Doctor Test")	NF M 07-029 ASTM D 325	Sodium plumbite test (coloration of interface)
Sulfur/propane	NF M 41-009 ASTM D 2784	Combustion in lamp and analysis of sulfur oxides formed
Total sulfur	NF EN 24260 ISO 4260 ASTM D 2785	Combustion in Wickbold burner and analysis
Total sulfur	NF M 07-025 ASTM D 1552	Combustion at high temperature (induction furnace) and analysis
Total sulfur	NF M 07-031 ISO 2192 ASTM D 1266	Combustion in lamp and analysis
Total sulfur	M 07-052	Combustion in furnace and colorimetric analysis

Characteristic	Method	Principle
Total sulfur	M 07-053 ISO 8754 ASTM D 4294	Non-dispersive X-ray fluorescence
Total sulfur	NF M 07-059	Combustion and analysis by UV fluorescence
Total sulfur	NF T 60-109 ASTM D 129	Oxidation in pressurized bomb and gravimetric analysis
Vanadium content	NF M 07-027 ASTM D 1548	Dissolving ash in acid and colorimetric analysis
Vapor pressure of LPG	NF M 41-010 ISO 4256 ASTM D 1267	Pressure in a sample bomb held at predetermined temperature
Viscosity – Brookfield	NF T 60-152 ASTM D 2983	Rotation of a bob in a sample
Viscosity index	NF T 60-136 ISO 2909 ASTM D 2270	Calculation based on kinematic viscosity
Volatility of LPG	NF M 41-012 ISO/DIS 6620 ASTM D 1837	Measurement of residue temperature after 95% evaporation
Water (content)	NF T 60-113 ISO 3733 ASTM D 95	Azeotropic distillation
Water (content)	NF T 60-154 ISO/DIS 6296 ASTM D 1744	Karl Fischer Method (electrometric, after addition of KF reagent)
Water in propane	NF M 41-004	Change in color of cobalt bromide
Water and sediments	NF M 07-020 ISO 3734 ASTM D 1796	Mix with toluene and centrifugation

References

Abbott, M.M., T.G. Kaufmann and L. Domash (1971), "A correlation for predicting liquid viscosities of petroleum fractions". *Can. J. Chem. Eng.,* Vol. 49, p. 379.

Anon. (1983), "Assessment of the energy balances and economic consequences of the reduction and elimination of lead in gasoline". Working Group ERGA (Evolutions of Regulations, Global Approach). *CONCAWE,* La Haye.

Anon. (1983), *Handbook of aviation fuel properties.* Coordinating Research Council, report No. 530, Atlanta, GA. Distributed by SAE, Inc., Warrendale, PA.

Bauer, C.R. and J.F. Middleton (1953), "Enthalpy of petroleum fractions". *Petroleum Refiner,* Vol. 1, p. 111.

Benson, J.D. et al. (1991), "Effects of gasoline sulfur level on mass exhaust emissions". *SAE paper* No. 91-2323, *International fuels and lubricants meeting,* Toronto, Ontario.

Bert, J.A., J.A. Gething, T.J. Hansel, H.K. Newhall, R.J. Peyla and D.A. Voss (1983), "A gasoline additive concentrate removes combustion chamber deposits and reduces vehicle octane requirement". *SAE paper* No. 83-1709, *Fuels and Lubricants meeting,* San Francisco, CA.

Biedermann, J.M., J.-P. Périès and J. Bousquet (1987), "SOLVAHL®: an attractive way to provide conversion units with high quality feedstocks". *National Petroleum Refiners Association (NPRA) paper* No. AM-87-41, *Annual meeting,* San Antonio, TX.

Billon, A., J. Bousquet and J. Rossarie (1988), "HYVAHL® F and T processes for high conversion and deep refining of residues". *National Petroleum Refiners Association (NPRA) paper* No. AM-88-62, *Annual meeting,* San Antonio, TX.

Bouquet, M. and A. Bailleul (1986), "Routine method for quantitative carbon 13 NMR spectra editing and providing structural patterns. Application to every kind of petroleum fraction including residues and asphaltenes". *Fuel,* Vol. 65, p. 1240.

Brandes, G. "Die Strukturgruppenanalyse von Erdölfraktionen mit Hilfe der ultrarot spektroskopie". *Brennstoff Chemie*, Vol. 37, No. 17-18, p. 263.

Brandes, G. (1958), "Die Strukturgruppenanalyse von Erdölfraktionen". *Erdöl und Kohle,* 11 Jahr, Vol. 10, p. 700.

Briant, J., J. Denis and G. Parc (1985), *Propriétés rhéologiques des lubrifiants.* Éditions Technip, Paris.

Bromley, L.A. and C.R. Wilke (1951), "Viscosity behavior of gases". *Ind. Eng. Chem.,* Vol. 43, No. 7, p. 1641.

Brown, J.K. and W.R. Ladner Jr (1960), "Distribution in coallike materials by high-resolution nuclear magnetic resonance spectroscopy". *Fuel,* Vol. 39, p. 87.

Burdett, R.A., L.W. Taylor and L.C. Jones Jr (1955), "Determination of aromatic hydrocarbons in lubricating oil fractions by far UV absorption spectroscopy", p. 30. In *Molecular Spectroscopy Report Conf.* Institute of Petroleum, London.

Burns V.R., M.C. Ingham and H.M. Doherty (1992), *Auto/Oil air quality improvement research program.* Special publication 920, SAE, Inc., Warrendale, PA, p. 468.

Cavanaugh, T.A., D.E. Blaser and R.A. Busch (1978), "Fluid coking/Flexi-coking, a flexible process for upgrading heavy crudes". *Japanese Petroleum Institute (JPI) Conference,* Tokyo.

Chevron (1988), *US Pat.* 4 753 611.

Clark, G.H. (1988), *Industrial and marine fuels.* Butterworths, London.

Convers, A. and M. Valais (1992), *World refining trends and prospects.* Publications of Indian Institute of Petroleum, Dehra Dun.

Coley, T.R. (1989), "Diesel fuel additives influencing flow and storage properties". In *Gasoline and diesel fuel additives* (Owen, K. Ed.). John Wiley.

Cox, F.W. (1979), *Physical properties of gasoline-alcohol blends.* US Department of Energy, Bartlesville, OK.

Damin, B., A. Faure, J. Denis, B. Sillion, P. Claudy and J.M. Letoffé (1986), "New additives for diesel fuels: cloud point depressents". *SAE paper* No. 86-1527, *International fuels and lubricants meeting and exposition,* Philadelphia, PA.

David, P., G.I. Brown and E.W. Lehman (1993), "SFPP – A new laboratory test for assessment of low temperature operability of modern diesel fuels". *CEC 4th International Symposium,* Birmingham.

Dean, D.E. and L.I. Stiel (1965), "The viscosity of nonpolar gas mixtures at moderate and high pressures". *AIChEJ*, Vol. 11, No. 3, p. 526.

Descales, B., D. Lambert and A. Martens (1989), "Détermination des nombres d'octane RON et MON des essences par la technique proche infrarouge". *Pétrole et Techniques (Revue de l'Association Française des Techniciens du Pétrole)*, No. 349.

Devos, A., G. Heinrich, P. Truffinet and F. Villette (1990), "La longue route vers la conversion profonde". *Pétrole et Techniques (Revue de l'Association Française des Techniciens du Pétrole)*, No. 357, p. 27.

Durand, J.P., Y. Boscher and N. Petroff (1987), "Automatic gas chromatographic determination of gasoline components. Application to octane number determination". *Journal of chromatography*, No. 395, p. 229.

Edmister, W.C. and D.H. Pollock (1948), "Phase relations for petroleum fractions". *Chem. Eng. Prog.*, Vol. 44, No. 12, p. 905.

Edmister, W.C. and K.K. Okamoto (1959), "Applied hydrocarbon thermodynamics. Part 12: equilibrium flash vaporization correlations for petroleum fractions". *Petroleum Refiner*, Vol. 38, No. 8, p. 117.

Edmister, W.C. and K.K. Okamoto (1959), "Applied hydrocarbon thermodynamics. Part 13: equilibrium flash vaporization for heavy oils under sub-atmospheric pressures". *Petroleum Refiner*, Vol. 38, No. 9, p. 271.

Eyzat, P. and J. Trapy (1982), "Caractérisation de l'accroissement des échanges thermiques dans les moteurs à allumage commandé fonctionnant avec cliquetis". *CR Acad. Sc.*, Paris, t. 295.

Feugier, A. and G. Martin (1985), "Influence de la nature des fuels lourds sur la qualité de leur combustion". *Rev. Inst. Franç. du Pétrole*, Vol. 40, No. 4, p. 787.

Fisher, I.P. and P. Fisher (1974), "Analysis of high boiling petroleum streams by high resolution mass spectrometry". *Talanta*, Vol. 21, p. 867.

Fiskaa, G., K. Langnes, O. Toff and G. Ostvold (1985), "Some aspects on utilizing modern marine Diesel fuels". *Congrès international sur les machines à combustion interne (CIMAC)*, Oslo.

Fuller, E.N., P.D. Schettler and J.C. Giddings (1966), "A new method for prediction of binary gas phase diffusion coefficients". *Ind. Eng. Chem.*, Vol. 58, No. 5, p. 19.

Girard, C., J.C. Guibet, A. Billon and X. Montagne (1993), "La désulfuration du gazole. Aspects technico-économiques. Impacts sur l'environnement". *Congrès international de la Société des Ingénieurs de l'Automobile (SIA) "Moteur diesel: actualité – potentialité"*, Lyon.

Glavincevski, B., O.L. Gulder and L. Gardner (1984), "Cetane number estimation of diesel fuels from carbon type structural composition". *SAE paper* No. 84-1341, *International fuels and lubricants meeting*, Baltimore, MD.

Goodacre, C. (1958), "Les antidétonants, en particulier le plomb tétraéthyle. Son passé, son présent, son avenir". *Congrès de la Société des Ingénieurs de l'Automobile,* Paris.

Groff, J. (1961), *ABC du graissage.* Éditions Technip, Paris.

Guibet, J.C. and B. Martin (1987), *Carburants and moteurs.* Éditions Technip, Paris.

Hall, D.W. and L.M. Gibbs (1976), "Carburetor deposits are clean throttle bodies enough?". *SAE paper* No. 76-0752, *Automobile Engineering Meeting,* Dearborn, MI.

Hamon, D. and B. Damin (1993), "New diesel low temperature operability. AGELFI filtration test". *CEC 4th International Symposium,* Birmingham.

Hankinson, R.W. and G.H. Thompson (1979), "A new correlation for saturated densities of liquids and their mixtures". *AIChEJ,* Vol. 25, No. 4, p. 653.

Hayward, P., L. Bekourian, E. Meheust, E. Lesimple, H. Benighi and A. Nouven (1980), "La reconnaissance quantitative et qualitative des cargaisons de pétrole brut". *Pétrole et Techniques (Revue de l'Association Française des Techniciens du Pétrole),* No. 269, p. 17.

Heinrich, G., M. Valais, M. Passot and B. Chapotel (1991), "Mutations of world refining: challenge and answers". *13th World Petroleum Congress,* Buenos Aires, Vol. 3, p. 189-198.

Heinrich, G., R. Bonnifay, J.-L. Mauleon, M. Demar and M.A. Silverman (1993), "Advances in FCC design, Parts I and II". *Refining process services Seminar,* Amsterdam.

Hennico, A., A. Billon, P.-H. Bigeard and J.-P. Périès (1993), "IFP's new flexible hydrocracking process combines maximum conversion with production of high viscosity, high VI lube stocks". *Rev. Inst. Franç. du Pétrole,* Vol. 48, No. 2, p. 127.

Hibbard, R.R. and R.L. Schalla (1952), "Solubility of water in hydrocarbons". RM E 52D24, *USA Nat. Adv. Comm. Aeron.*

Hildebrand, J.H. and R.L. Scott (1950), *The solubility of non electrolytes,* 3rd edition. Reinhold Pub. Corp., New York.

Hildebrand, J.H. and R.L. Scott (1962), *Regular solutions.* Prentice Hall, Engelwood Cliffs, NJ.

Hublin, M. and J.C. Griesemann (1993), "Le GPL: un carburant propre pour des automobiles propres". *Gaz d'Aujourd'hui.*

IFP (1989), "Gasoline production in the next decade: new trends". *Booklet from Direction Industrielle de l'Institut Français du Pétrole, symposium Venezia.*

Johnson, R.L. and H.G. Grayson (1961), "Enthalpy of petroleum fractions". *Petroleum Refiner,* Vol. 40, No. 2, p. 123.

Kabadi, V.N. and R.P. Danner (1985), "A modified SRK equation of state for water-hydrocarbon phase equilibria". *Ind. Eng. Chem. Proc. Des. Dev.,* Vol. 24, No. 3, p. 537.

Kesler, M.G. and B.I. Lee (1976), "Improve prediction of enthalpy of fractions". *Hydrocarbon Processing,* Vol. 55, No. 3, p. 153.

Kouzel, B. (1965), "How pressure affects liquid viscosity". *Hydrocarbon Process. Petrol. Refiner,* Vol. 44, No. 3, p. 120.

Kovats, E. and A. Wehrli (1959), "Gas-chromatography characterization of organic compounds. III. Calculation of the retention indexes of aliphatic, alicyclic and aromatic compounds". *Helv. Chim. Acta,* Vol. 42, p. 2709.

Le Breton, M.D. (1984), "Hot and cold fuel volatility indexes of french cars. A cooperative study by the GFC volatility group". *SAE paper* No. 84-1386, *International fuels and lubricants meeting,* Baltimore, MD.

Lee, B.I. and M.G. Kesler (1975), "A generalized thermodynamic correlation based on three-parameter corresponding states". *AIChEJ,* Vol. 21, No. 3, p. 510.

Lefebvre, G. (1978), *Chimie des hydrocarbures.* Éditions Technip, Paris.

Lenoir, J.M. (1957), "Effect of pressure on thermal conductivity of liquids". *Petroleum Refiner,* Vol. 36, No. 8, p. 162.

Letsou, A. and L.I. Stiel (1973), "Viscosity of saturated nonpolar liquids at elevated pressures". *AIChEJ,* Vol. 19, No. 2, p. 409.

Li, C.C. (1976), "Thermal conductivity of liquid mixtures". *AIChEJ,* Vol. 22, No. 5, p. 927.

Lindsay, A.L. and L.A. Bromley (1950), "Thermal conductivity of gas mixtures". *Ind. Eng. Chem.,* Vol. 42, No. 8, p. 1508.

Lowi, A. and W.P.L. Carter (1990), "A method for evaluating the atmosphere ozone impact of actual vehicle emissions". *SAE paper* No. 90-0710, *International congress and exposition,* Detroit, MI.

Lubrizol (1980), *US Pat.* 4 191 659.

Lydersen, A.L. (1955), "Estimation of critical properties of organic compounds by the method of group contributions". *Univ. Wisconsin Coll., Eng. Exp. Stn. report* No. 4, Madison, WI.

Maier, C.E., P.-H. Bigeard, A. Billon and P. Dufresne (1988), "Boost middle distillate yield and quality with a new generation of hydrocracking catalyst". *NPRA paper* No. AM-88-76, *Annual meeting,* San Antonio, TX.

Mariette, L., A. Billon and T. Descourières (1988), "Hyvahl® process for high conversion of resids". *Japanese Petroleum Institute Conference,* Tokyo.

Martin, B. and P.-H. Bigeard (1992), "Hydrotreatment of diesel fuels –its impact on light– duty diesel engine pollutants". *SAE paper* No. 92-2268, *International fuels and lubricants meeting,* San Francisco, CA.

Maxwell, J.B. and L.S. Bonnel (1955), *Vapor pressure charts for petroleum engineers.* Esso Research and Engineering Co., NJ.

Mc Arragher, J.S. and al. (1990), "The effects of temperature and fuel volatility on evaporative emissions from european cars". *I. Mech. Eng.*, No. 394/028, London.

Mehrotra, A.K. (1990), "Development of mixing rules for predicting the viscosity of bitumen and its fractions blended with toluene". *Can. J. Chem. Eng.*, Vol. 68, p. 839.

Misic, D. and G. Thodos (1961), "The thermal conductivity of hydrocarbon gases at normal pressures". *AIChEJ*, Vol. 7, No. 2, p. 264.

Mobil Oil Corp. (1980), *US Pat.* 4 208 292.

Montagne, X., D. Herrier and J.-C. Guibet (1987), "Fouling of automotive diesel injectors. Test procedure, influence of composition of diesel oil and additives". *SAE paper* No. 87-2118, *International fuels and lubricants meeting*, Toronto, Ontario.

Nelson, O.L., R.W. Krumm, R.S. Fein, D.D. Fuller, G.K. Rightmire and G.E. Ducker (1989), "A broad spectrum, non metallic additive for gasoline and diesel fuels: performance in gasoline engines". *SAE paper* No. 89-0214, *Int. Congress*, Detroit, MI.

Nelson, W.L. (1958), *Petroleum Refinery Engineering*, 4th edition. Mc Graw-Hill.

Odgers, J. and D. Kretschmer (1986), *Gas turbine fuels and their influence on combustion.* Abacus Press, Cambridge, USA.

Oelert, H.H. (1971) "Entwicklung und Anwendung einer auf IR-NMR Spektroskopie und elemental Analyse beruhenden Strukturgruppenanalyse für höhersiedende Kohlenwasserstoffgemische und Mineralölanteile". *Z. Anal. Chem.*, Vol. 255, p. 177.

Pande, S.G. and D.R. Hardy (1990), "A practical evaluation of published cetane indices". *Fuel*, Vol. 69.

Parrish, W.R. and J.M. Prausnitz (1972), "Dissociation pressures of gas hydrates formed by gas mixtures". *Ind. Eng. Chem. Proc. Des. Dev.*, Vol. 11, No. 1, p. 26.

Peng, D.Y. and D.B. Robinson (1976), "A new two-constant equation of state". *Ind. Eng. Chem. Fund*, Vol. 15, No. 1, p. 59.

Périès, J.-P., A. Quignard, C. Farjon and M. Laborde (1988), "Les procédés ASVAHL thermiques et catalytiques sous pression d'hydrogène pour la conversion des bruts lourds et des résidus de bruts classiques". *Rev. Inst. Franç. du Pétrole*, Vol. 43, No. 6, p. 847.

Pitzer, K.S. and al. (1955), "The volumetric and thermodynamic properties of fluids. I: theoretical basis and virial coefficients". *J. Am. Ch. Soc.*, Vol. 77, No. 13, p. 3427.

Quayle, O.R. (1953), "The parachors of organic compounds – an interpretation and catalogue". *Chem. Rev.,* Vol. 53, p. 439.

Quivoron, C. (1978), "Thermodynamique des solutions de polymères". In *Initiation à la physicochimie macromoléculaire.* Groupement Français des Polymères, Strasbourg.

Rackett, H.G. (1970), "Equation of state for saturated liquids". *J. Chem. Eng. Data,* Vol. 15, No. 4, p. 514.

Rall, H.T., C.J. Thompson, H.J. Coleman and R.L. Hopkins (1972), "Sulfur compounds in crude oil". *Bureau of Mines Bull.* No. 659. Distributed by *National Technical Information Service (NTIS),* US Dpt of Commerce, Springfield, VA.

Ranney, M.W. (1974), "Fuel additives". *Chemical Technology Review* No. 26, Noyes Data Corporation, Park Ridge, NJ.

Riazi, M.R. (1979), "Prediction of thermophysical properties of petroleum fractions". *Ph. D. Thesis,* Dpt of Chem. Eng., The Pennsylvania State University, PA.

Rihani, D.N. and L.K. Doraiswamy (1965), "Estimation of heat capacity of organic compounds from group contributions". *Ind. Eng. Chem. Fund.,* Vol. 4, No. 1, p. 17.

Satriana, M.J. (1982), "Synthetic oils and lubricant additives (advances since 1970)". *Chemical Technology Review* No. 207, Noyes Data Corporation, Park Ridge, NJ.

Shell Oil Comp. (1980), *US Pat.* 4 212 751.

Sirtori, S., P. Garibaldi and F.A. Vicenzetto (1974), "Prediction of the combustion properties of gasolines from the analysis of their composition". *SAE paper* No. 74-1058, *International Automobile Engineering and Manufacturing Meeting,* Toronto, Ontario.

Smith, H.M. (1940), "Correlation index to aid interpreting crude oil analysis". *US Bureau of Mines,* techn. paper No. 610, p. 54.

Soave, G. (1972), "Equilibrium constants from a modified Redlich-Kwong equation of state". *Chem. Eng. Sci.,* Vol. 27, p. 1197.

Speight, J.G. (1991), *The chemistry and technology of petroleum,* 2nd edition. Ed. Marcel Dekker, New York.

Stiel, L.I. and G. Thodos (1963), "Viscosity of hydrogen in the gaseous and liquid states for temperatures up to 5000 K". *Ind. Eng. Chem. Fund.,* Vol. 2, p. 283.

Strelzoff, S. (1974), "Partial oxidation for syngas and fuel (comparaison des procédés Texaco, Shell et Union Carbide)". *Hydrocarbon processing,* Vol. 53, No. 12, p. 79.

Sugden, S. (1924), "The variation of surface tension. VI. The variation of surface tension with temperature and some related functions". *J. Chem. Soc.,* Vol. 125, p. 32.

Thomson, G.H., K.R. Brobst and R.W. Hankinson (1982), "An improved correlation for densities of compressed liquids and liquid mixtures". *AIChEJ,* Vol. 28, No. 4, p. 671.

Tims, J.M. (1983), "Benzene emissions from passenger cars". *CONCAWE,* report No. 12/83, La Haye.

Umesi, N.O. (1980), "Diffusion coefficients of dissolved gases in liquids – Radius of gyration of solvent and solute". *M.S. Thesis,* The Pennsylvania State University, PA.

Unterzaucher, J. (1940), "Die mikroanalytische Bestimmung des Sauerstoffes". *Ber. Deut. Chem. Ges.,* Vol. 73, p. 391.

Unzelman, G.H. (1984), "Diesel fuel demand. A challenge to quality". *Petroleum Review.*

Unzelman, G.H. (1989), "Future role of esters in US gasoline". *National Petroleum Refiners Association (NPRA), Annual meeting,* San Francisco, CA.

Van Den Dool, H. and P.D. Kratz (1963), "Generalization of the retention index system including linear temperature programmed gas-liquid partition chromatography". *J. Chromatogr.,* Vol. 11, p. 463.

Vidal, J. (1973), *Thermodynamique. Méthodes appliquées au raffinage et au génie chimique.* Éditions Technip, Paris.

Watson, K.M. and E.F. Nelson (1933), "Improved methods for approximating critical and thermal properties of petroleum fractions". *Ind. Eng. Chem.,* Vol. 25, No. 8, p. 880.

Whitcomb, R.M. (1975), "Non-lead antiknock agents for motor fuels". *Chemical Technology Review* No. 49, Noyes Data Corporation, Park Ridge, NJ.

Wilke, C.R. and P. Chang (1955), "Correlation of diffusion coefficients in dilute solutions". *AIChEJ,* Vol. 1, No. 2, p. 264.

Wuithier, P. (1972), *Le pétrole. Raffinage et Génie Chimique,* 2nd edition. Éditions Technip, Paris.

Yoon, P. and G. Thodos (1970), "Viscosity of nonpolar gaseous mixtures at normal pressure". *AIChEJ,* Vol. 16, No. 2, p. 300.

Index

ACHEVÉ D'IMPRIMER
SUR LES PRESSES DE
L'IMPRIMERIE CHIRAT
42540 ST-JUST-LA-PENDUE
EN JUILLET 1995
DÉPÔT LÉGAL 1995 N° 1288
N° D'ÉDITEUR : 919

IMPRIMÉ EN FRANCE

Typesetting and diagram
MACH3, 60720 BURY